Insects for Human Consumption

Considering their growing popularity as a food source, this book examines the main nutrients and bioactive compounds present in a range of different edible insects. Focusing on the fundamentals of nutrient obtention and rearing management, it dives into the latest developments in the science behind an industry niche that continues to grow as the international community recognizes the uses of insects as a natural, sustainable source of important proteins and fats. It also features accessible discussions of safety concerns, regulatory limitations, and other key aspects, including food neophobia and consumer acceptance. This book is ideal for researchers, students, and professionals keen to understand the potential benefits of edible insects and how to navigate potential industry challenges.

Contemporary Food Engineering

Series Editor
Prof. Dr. Theodoros Varzakas
Full Professor in Food Technology and Food Safety Management Systems
University of Peloponnese
Dept. Food Science and Technology
Kalamata, Hellas, Greece

Founding Editor
Professor Da-Wen Sun, Director
Food Refrigeration & Computerized Food Technology
National University of Ireland, Dublin
(University College Dublin)
Dublin, Ireland

The **Contemporary Food Engineering Series** will consist of edited books addressing engineering processes related to food production. Classical unit operations in engineering applied to food manufacturing will be covered as well as such topics including the transport and storage of liquid and solid foods; heating, chilling and freezing of foods; mass transfer in foods; dehydration, thermal processing, non-thermal processing, extrusion, liquid food concentration, membrane processes and applications of membranes in food processing; packaging, cleaning, and sanitation. The books will be aimed at professional food scientists, academicians researching food engineering problems, and graduate level students.

Bioreactor Technology in Food Processing
edited by Rosane F. Schwan, V. K. Joshi, and Disney R. Dias

Innovative Technologies for Meat Processing
edited by Slim Smaoui and Tanmay Sarkar

Engineering Innovations in Sensory Science
edited by Mahendran Radhakrishnan and Anbarasan Rajan

Food Toxicology and Safety
edited by Sofia Agriopoulou and Maria Tarapoulouzi

Non-Thermal Approaches in Extraction of Bioactive Compounds from Diverse Food Sources
edited by Mahendran Radhakrishnan and Kamalapreetha Baskaran

Insects for Human Consumption: A Safe and Sustainable Food Source
edited by José Manuel Lorenzo

For more information about this series, please visit: https://www.crcpress.com/Contemporary-Food-Engineering/book-series/CRCCONFOOENG

Insects for Human Consumption

A Safe and Sustainable Food Source

Edited by José Manuel Lorenzo

CRC Press
Taylor & Francis Group
Boca Raton London New York

CRC Press is an imprint of the
Taylor & Francis Group, an **informa** business

Designed cover image: Shutterstock

First edition published 2026
by CRC Press
2385 NW Executive Center Drive, Suite 320, Boca Raton FL 33431

and by CRC Press
4 Park Square, Milton Park, Abingdon, Oxon, OX14 4RN

CRC Press is an imprint of Taylor & Francis Group, LLC

ISBN: 978-1-032-78495-3 (hbk)
ISBN: 978-1-032-77861-7 (pbk)
ISBN: 978-1-003-48815-6 (ebk)

DOI: 10.1201/9781003488156

Typeset in Times LT Std
by Apex CoVantage, LLC

Contents

Preface

The use of insects as food for humans is ancient. Edible insects have been widely used in the culinary recipes of many countries throughout the world, including America, Africa, Europe, Australia, and Asia. However, although certain insects are eaten, their importance in the human diet is residual compared to other types of food. In recent years, the international community has focused on the use of insects to obtain a natural and sustainable source of important ingredients, including protein and fat. This, together with the increase in the world population, and the need to find new food sources, specifically protein, makes insect farming a real and viable alternative. The rapid growth and the low need for resources and water for their production make them a viable alternative to create new foods that reduce the pressure on the environment and manage to feed the entire population, with which it is also possible to combat problems such as malnutrition.

However, it is important to make a correct characterization of edible insects, as well as to know the rearing possibilities, see their nutritional value, or their application in food reformulation. Another important aspect is the safety of insect consumption, and it is vital to study possible allergies or antinutritional compounds that they may have. Finally, to make insects a reality in the human diet, certain limitations must be studied and resolved, which include consumer acceptance, or certain regulatory implications. In the present book, all these aspects are studied and discussed comprehensively.

This book will serve as a comprehensive source of information about the main nutrients and bioactive compounds from different edible insects. The fundaments and the latest developments for nutrient obtention and rearing management are the main focus of this book. Also, safety concerns, regulatory limitations, and other aspects such as food neophobia, or consumer acceptance are discussed. All this information is presented in a way that is comprehensive for researchers, students, and professionals in the food industry. The rearing, processing, use, and characterization of insects are subjects that have been relatively little studied. That is why it is still a major challenge for both researchers and the industry. Thus, it is necessary to offer information to the entire community that promotes their study and responds to the concerns of international authorities with the intention of approving regulations that allow more extensive studies and allow the use of insects as food. However, as we have stated, in order to accept changes in the regulations, it is essential to ensure the safety and health of edible insects.

Therefore, as commented above, this book will provide a comprehensive view of edible insects and their use in the food industry by discussing in detail the aspects of rearing, sustainable production, composition, nutritional value, safety, health, acceptance, and regulatory implications of insect consumption, which are relevant in the development of new and functional foods based on edible insects. More specifically, academics and researchers in the area of Food Science and Technology and related areas will find the latest developments about all aspects of edible insects, as well as the current needs and opportunities to advance the knowledge in this field of research. Master and Ph.D. candidates will be able to use this source of information as a textbook due to the comprehensive approach to the fundaments and the progressions in the use of insects to develop new foods and food ingredients.

About the Editor

José Manuel Lorenzo is Head of Research at the *Centro Tecnológico de la Carne* (CTC), Ourense, Spain. He has a degree in food science and technology from the University of Vigo (1995–2000). From October 2005 to March 2006, he completed his training with a predoctoral stay at the *Stazione Sperimentale per L'Industria delle Conserve Alimentaria* (Parma, Italy). He obtained his PhD in food science and technology (University of Vigo) in 2006. He started his scientific career in the Department of Food Science and Technology at the University of Vigo, first as a research scholar and then, since April 2006, as an academic researcher. In 2007, he joined the CTC staff, developing since then an intense collaboration with the industrial sector, private companies, and research institutions as a PI of several projects of R&D and innovation in food science area, with emphasis on healthier and functional food development. He was a principal researcher in *four European projects, nine national projects,* and *58 regional projects* (Galicia, NW Spain) with meat industries. Moreover, he has participated in more than 110 projects as a research collaborator. He is currently leading the Network of Sustainable Regional Protein-Rich Foods (AlProSos; CYTED 125RT0165). It is a robust network comprising 229 researchers from 14 Ibero-American countries. It boasts the support of 29 research centers and universities, along with 21 companies. The network's core mission is to identify, characterize, extract, purify, functionalize, and assess the allergenicity and safety of regional alternative proteins for developing novel food products. Regarding his research experience, the quality and impact of his scientific production stand out:

1. His scientific production includes *more than 1,000 research papers* in well-recognized peer-reviewed international journals (SCI), with *56% of them in the first quartile* (the number of publications in Q1 is higher than 500) and *283 communications to national and international congresses.* His *h-index is 111* with *49,620 citations* in Scopus.
2. He has *edited 22 international books* and one national book. Moreover, he has written *125 chapters in international and national books.*
3. He has five *national patents* as an inventor.
4. He is *Chief Editor* of *Current Food Science and Technology Reports* and *Associate Editor* of several prestigious journals: *Quality Assurance and Safety of Crops & Foods*, *Food Frontiers*, *Food Analytical Methods*, *Journal of the Science of Food and Agriculture*, *Animal Science Journal*, *eFoods*, *Measurement: Food*, and *Canadian Journal of Animal Science.* He has also edited several special issues for these high-impact journals to create more discussion around key aspects of functional food development: innovative technologies, functional ingredients, and food safety and quality.

Contributors

Rubén Agregán
Centro Tecnológico de la Carne de Galicia
Ourense, Spain

José Eleazar Aguilar-Toalá
Purdue University
Indianapolis, Indiana, USA

Simone Belluco
Istituto Zooprofilattico Sperimentale delle Venezie
Legnaro, Italy

Ji Yoon Cha
Korea Food Research Institute
Wanju, Korea

Yun-Sang Choi
Korea Food Research Institute
Wanju, Korea

Rubén Domínguez-Valencia
Centro Tecnológico de la Carne de Galicia
Ourense, Spain

Noemí Echegaray
Centro Tecnológico de la Carne de Galicia
Ourense, Spain

Tiziana Fornari
Universidad Autónoma de Madrid
Madrid, Spain
and
Instituto de Investigación en Ciencias de la Alimentación
Madrid, Spain

Mónica R. García-Risco
Universidad Autónoma de Madrid
Madrid, Spain
and
Instituto de Investigación en Ciencias de la Alimentación
Madrid, Spain

Rodrigo Jiménez-Pichardo
Universidad Autónoma del Estado de Hidalgo
Pachuca, Mexico

Yea-Ji Kim
Korea Food Research Institute
Wanju, Korea

Beatrice La Porta
Università degli Studi di Padova - TESAF
Padova, Italy

Jae Hoon Lee
Jeonbuk National University
Jeonju, Korea

Andrea M. Liceaga
Purdue University
Indianapolis, Indiana, USA

José Manuel Lorenzo
Centro Tecnológico de la Carne de Galicia
Ourense, Spain
and
Universidade de Vigo
Vigo, Spain

Diana Martin
Universidad Autónoma de Madrid
Madrid, Spain
and
Instituto de Investigación en Ciencias de la Alimentación
Madrid, Spain

Anthia Matsakidou
Aristotle University of Thessaloniki
Thessaloniki, Greece

Heimo Mikkola
Eastern Finland University
Kuopio, Finland

Adamantini Paraskevopoulou
Aristotle University of Thessaloniki
Thessaloniki, Greece

Mirian Pateiro
Centro Tecnológico de la Carne de Galicia
Ourense, Spain

Jose Angel Perez-Alvarez
Miguel Hernández University
Alicante, Spain

Juliane Pinarelli Fazion
Istituto Zooprofilattico Sperimentale delle Venezie
Legnaro, Italy

Fernando Rivero-Pino
European Food Safety Authority
Parma, Italy
and
University of Seville
Seville, Spain

José Antonio Rodríguez Ávila
Universidad Autónoma del Estado de Hidalgo
Pachuca, Mexico

Irais Sánchez-Ortega
Universidad Autónoma del Estado de Hidalgo
Pachuca, Mexico

Eva María Santos López
Universidad Autónoma del Estado de Hidalgo
Pachuca, Mexico

Luis Vázquez
Universidad Autónoma de Madrid
Madrid, Spain
and
Instituto de Investigación en Ciencias de la Alimentación
Madrid, Spain

Ermolaos Ververis
European Food Safety Authority
Parma, Italy

David Villanueva-Bermejo
Universidad Autónoma de Madrid
Madrid, Spain
and
Instituto de Investigación en Ciencias de la Alimentación
Madrid, Spain

1 Introductory Aspects of Insects as Food: Their Historical and Cultural Importance

Heimo Mikkola

1.1 BACKGROUND

The world population is expected to reach 8.6 billion in 2030 [1]. The main population growth will take place only in nine countries (given in the order of expected growth): India, Nigeria, the Democratic Republic of the Congo, Pakistan, Ethiopia, Tanzania, the United States of America, Uganda, and Indonesia [1]. It is anticipated that despite declining fertility levels, the upward trend in the world population will continue, so by 2050, we will have more than nine billion people to feed – how can we manage? From our globe, only 40% of the ice-free land surface is suitable for agriculture, while forests take 30%, deserts and mountains take 25%, and the urban and wetlands take the rest [2]. Of the total agricultural land, 70% is used for feeding livestock, and 30% of all grains produced are fed to animals (80% of soy and half of the corn production).

Similarly, 70% of the total freshwater is used for agriculture with 110 million tons of chemical fertilizers and 2.3 million tons of pesticides (like 1/3 glyphosates–Round-Up), and agriculture is causing 14–17% of total greenhouse gas emissions and producing 6 billion tons of manure [2]. By 2050, if not before, more than 60% of people will be living

DOI: 10.1201/9781003488156-1

in urban areas, and they will have higher demands for animal-derived proteins up to 60 kg of meat and 30 kg of fish. When it comes to farmed animals for human consumption, there is a need for a substantial annual increase in chicken and aquaculture to produce 60 billion (vs. present 20 billion) chickens and 140 million tons (vs. present 70 million) of aquaculture products [2].

Other protein alternatives will be needed and could come from [2, 3]:

- Capturing more out of oceans: medusa, jellyfish, krill. . . .
- Farming the sea: macro- and microalgae (Spirulina–High tech)
- Producing artificial proteins but now vitro meat would cost 120,000 USD/kg
- Getting more out of agro-industry processing by-products: corn gluten, brewers' grains, yeasts, potato protein concentrate, distiller's dried grain with soluble. . . .
- Farming more plant protein sources: oil seeds, legumes, forages, trees (like *Moringa* leaves). . . .

We also should worry about how to feed our animals: in 2013, it was estimated that 795 million tons of cereals, which was one-third of the global cereal production, went for feed [2]. By 2050, we would need an additional 520 million tons for our animals, that is, half of the estimated cereal production. In 2013, aquaculture, chickens, and pigs consumed 155 million tons of mainly soy feed protein. In 2050, an additional 110 million tons of feed protein are needed, this being half from cereals/soy protein sources. In 2013, some 110 million tons of coarse grains were used for bioethanol production. As scarcities of agricultural land, water, forest, fishery, and biodiversity resources, as well as nutrients and nonrenewable energy, are foreseen, insect rearing as "Mini-livestock" could be a solution for food and feed security [2].

1.2 HISTORY OF INSECT EATING AND FARMING

Insects have occupied a prominent position in the human diet since ancient times. Countries around the world used to have the habit of eating insects [4]. The Middle East people ate desert locusts, *Schistocerca gregaria*, before the Eighth Century BC [5–7]; in ancient Rome (before 77 AD), people liked to eat the larva of the great Capricorn beetle *Cerambyx cerdo* after it was fed with flour [7, 8]. In addition, the ancient Hebrew people hunted locusts for eating, American Indians ate the locust pest braised in soy sauce, Australian people loved eating cutworms *Agrotis* spp., Africans fed termites, hoppy of South Americans were to eat red ants, and the Spanish made worm bean sauce from ant eggs, etc. Even in old Europe, the locusts were dried and ground into flour, baking cookies or bread. African residents in many countries eat ants and termites due to their delicious taste. American Indians, in southern California and Mexico, collected a vast number of white oblong eggs of the backswimmer *Notonecta glauca* in the water for consumption and sale. The blue butterfly *Maculinea arion* larvae were popular in the United States and Mexico; despite the expensive price, they were the famous, rare dishes in restaurants or hotels. The edible insects also have a long history in China. The emperor's banquet, in the Third Century BC, was made up of ant eggs, bees, cicadas, and locusts such as Chan (cicada), Huangchong (locust pest), Mifeng (bee), and Yiluan (ant egg sauce), which were listed as the emperor's favorite meat dish and the banquet delicacy during the aristocratic gatherings. There is a famous, traditional Chinese speciality named Eight Jane Cakes baked from fly maggots [6, 9].

Insect farming has been practiced past 5000 years by cultivating silk moth *Bombyx mori*, bees, Polish cochineal *Porphyrophora polonica*, formerly used to produce a crimson dye, and lac beetle *Kerria lacca* being a main source of lac. Insect rearing for pet food, fish bait, zoos, medicinal purposes, and ornamentals started in the early 1900s. These insects included some 20 species of mealworms, fly larvae, butterflies, crickets, and beetles. Since the 1960s, some 100 insect species have been used for integrated pest management and plant pollination. Insect farms for human food started around 2000 by using less than ten species like crickets and mealworms. By 2010, mass rearing of insects began as feed for farmed animals with less than five species like fly larvae–black soldier flies *Hermetia illucens* [2, 3].

1.3 A VARIETY OF INSECTS EATEN NOWADAYS

Entomophagy is prevalent in many regions, and some 1500–2000 species of insects and other invertebrates are consumed by 3000 ethnic groups across 113 countries in Africa, Asia, Australia, and Central and South America [10].

Africa is a hotspot of edible insect biodiversity and more than 500 species are consumed daily [10]. These insects include a large number of caterpillar species, out of which a list of 97 species has been compiled [11]. Often, the caterpillars are sold alive because there are no ways to conserve them. For this reason, the animals are collected very early in the morning from leaves or pitfall traps around the bases of trees. Then, caterpillars are starved to eliminate the gut contents, boiled or smoked for a few hours, sun-dried on mats, and then cooked in a stew of vegetables. In southern Africa, one of the most frequently eaten is the mopane caterpillar from the emperor moth *Gonimbrasia belina*, and in West Central Africa, the shea caterpillar of *Cirina butyrospermi* butterfly is a species with high nutritional value [12]. The caterpillar (Figure 1.1) can be twice as expensive as beef probably giving the reason that in Burkina Faso, this species is even called "caviar of the savannas" [11].

In Cameroon, people have developed a great knowledge about the bioecology of the most marketed insect species. About 200 species have been reported, and many are still to be unveiled. Some are used by limited communities, making their identification a real challenge. Tropical forests provide shelter and a perfect milieu for numerous of them [13]. An important group of consumed insects is made up of termites with two *Macrotermes* species, being the most widely exploited all over the country. The common recipe is to grill them after removing the wings with onions. Crunchy and crispy insects are highly regarded as a delicacy, mainly of Coleoptera order. Adult palm weevils *Rhynchophorus phoenicis* are known as pests, but adults and grubs are "domesticated" by raffia palm

FIGURE 1.1 *Cirina butyrospermi* caterpillars served on white bread in West Africa. This could be a new candidate for insect farming. Photo from Wikimedia.

harvesters [13]. The migratory locusts are commonly exploited in drier regions of Cameroon, whereas long-horned grasshopper *Ruspolia differens*, great green bush cricket *Tettigonia viridissima,* and variegated grasshopper *Zonocerus variegatus* are exploited in humid environments in the southern parts of the country. Painted or variegated grasshoppers get special treatment because their eating can in some cases cause muscular and articular paralysis if not well prepared before ingestion. Some people (3% of groups) do not eat them at all because of bad body odor [14].

Interestingly, it has been calculated that only 135 (0.2%) of the 60,000 described insect species are known to be edible in the Amazon basin [15]. Mainly, Hymenoptera insects (ants, termites, wasps, and bees) are consumed in the Amazon region. Also, the South American palm weevil *Rynchophorus palmarum* and bottlebrush or bearded weevil *Rhinostomus barbirostris* larvae are an important source of proteins and fats used in native Amazonian indigenous communities to supplement their diet [15].

The results of the edible insect survey in China reported 283 species with subspecies, involving 13 orders and 71 families. Among them, the Orthoptera, Hemiptera, Coleoptera, Lepidoptera, and Hymenoptera accounted for the majority number. The quantity of edible species is still growing because many of the species have not been identified and reported in this survey [9].

In Thailand, entomophagy has spread to the south from the northeast as people migrate toward city centers. It has become so popular that >150 species are sold in the markets of Bangkok. The most common edible insects are moths, cicadas, beetles, mealworms, flies, grasshoppers, and ants [10].

Globally, the most commonly consumed insects are beetles (Coleoptera, 31%), caterpillars (Lepidoptera, 18%) and bees, wasps, and ants (Hymenoptera, 14%), grasshoppers, locusts and crickets (Orthoptera, 13%), cicadas, leafhoppers, planthoppers, scale insects and true bugs (Hemiptera, 10%), termites (Isoptera, 3%), dragonflies (Odonata, 3%), flies (Diptera, 2%), and other orders (5%) [16, 17].

1.4 WHAT ARE THE NUTRITION VALUES OF INSECTS?

Many insects have good but highly variable nutritional values depending on species, metamorphic stage, habitat, and diet. Edible insects contain various nutrition substances, such as essential proteins, amino acids, fatty acids, carbohydrates, mineral elements, and other functional ingredients with the insect secondary metabolite, including the flavonoids, alkaloids, polysaccharides, hormones, and phospholipids, which have high economic value for development and utilization [9]. There are numerous vitamins in the insect body, mainly including vitamin B1 (thiamine), B2 (riboflavin), B3 (niacin), B6 (pyridoxine), C, D, E, K, and carotene. Termites like *Macrotermes annandalei* contain vitamin A 25.0, vitamin D 85.4, and vitamin E 11.7 International Unit/g. Vitamins are essential substances for maintaining the normal physiological function of the human body [9]. Interestingly, traditional medicine in Africa includes the use of termites Hodotermitidae and Termitidae as popular medicine. They are used in the treatment of various diseases that affect humans such as influenza, asthma, bronchitis, whooping cough, sinusitis, tonsillitis, and hoarseness [16]. Processing of the insects plays a significant role in determining nutritional content. Insects are particularly important as food and feed supplements, as they provide satisfactory amounts of energy and protein and meet amino acid requirements. Insects have plenty of monounsaturated and/or polyunsaturated fatty acids and are high in micronutrients like biotin, copper, folic acid (in some cases), iron, magnesium, manganese, pantothenic acid, phosphorus, riboflavin, selenium, and zinc [2, 3]. Insects have been utilized as medicines in diverse cultures, especially in traditional Chinese medicines. China has ongoing investigations in searching for new antibacterial structures from insects' natural products. More studies designed to determine the medicinal properties of isolated chemical components from insects, and other arthropods should be performed for the development of useful drugs [9].

Unfortunately, allergic reactions to edible insect eating have been reported, especially in countries where edible insects are traditionally consumed [18]. The allergenic character from the consumed yellow mealworm *Tenebrio molitor*, silkworm *Bombyx mori*, house fly *Musca domestica* maggot, migratory locust *Locusta migratoria*, house cricket *Acheta domesticus*, greater wax moth *Galleria mellonella*, black soldier fly *Hermetia illucens*, American grasshopper *Schistocerca a. americana*, and Indian meal moth *Plodia interpunctella* was studied. Most of the positive hits consisted of allergens from arthropods such as dust mites, crustaceans, and insects, and more rarely, of allergens from molluscs, nematodes, and fungi. The allergens of edible insects form clusters closely related to the crustacean, mollusk, and nematode clusters. Edible insects seem to possess a large repertoire of IgE-binding allergens they share with related groups of arthropods, molluscs, and nematodes. These cross-reacting allergens can trigger allergic reactions in individuals previously sensitized to shellfish or mollusks. Allergies induced through insects' ingestion should always be considered, and all insect food and insect-containing food products should mention this allergy possibility very clearly in the product labels [19]. The processing and storage of insects and their products should follow the same health and sanitation regulations as for any other food or feed to ensure their microbiological and chemical safety [20].

In a study of the calorific value of 94 insect species, 50% were higher than soybeans, 87% higher than maize, 63% higher than beef, and 70% higher than fish [10]. The composition of omega-3 and omega-6 fatty acids in mealworms is comparable to that of fish, and other insects with ideal fatty acid ratios are house crickets, short-tailed crickets, Bombay locusts, and scarab beetles. Some insect species have micronutrients not found in some conventional animal proteins, such as riboflavin in termites and high concentrations of thiamine in silk moth larvae (225% daily human requirement) and palm weevils (201%) compared to chicken (5.4%). Mealworms have a higher content of protein (all essential amino acids), calcium, vitamin C, thiamine, vitamin A, and riboflavin per kg than beef. Although the nutritional content of many insects is well-described in the literature, there is a variation depending on diet, sex, life stage, origin, and environmental factors, and the realized nutritional content also depends on preparation and cooking [10].

Depending on the source of insects and type of animals, the insect meals can replace 25–100% of soymeal or fish meal in the feed of our domesticated animals or fish. Aroma, taste, and carcass quality will not be different with the use of insect meals. Overall levels of essential amino acids are good in insect meals, over 42% in black soldier fly larvae, 61% in silkworm meal, and over 57–82% in defatted insect meals, and the protein digestibility is good. Insect meal energy value is higher than those of both soymeal and fish meal, and the oil content is as high as 36%. Insect meals are low in calcium, except for black soldier fly larvae, which is 7.6% [2, 3].

1.5 WHAT ARE THE BENEFITS OF INSECT FARMING?

Insect farming contributes to a closer, local circular economy in livestock rearing, as locally produced sidestream from agriculture, agro-industries, food, and waste management is available to local insect farmers to produce proteins and fats. These can be used as ingredients for livestock, meat, and fish producers in the same region. This will shorten the chain for feed producers by incorporating better locally produced ingredients. Insect rearing can improve local farming economies, including small farm operators [2, 3].

The benefits of insect farming can be summarized as follows:

- It can provide food for two billion people, plus natural feed for chickens, carnivore fish in aquaculture, and pigs;
- Insects are a good source of protein, fats, minerals, plus micronutrients;
- Feed conversion efficiency is four times better than in cattle rearing;
- Insects can be fed with organic waste, so not competing with foods for humans;
- They have a fast-breeding cycle (10 days to 1 month);

- Insect farming requires less water, land, and produces less greenhouse gas emissions;
- Insects may help reduce the use of antibiotics in feed;
- Insect farming is socially more inclusive – easy to farm, no need for huge investment, or "high-tech" skills;
- It opens opportunities for business and jobs at any level, everywhere in the world;
- It supports local-circular agro-economy; and
- Insect farming is interdisciplinary: food-feed-health-waste disposal-climate change and jobs [2, 3].

Insects have attracted the attention of mankind since ancient times for both negative and positive reasons: Negative, related to their destructive effects on agricultural and industrial crops, causing large economic losses and their harmful effects on human health causing huge human losses by transmitting diseases such as Chagas disease, dengue, malaria, yellow fever, chikungunya, leishmaniasis, and others. Positive, related to their use as a human food source, of particular importance to help mitigate, in the medium term, critical cases of food insecurity and famine, and feeding other animal organisms [21].

However, insect eating is declining in many traditional markets and is often seen as taboo in Western cultures. Its recent introduction has been met by a range of barriers. Nutritional arguments are not thought to be enough to overcome the "disgust factor" and convert Westerners to insect-based dishes [22]. This chapter reviews several of these barriers and seeks new concepts in promoting the wider acceptance of insect-eating.

1.6 FINLAND AS A CASE STUDY

1.6.1 Introduction

Several newspaper articles from Finland have been collected (Finnish names of the articles translated into English) on the insect food business development between 2016 and 2024. These papers show at first an enormous boom in insect food production and sales for human consumption after that business was finally legalized in September 2017. However, already in late 2018, many insect farmers and market people expressed their concerns that the business development has not been as lucrative as anticipated. The year 2019 brought more not-so-positive evaluations of insect food markets in Finland but concluded that the insect food boom will come one day, latest with the next generation of people. And lately, a good number of academic theses have been published on various topics in the insect farming and eating business giving some hope for the future.

1.6.2 Illegal Marketing

Due to European Union regulations, insect food marketing was at first illegal in Finland, although some production existed. It was also interesting that for highly valued free-roaming, chicken insects are part of their feed, but insects were not legally allowed in chicken feed in the European Union countries. Enthusiastic farmers and researchers started, however, in 2016 to organize insect food-tasting events in some restaurants and schools around the country. One restaurant in Helsinki (Restaurant Rupla) had a three-course insect menu one evening every 2 months (21.07, 21.09, etc.). None of the participants refused to eat the insect food [23]. Many people had to queue to get to taste the insect food.

In October 2016, Natural Research Institute informed people about the insect business and told them that insect producers were invited to attend a study trip to Holland where insect production and marketing are far ahead of Finland [24]. In February 2017, newspapers wrote that many start-up companies on the Otaniemi University campus believe that the insect food business could bring millions in income to the participants [25]. A bit later, it was written that people in North Karelia are well prepared to start insect food production and that 70% of the Finnish population is ready to taste insect food products [26, 27]. A good introduction articles were written next in March–August 2017 referring to the Food and Agriculture Organization of the United Nations (FAO) report on the benefits of insect farming but warning that some consumers may have some reservations about eating insects [28–30].

1.6.3 Insect Food Was Legalized

Finally, in September 2017, Finland adopted the European Union (EU) regulations so that it was acceptable to start selling insect food as human food products [31, 32]. This sales permit allows the selling of insect products made out of the following insects [33]:

- Black soldier fly larvae *Hermetia illucens,*
- Cricket *Acheta domesticus,*
- Desert locust *Schistocerca gregaria,*
- Drone bee larvae *Apis mellifera,*
- Lesser mealworm (= chicken hog larva) *Alphitobius diaperinus,*
- Migratory locust *Locusta migratoria,*
- Tropical domestic cricket *Gryllodes sigillatus,* and
- Yellow mealworm *Tenebrio molitor.*

Several farmers modified their farms by giving up pig farming and starting to grow insects instead. Some tens of restaurants started to offer insect food on their menu. Many new insect products were introduced to the market when the sale of insect food became legal. Some producers stated, "We will start slowly, learning the markets and markets getting to know us." When asked about the taste, people stated: "Taste is mild, something between chicken and shrimps. Best crickets are when well fried" or "Crickets taste like the Finnish Rye Crispbread [30, 34].

1.6.4 Fazer Cricket Bread

Huge Food industry company Fazer with 15,000 employees decided to take the lead in the industrial use of edible insects in Finland. Already on 24 November 2017, the

FIGURE 1.2 Fazer cricket bread costs 3.99 Euros per piece. Free photo from the company.

company launched its cricket bread (Figure 1.2) in five supermarket bakeries in the Helsinki region [35]. After that launching, cricket bread was meant to be available in all 47 Fazer bakeries. Soon it came through that Finnish production of cricket flour was not large enough, so Fazer had to buy the cricket flour from Holland at a very high price (almost 100 Euros/kilo) [36]. Newspaper people were invited to taste the bread the night before the launch. The bread is 250 g, and it contains 70 crickets in the form of flour. Bread has 18 g of protein, of which 5 g comes from the cricket flour. The price of one piece of bread is exorbitant 3.99 Euros (almost 16 Euros/kg). The taste group was somehow disappointed that the crickets could not be seen or tasted from the bread, but the bread was as good and crispy as any other full-grain product. Many felt that there could have been more crickets [37].

Leader Foods Oy, for instance, started to sell cricket protein bars. One Zircca bar contains 15 crickets and 34% protein, and it is a gluten-free product. The company says that cricket bars meet all nutrition requirements equally well as any meat or fish products. And the iron content of crickets is higher than that of spinach [38]. On 11 June 2018, Oy Halva Ab brought to the markets cricket liquorice bars first time in Finland and in the world [39]. One of the largest insect food companies in Finland is Finsect, which also exports insect products under the name "Griidy." They have 26 contract farmers, mainly in Western Finland. They produce cricket bread, cricket chocolate, cricket liquorice, roasted and seasoned crickets, and cricket flour for 150–450 Euros/kg for the consumers depending on the package size [40].

Then, on 22 June 2018, Fazer hit the jackpot with the cricket bread after winning a third prize in the Cannes Lions International Advertising Festival in France [41, 42].

1.6.5 Problems Started to Build Up

After Fazer cricket bread celebrations, the problems started in all fronts of the insect farming. The first major problem many commercial insect farmers faced when starting large-scale production was the lack of commercial feed for the insects. In 2018, the Natural Resources Institute Finland and Eastern Finland University started a project to produce plant-based pelleted feed for the insect farmers. This "Hyvä Rehu" (= Good Feed) project lasts 2 years and is funded by the Ministry of Agriculture of Finland. One of the most amazing problems in cricket farms in Finland has been the extremely warm and dry summer weather we have had in recent years [43]. Even the tropical species stopped eating due to the heath or rather due to the dry air. In the tropics, the weather is always humid even in high temperatures. So the farmers were forced to invest in expensive air humidifiers [44]. In Loviisa town, there was a huge insect farm that aimed to be the largest in Europe but had to be closed down in 2019 as not profitable enough [42]. At the end of 2018, newspapers started to write negative news about the farmers who had invested money to make millions with insect food [44, 45].

1.6.6 End of the Cricket Bread Production

In early 2019, the most shocking news came from Fazer. They decided to stop making cricket bread and bread rolls [46]. The boom for the bread lasted 1 year and for the bread rolls only 4 months. Despite that, Fazer stated that they were very happy with this project giving them positive visibility in all other continents than Antarctica. They also felt that Finnish insect producers got the door open to sell their insects to Germany, etc. However, Fazer had to stop because people were not ready for these products – customers bought the products just once to try and to form their own opinion on insect food. Very few if any purchased the bread again [46]. After the sad end of Fazer cricket bread, the newspapers started to keep up the insect food by dreaming about some new products, like cricket snacks with garlic and sports drinks with insect flour addition [47, 48] or by spirit-lifting articles: "How the insect boom is only starting [49]. Sad but more realistic stories started to come at the end of 2019 when it was estimated that only 15 insect farms had survived the hard times [50]. Long silence started in the newspaper on insect farming, but two very good summarizing papers were published in January 2021 on how the cricket bread boom started in 2017 and ended in 2018 [51, 52]. The conclusion was that the cricket business did not collapse, but it just slowly vanished without people noticing it.

1.6.7 Last But Not Least Insect Farmer

End of January 2021, one of our most dedicated insect farmers, Hermanni Nieminen, told us what was happening in his vegetable and insect farm near Joensuu. He was one of the first who started to farm the insects in 2016, even before it was legal to sell them for food. Now he believes that he is one of the few remaining in the business. His wife got fed up with the extra work insect farming required, but he continues with a 10% production capacity. The only customers are the lizard pet owners in Joensuu, who feed their pets with crickets. Nieminen is one of those farmers saying that the marketing of insects was better when they were sold illegally as human food. The products were often labelled and sold as kitchen and food decoration items. However, he is still optimistic that if Finland were to produce insects for European markets, then we could get rid of soy feed used as the main domestic animal protein source. Eventually, insect farming would save or at least reduce the rainforest destruction in South America [52]. In 2020, very few papers were published on insect food but one was very much positively repeating Finnish failures but also what FAO and others said at the beginning of the insect farming in Finland [53].

1.6.8 Hopes Up for the Pet Food

In 2021, papers concentrated much on the importance the insect farming in the production of good feed alternatives for our home pets, of which we have many: cats, 600,000, and dogs, 700,000. The value of pet feed sales is estimated at 400 million Euros/year in Finland [54, 55].

1.6.9 Cruel Internet Attack

Already beaten should not be beaten again, but some insect food opponents did that on the internet in April 2023 against the Fazer and other insect food companies. These unknown attackers falsified the Fazer information on the cricket bread, as it would have been something new although the bread was not made after 2018. Facebook attack stated that insect bread has chitin, parasites, and cyanide, making the product unfit for human consumption [56]. Even some EU sceptic politicians were said to spread this false news. Fazer and the Finnish Food Authority reacted immediately, making it clear that no cricket bread was in the supermarkets, but even if there was, it would be perfectly safe to eat. Insects cannot carry cyanide as it is equally mortal to them as it is to humans. Chitin is a normal food fiber, which would not cause any harm to the food. Correctly produced food does not contain parasites of any kind. Corrected news was widely published, but as we know, the false information will also remain on the internet [56].

A large part of the population still finds the insect food too exotic, but the producers and salespeople hope that the next generation will be a real insect eating generation. Similarly, supermarket chains, S-group & K-markets, will keep some of the insect food available hoping that the sales will eventually pick up. When that will happen is still unknown [33, 46, etc.].

1.6.10 Was COVID-19 the Last Nail in the Insect Boom Coffer?

The outbreak of COVID-19 changed everyone's lifestyle with people staying at home and practicing social distancing. The lockdown caused a significant decrease in the promotion of insect food, as it was illegal to organize any restaurant taste or public relations (PR) gatherings. The popularity of the insect food went down clearly as did the demand and availability of the products because many producers decided to give up the production [57]. Findings also revealed that consumers who contracted COVID-19 strictly followed the regulations during the confinement. Although the Finnish Food Authority made it very clear that not a single case of COVID-19 had any relation with food, people were more likely to reject the consumption of insect-based products [58]. A similar reaction was noted in Spain where a more detailed study was conducted to determine the effect of the COVID-19 lockdown on consumers' expected acceptance of insect-based products [59].

1.6.11 Academic Research Gives Some New Hope

In **2018**, an interesting MSc thesis was presented at the University of Eastern Finland (hereafter UEF) after interviewing students on their insect food relation [60]. It was clear that students were interested in eating insects and saw a future for it in Finland. Most interviewed had never tasted the insect food, but most were eager to try it. Students felt that the university lunch restaurant could be the place to introduce insect food, especially if the insects would be produced at the university by using the food waste from the restaurant. In the postscript of that article, it was stated that 2/3 of Finnish people have a positive attitude toward insect food. The price of crickets has gone down in one year by some 60% being now equal to that of whitefish *Coregonus lavaretus* price [61].

In **2019**, the biology of the long-horned grasshopper *Ruspolia differens* was studied in PhD thesis at the UEF [62] to establish the basis for a mass production method for it, as the information on the basic biological parameters for rearing this species is still largely missing. It is an economically important species in East Africa where it has existing markets. It is rich in protein, essential fatty acids, vitamins, and minerals. The consumption is currently based on wild harvest during two unpredictable annual swarming seasons. The intensification of harvesting, together with a decrease in the cover of grasslands due to urbanization, land use changes, and changing precipitation patterns, might cause the overexploitation of wild populations in the future. Farming of the grasshopper could help to improve food security in this region, create income for locals, and help preserve the wild populations in the long term. For insect production to be economical, insects should be reared at high densities with low mortality and a short development time, with feeds that maximize performance [62].

Furthermore, to maintain the cultivation throughout many generations, the insects should breed and lay eggs, and larvae should be able to hatch effectively. In this study, the optimal temperature range for development, growth, and survival was experimentally explored at a temperature range of 18–32°C, and the effect of rearing density on young nymphs was examined in three different rearing environments. The potential of manipulating the fatty acid content and composition of *R. differens* was explored in a short-term feeding experiment with seven artificial feeds. Finally, different egg-laying media were compared to find an efficient and economical material that allows high hatchability [62].

The results show that this species can be reared from egg to adult in a relatively short time, on artificial diets, and at relatively high densities. It was also found a promising egg-laying medium for large-scale production. The most suitable rearing temperature for *R. differens* is 28–30°C. The development rate and survival were maximized at 30°C, whereas the highest weight gain was observed at 28°C. The development rate slows down at temperatures below the optimum, so temperature could be used to synchronize the development of different batches of nymphs. The weight of insects can also be modified with rearing duration after the final molt. Up to a 50% increase in weight could be achieved, and this weight gain is stored as fat by rearing the insect for up to a week after the final molting. Young grasshopper nymphs can be reared at densities ≤ 36 nymphs/liter with a 60% survival up to 28 days of rearing. Highly variable feed ingredients can be used, and the results confirm that dietary fatty acids significantly alter the fatty acid composition and content of the insects. Polyunsaturated fatty acids accumulate in the fat body. Linseed can be used to lower the omega-6/omega-3 (n-6/n-3) ratio of the insect, highlighting the potential to enhance the nutritional value of (at least some) edible insects to make them more suitable for humans. Even a slight modification to the feed can drastically alter the whole fatty acid composition of the insect. Finally, the most promising egg-laying medium with high hatchability was found to be a folded plastic cloth with a short fleece cover on one side [62].

It remains open to how important part grasshopper farming can have in the East African food security, but there are vast markets for this species. The current wild-harvesting methods might not be sustainable in the long term and can only meet the demand seasonally with highly unpredictable yields. Further developing sustainable technologies that enable the production of insects at small-scale farms or on an industrial scale is one focal area of edible insect research [62].

In **2019**, the Eastern Finland University team also studied the insect consumption attitudes in Joensuu because consumption of foods of insect origin is encouraged since insect consumption is seen as one of the responses to the environmental impact of meat production [63]. However, it has been noted that consumers' willingness to eat foods of insect origin is weak and varies between different insects. The negative attitude is explained by Western food culture traditions, in which insects have not been classified as a source of food. Instead, the majority of consumers regard them as rather disgusting and inappropriate for human nutrition. Respondents ($n = 567$) were mostly women, highly educated, and city dwellers, a demographic profile known to impact food choice. However, women are generally less eager compared to men to consume foods of insect origin [63].

This study examined the attitude, subjective norm, perceived behavioral control, and food neophobia toward the consumption of foods of insect origin, as well as the conditions for eating insect-based foods among vegans, vegetarians, and omnivores. The data were obtained by using an online survey with the respondents of whom omnivores represented 74%, vegans 5%, and nonvegan vegetarians 22%. These three dietary groups exhibited significantly different intentions to eat foods of insect origin. Vegans held the most rigid negative attitude, and their subjective norm to eat insects was weaker compared to that of omnivores and nonvegan vegetarians. Vegans' perceived behavioral control over their eating of insects was stronger compared to that of omnivores and nonvegan vegetarians, and they were more neophobic than omnivores and nonvegan vegetarians. Nonvegan vegetarians held the most positive attitude toward eating insects, and both nonvegan vegetarians and omnivores thought that insect consumption is wise and offers a solution to the world's nutrition problems. In contrast, vegans regarded insect consumption as immoral and irresponsible. The study concluded that vegans' weak intention, negative attitude, and low willingness to eat insects in the future exhibit their different dietarian identity compared to that of omnivores and nonvegan vegetarians [63].

In **2020**, a good Bachelor's thesis was published on the insect-based food market at the Tampere University of Applied Science [64]. The study surveyed the reasons why consumers used or did not use groceries including insects. An online survey consisted of 271 people with the age varying from 14 to 77 years. The most important reason for not buying the offered groceries was that insects were considered unappealing. Price, nutrition values, and availability were also important for the consumers who hoped to see insect products that could replace common meat products. The study recommended to include recipes in the packaging since insects have not been a traditional part of the Finnish cuisine. Thus, the insect cooking methods have not been transferred from one generation to the next. Despite slow down of the early insect boom, the author believes that the Finnish insect market could be revived by offering the customers science-based information on insect food through PR events [64].

In *2021*, an interesting PhD thesis of the Eastern Finland University (EFU) examined the potential of food industry by-products as feed ingredients for three edible insect species [65]. Three species studied were house cricket *Acheta domesticus*, two-spotted cricket *Gryllus bimaculatus*, and long-horned grasshopper *Ruspolia differens*. All are economically valuable as human food, and crickets are already widely reared for human consumption. Long-horned grasshopper is mainly harvested from nature in Eastern Africa, but mass-rearing technologies are currently being developed. The study goal was to develop artificial diets by reusing materials from the food industry that are not consumed by humans. Specific study objectives were first to investigate how long-horned grasshopper grows and develop on by-product diets and to determine if the diets can alter the fatty acid content and composition of that insect; second to identify the protein-level requirements for the feed and to investigate how house cricket and two-spotted cricket grow and develop on diets including food industry by-products; third to examine the effect of by-product diets on the fatty acid composition and content of these crickets; and fourth to evaluate how the artificial diets for crickets with the most potential perform under small-scale farming conditions [65].

The by-products were selected among those readily available from the Finnish food industry including potato protein, barley mash (brewer's spent grain), barley feed from the ethanol industry, compressed leftover turnip rape, a mix of broad bean and pea, and a mix of potato, carrot, and apple. The fatty acid composition and content of the produced insects were analyzed, as well as the glycoalkaloid content of insects feeding on potato protein. Finally, the feasibility of the most suitable by-product feeds was examined in a small-scale cricket farm using current rearing technology [65]

The studied insect species accepted several food industry by-products in their feed and could successfully grow and develop when fed on by-product diets. For long-horned grasshoppers, the suitable by-product diets included potato protein, barley feed, and barley mash, which resulted in a good survival rate, large individual weight, fast development, and the highest feed conversion efficiency. Diets containing 17–22% protein significantly improved the growth performance of grasshoppers. Furthermore, diets containing barley increased their polyunsaturated fatty acid content, these being the essential fatty acids for humans. For two cricket species, several suitable by-product feeds were identified, for example, compressed leftovers of turnip rape from rape seed oil production, barley mash from the beer industry, and barley feed from the ethanol industry. The crickets showed high growth performance on these diets when the protein content of the diet ranged between 22.5% and 30.5%. The total fatty acid content of crickets can be modified by the diet. The small amounts of glycoalkaloids in potato protein by-products did not affect the growth performance of the studied insects, and both crickets and grasshoppers feeding on potato by-products were safe for humans to eat [65].

The study concluded that insect rearing creates good opportunities for a circular economy. It is viable to use food and agricultural by-products as insect feeds because many edible insect species can utilize a wide variety of plant-based materials. Developing insect feeds from food industry by-products could be a way to reuse, recycle, and valorize materials that are not consumed by humans. However, the bare use of by-products, unprocessed or low-quality materials, is not sufficient to maximize rearing yields or to guarantee insect health and proper development. Insect feeds need to be designed specifically for each species to meet their nutritional needs, to ensure successful growth and development, and to ensure they are nutritious and safe for human consumption [65].

1.7 DISCUSSION AND CONCLUSION

Insects have attracted the attention of mankind since ancient times, and countries in Africa, Asia, and South America have a long history of using insects in their cuisine [16]. My interest in insect food started when living some 22 years in Africa, east, south, and west starting in 1977 in Kenya and ending in 2004 in Nigeria. When living in Malawi (1993–1997), I got even more involved in the use of insects in the diet of local people. In Malawi, population growth outstrips food production, especially during times of drought. According to a small survey [66], the use of wild foods is often the only coping mechanism during malnutrition and famine. Wild fruits, mushrooms, nuts, seeds, and leaves were eaten by 83–97% of the interviewed families (110 in total). Insect food was almost as popular: larvae/caterpillars 68%, grasshoppers/locusts 46%, flying ants/termites 86%, lake flies *Chaoborus edulis* 17%, crickets/cicadas 58%, and other insects 58%. Interestingly, this insect-eating is not shown in the official food statistics of the country, which state that almost 80% of protein intake in the poorest households in Malawi is from cereal crops, predominantly maize [67]. Does it mean that insect-eating is declining in Malawi as is happening in many traditional markets? In many regions that have traditionally eaten insects, the practice is declining due to globalization, and their consumption has decreased over the last decade as agriculture and living standards change, and the availability of wild-caught insects has decreased [68].

It is not agreed in the literature how many insect species are eaten in the world, with the number varying from 1,500 to 3,650 [9, 10], but one thing is sure: we are not farming even a fraction of potential species. One good example is the shea caterpillar *Cirina butyrospermi*, which is an important wild collected insect, highly valued as a human food item in West Africa [12]. Caterpillars can be dried, fried, or boiled and used in various meals. The insect has exceptional nutritional characteristics, with 63% protein and 15% fat, as well as vitamins and minerals [69]. It also has potential as an ingredient in fish feed [70] and even as a bioconverter [71]. Despite its importance as food or feed insect, thus far no attempts have been made for farming production. The season of the moths occurs once per year in West Africa, and the waiting period for fresh caterpillars is 11 months. This new study suggests the possibility of the continuous rearing and supply of fresh shea caterpillars at any point in time in the year [12]. The dipping technique on 20-hydroxyecdysone solution seems to be effective both for larvae and pupae with the same emergence rates. The technique used had no visible effect on the fecundity of the moth. These results open up new perspectives in the insect food sector and for entrepreneurship in the region, if not globally.

Although human insectivory is an ancient practice and 80% of the world's population consumes insects, it is relatively uncommon in contemporary Western culture. In the Western world, insects are largely unfamiliar and mostly viewed as holiday novelty or "yuk;" thus, awareness of local taboos, cultural preferences, and the population's exposure to insects as food are crucial for the successful promotion of insect farming for food. In many urban and developed populations, a central issue is food neophobia, but after taking the first step in trying an insect, continued exposure correlates with increased acceptance. Processed insect products such as cookies, snack bars, or powders further normalize the protein source. Conventional meat has a special status in society, both culturally and structurally in meals, so a sustainable culinary culture must be promoted to associate insect protein with pleasurable food [72]. Nutritional arguments are not thought to be enough to overcome the "disgust factor" and convert Westerners to insect-based dishes [23]. This just shows how sensitive issues with any new food items can be in diverse cultures.

In Finland, several of these barriers and concerns were known but not noted before the boom started. The first forgotten rule was that the consumers must want to buy the insect food [73] and that a good nutritional value will not help if insect products do not please the consumer [74]. Consumers should have been much more involved in product development [64]. Naturally, the very high consumer price of any insect products may partly explain the marketing problems in Finland [35–37]. The whole point of the ecological and environmental value of insect farming was lost in Finland when the insects were fed with quality beans, legumes, and grain. No wonder the cricket bread became far too expensive for the customers to appreciate and buy after the curiosity trials. Luckily, academic research has now demonstrated new concepts in promoting the wider acceptance of insect eating and how industrial waste products can safely be used for future insect farming. Surely, future research will provide necessary advice and standards in appropriate production methods and for marketing hygienic insect products.

It is interesting that the World Resources Institute (WRI) in their "How to Sustainably Feed 10 Billion People by 2050" does not even mention insect farming in any of their 21 charts [75]. Let us hope that WRI is wrong, and insects will feed the people and production animals due to their nutritive and medicinal value in the future more than today.

I would like to share the Chinese belief that edible insects can be converted into a steady stream of elegant food dedicated to all humanity.

REFERENCES

[1] UN Desa. World population projected to reach 9.8 billion in 2050, and 11.2 billion in 2100. https://www.un.org/en/desa/world-population-projected. Accessed 7 April 2024.

[2] Vantomme P. World population growth and the search for new protein alternatives. Conference Paper. 2015. https://www.researchgate.net/publication/282753885

[3] Mikkola H (Ed.). *Future foods.* InTech Open Access, University Campus, Rijeka, Croatia; 2017. http://dx.doi.org/10.5772/65132
[4] José M, Murosa S, Barroso FG, et al. Insect meal as renewable source of food for animal feeding: A review. Journal of Cleaner Production. 2013;4:1–43.
[5] Defoliart GR. Insects as food: Why the western attitude is important. Annual Review of Entomology. 1999;44:21–50.
[6] Zhou ZH, Yang W, Xu DP, et al. Analysis of resource value of edible insect. Science and Technology of Food Industry. 2014;17:390–394. (In Chinese)
[7] Bodenheimer FS. Insects *as human food.* Junk Publishers, The Hague, Netherlands; 1951.
[8] Plinius C Secundus. The Historie of the world: Commonly called the natural historie. Originally 77 AD; London edition printed 1635.
[9] Yin W, Liu J, Liu H, Lv B. *Nutritional value, food ingredients, chemical and species composition of edible insects in China.* IntechOpen, Rijeka, Croatia; 2016. https://doi.org/10.5772/intechopen.70085
[10] Dickie F, Miyamoto M, Collins CM. *The potential of insect farming to increase food security.* IntechOpen, London; 2020. https://doi.org/10.5772/intechopen.88106
[11] Van Huis A. Cultural significance of Lepidoptera in sub-Saharan Africa. Journal of Ethnobiology and Ethnomedicine. 2019;15:26. https://doi.org/10.1186/s13002-019-0306-3
[12] Bama HB, Dabire RA, Ouattara D, Niassy S, Ba MN, Dakouo D. Diapause disruption in Cirina butyrospermi Vuillet (Lepidoptera: Attacidae), the shea caterpillar, in Burkina Faso. Journal of Insects as Food and Feed. 2018;4(4):239–245. https://doi.org/10.3920/JIFF2017.0068
[13] Félix M. *Edible insects diversity and their importance in Cameroon.* IntechOpen, London; 2020. https://doi.org/10.5772/intechopen.88109
[14] Kekeunou S, Weise S, Messi J, Tamò M. Farmer's perception of the importance of variegated grasshoppers (Zonocerus variegatus (L.)) in the agricultural production system of the humid forest zone of Southern Cameroon. Journal of Ethnobiology and Ethnomedicine. 2006;2:17. https://doi.org/10.1186/1746-4269-2-17
[15] Cartay R, Dimitrov V, Feldman M. *An insect bad for agriculture but good for human consumption: The case of Rhynchophorus palmarum: A social science perspective.* IntechOpen, London; 2020. https://doi.org/10.5772/intechopen.87165
[16] Jideani AIO, Netshiheni RK. *Selected edible insects and their products in traditional medicine, food and pharmaceutical industries in Africa: Utilisation and prospects.* IntechOpen, Rijeka, Croatia; 2016. https://doi.org/10.5772/intechopen.68330
[17] de Figueirêdo RECR, Vasconcellos A, Policarpo IS, Alves RRN. Edible and medicinal termites: A global overview. Journal of Ethnobiology and Ethnomedicine. 2015;11:29. https://doi.org/10.1186/s13002-015-0016-4
[18] Rougé P, Barre A. *Allergy to edible insects: A computational identification of the IgE-binding cross-reacting allergen repertoire of edible insects.* IntechOpen, Rijeka, Croatia; 2016. https://doi.org/10.5772/intechopen.68124
[19] Fraqueza MJ, Patarata L. *Constraints of HACCP application on edible insect for food and feed.* IntechOpen, Rijeka, Croatia; 2016. https://doi.org/10.5772/intechopen.69300
[20] Hardouin J. Production d'insectes à des fins économiques ou alimentaires: Mini-élevage et BEDIM. Notes Fauniques de Gembloux. 2003;50:15–25.
[21] Iannacone J, Alvariño L. Diversidad de la artropofauna terrestre en la Reserva Nacional de Junín, Perú. Ecología Aplicada. 2006;5(1–2):171–174.
[22] Deroy O, Reade B, Spencer C. The insectivore's dilemma, and how to take West out of it. Food Quality and Preference. 2015;44:44–55. https://doi.org/10.1016/j.foodqual.2015.02.007
[23] Tarvonen H-M. Insects creeping to the plates. Helsingin Sanomat. 30 June 2016.
[24] Kurki E. Mealworms should have a legal way to the food table. Karjalainen. 5 October 2016.
[25] Moilanen K. Cricket beef to the food table. Helsingin Sanomat. 4 February 2017.
[26] Kurki E. Ackwardness is only a vision error: North Karelia is ready to start the production of insect food. Karjalainen. 16 February 2017.
[27] Rouvinen M. Tomorrow's dinner – many Finns would buy edible insects already. Karjalainen. 5 March 2017.

[28] Savolainen S. Would you like to eat crickets? Apu. 16 March 2017.
[29] Merimaa J. Food security from insects. Helsingin Sanomat. 8 May 2017.
[30] Kojo H. Insect food: Crickets are already chirping. Karjalainen. 17 August 2017.
[31] Lehtinen T. Now insects will be made into food products. Helsingin Sanomat. 30 September 2017.
[32] Nieminen K. Insects can now be sold as food products. Karjalainen. 10 November 2017.
[33] Mikkola H. Introductory chapter: Is the insect food boom over or when it will start? In: Mikkola H (Ed.), *Edible insects*. InTech Open Access, London; 2020, pp. 1–4. https://doi.org/10.5772/intechopen.77835
[34] Salminen J. Crickets feel in the mouth like the Finnish rye crispbread. Helsingin Sanomat. 30 September 2017.
[35] K-Ruoka. Fazer brings the world first cricket bread in the K-supermarkets. K-Ruoka. https://www.k-ruoka.fi/artikkelit/k-kaupassa/fazer-tuo-maailman-ensimmäisen . . . Accessed 8 April 2024.
[36] Mustonen R. Fazer brings cricket flour from Holland – the price per kilo is almost 100 Euros. Maaseudun Tulevaisuus. 23 November 2017.
[37] Baraka J. We tasted the new Fazer cricket bread – but where are the crickets? Iltasanomat. 23 November 2017.
[38] Leader Foods Oy. Crickets conquered the protein bars. Apu. 2017; Issue 6.
[39] Oy Halva Ab. Halva. https://www.halva.fi. Accessed 17 September 2019.
[40] Reen A. Finsect. Helsingin Sanomat. 30 January 2019.
[41] Fazergroup. Fazer cricket bread won bronze in the Cannes Lions competition. Fazergroup. https://www.fazergroup.com/fi/medialle/uutiset/fazer-sirkkaleipa-voitti . . . Accessed 8 April 2024.
[42] Ala-Siurua M. Fazer cricket bread popped up to the podium. Maaseudun Tulevaisuus. 23 June 2018.
[43] Koistinen A. Natural resources institute and Eastern Finland University developing feed for the insects. Karjalainen. 10 January 2018.
[44] Liiten M. Growing insects for food ended before starting. Helsingin Sanomat. 30 January 2019.
[45] Latvala J. Time was not ripe for the insect food, yet. Karjalainen. 23 December 2018.
[46] Tapio K. Large part of the population finds that insect food products are too exotic, yet. Karjalainen. 23 December 2018.
[47] Paljakka E. Sensation bread came with a rush to food stores – but insects did not become mainstream. Iltalehti. 22 January 2019.
[48] Olli T. Cricket bread flopped – now became fried snack cricket with garlic. Iltalehti. 22 March 2019.
[49] Massinen T. Crickets for hunger: Insects scurrying to the plate/insect boom is only starting. Karjalainen. 4 April 2019.
[50] Lehtinen T, Taskila S. New sports drink with insect flour addition. Helsingin Sanomat. 4 December 2019.
[51] Laitinen J. Fast end to cricket chirping. Karjalainen. 1 December 2019.
[52] Alanne J. Start and end of the cricket bread. Kotimaa. 24 January 2021.
[53] Alanne J. Where the edible crickets disappeared. Tekniikka & Talous. 31 January 2021.
[54] Alvar Pet. Black fly larvae to dogs. Apu. 4 February 2021.
[55] Brandt L. Pets eat insect food, and that is now produced in Finland. Verde. 11 August 2021.
[56] Hollingsworth A. Social media spreading disinformation on cricket bread by using five-year-old ad material. AFP Faktantarkistus. 24 April 2033.
[57] Hakala I, Tuominen V. *House cricket popularity growth and its future*. AMK-opinnäytetyö, Haaga-Helia; 2022.
[58] Finnish Food Authority. Often asked questions on COVID-19 and food safety to customers. Ruokavirasto. 16 April 2020.

[59] Khalil R, Kallas Z, Haddarah A, El Omar F, Pujolà M. Impact of COVID-19 pandemic on willingness to consume insect-based food products in Catalonia. Foods. 2021;10:805. https://doi.org/10.3390/foods10040805
[60] Pekkinen J. Insect food short-living boom or part of our future? Versus. 22 January 2019.
[61] Aaltio J. Insect food will hit through in the 2020-decade. Versus. 22 January 2019.
[62] Lehtovaara VJ. *Mass production of the edible long-horned grasshopper Ruspolia differens (Orthoptera: Tettigoniidae) – investigation of the basic parameters of rearing*. Publications of the UEF Dissertations in Forestry and Natural Sciences No. 350. University of Eastern Finland, Joensuu; 2019.
[63] Elorinne A-L, Niva M, Vartiainen O, Väisänen P. Insect consumption attitudes among vegans, non-vegan vegetarians, and omnivores. Nutrients. 2019;11:292. https://doi.org/10.3390/nu11020292
[64] Heina J. *The launch of insect-based foods to the Finnish market and changes in the market*. Bachelor's Thesis. Tampere University of Applied Sciences, Tampere; 2020.
[65] Sorjonen JM. *Could edible insects be one solution for a circular economy and food production? Designing feeds for insects using food industry by-products*. Publications of the UEF Dissertations in Forestry and Natural Sciences No. 444. University of Eastern Finland, Joensuu; 2021.
[66] Mikkola H. The use of wild foods in Malawi. The Society of Malawi – Historical and Scientific. 1997;50(2):40–53.
[67] Muleya M, Tang K, Broadley MR, Salter AM, Joy EJM. Limited supply of protein and lysine is prevalent among the poorest households in Malawi and exacerbated by low protein quality. Nutrients. 2022;14:1–12. https://doi.org/10.3390/nu14122430
[68] Commander NT, Anankware PJ, Royal OO, Obeng-Ofori D. *Econometrics of domestication of the African palm weevil (Rhynchophorus phoenicis F.) production as small-scale business in Ghana*. InTech Open Access, London; 2020. https://doi.org/10.5772/intechopen.90259
[69] Anvo MPM, Toguyéni A, Otchoumou AK, Zoungrana-Kaboré CY, Paul KE. Nutritional qualities of edible caterpillars *Cirina butyrospermi* in southwestern Burkina Faso. International Journal of Innovation and Applied Studies. 2016;18:639–645.
[70] Anvo MPM, Toguyéni A, Otchoumou AK, Zoungrana-Kaboré CY, Paul KE. Evaluation of *Cirina butyrospermi* caterpillar meal as an alternative protein source in *Clarias gariepinus* (Burchell, 1822) larvae feeding. International Journal of Fisheries and Aquatic Studies. 2016;4:88–94.
[71] Coulibaly K, Gnissien M, Yaméogo TJ, Traoré M, Nacro BHJ. Perception des producteurs sur l'utilisation des déjections de chenilles dans la gestion de la fertilité des sols dans la région des Hauts-Bassins au Burkina Faso. Journal of Applied Biosciences. 2016;108:10531–10542.
[72] McDade H, Collins CM. *How might we overcome 'western' resistance to eating insects?* InTech Open Access, London; 2020. https://doi.org/10.5772/intechopen.88245
[73] Piha S. Insect food and consumer psychology. Tieteessä tapahtuu. 2017;4.
[74] Mutanen M, Tuorila H. Tomorrow we will eat insect food – or do we eat? Good nutritional value will not help if insect products do not please the consumer. Duodecim. 2016;132(33):1204–1207. (In Finnish)
[75] Ranganathan J, Waite R, Searchinger T, Hanson C. *How to sustainably feed 10 billion people by 2050, in 21 charts*. World Resources Institute; 2018. https://www.wri.org/insights/how-sustainably-feed-10-billion-people.
[76] Rumpold BA, Schlüter OK. Potential and challenges of insects as an innovative source for food and feed production. Innovative Food Science and Emerging Technologies. 2013;17:1–11. https://doi.org/10.1016/j.ifset.2012.11.005

Insects of the World: Classification of the Main Edible Insects

2

Rubén Agregán, Noemí Echegaray, Eva María Santos López, Jose Angel Perez-Alvarez, Rubén Domínguez-Valencia, Mirian Pateiro, and José Manuel Lorenzo

2.1 INTRODUCTION

Insects are ubiquitous animals that are easy to find everywhere with their activities constantly colliding with those of humans. In this sense, insects have been the reason of multiple nuisances, causing agricultural losses and spoiling stored products, and are also carriers of diseases such as malaria. However, at the same time, they can provide pollination, control pests, and, of course, serve as food [1]. At this point, insects are an important food source for many societies around the world. The habit of eating insects probably dates back to prehistoric times, as suggested by fossilized feces of ancient humans and cave paintings [2]. The truth is that there is a custom of eating insects in certain populations, forming part of their gastronomic cultural heritage. In contrast, in Western societies, there is a widespread rejection to eat insects.

DOI: 10.1201/9781003488156-2

On the other hand, due to the current environmental context, with less land for agriculture, overexploitation of fishing practices, and adverse climate factors, all derived from the existing demographic pressure and the consequent increase in demand, the inclusion of insects in the human diet, including that of Western countries, as an alternative to other conventional protein sources such as meat, is gaining momentum. This way is expected to alleviate pressure on natural resources, reduce food insecurity, and improve health [3]. Therefore, edible insects seem good candidates for replacing livestock protein. The content of this macronutrient in these small invertebrate animals may range between 40% and 75%, exceeding that of farm animals (12–34.5%) and plants (7–50%). On the other hand, the body of insects is almost entirely edible (around 80%), while that of livestock has a smaller proportion of edible parts (e.g., 55% in chicken and pork and 40% in cattle). In addition, edible insects also apport other essential nutrients, as well as important compounds linked to a well body functioning, such as antimicrobials and enzymes. The presence of the carbohydrate chitin also ensures a valuable source of fiber [4].

More than 1,900 insect species have been classified as edible and are part of the traditional diet of at least two billion people worldwide, representing around 30% of the world's population, distributed in more than 100 countries [4]. Given the great diversity and number of insects considered edible by many societies and the growing interest that this topic arouses, the objective of this chapter was to compile the main orders of edible insects, describing both morphological and nutritional differences, as well as presenting the most representative species within each group and their worldwide distribution.

2.2 DEFINITION OF INSECT AND TAXONOMY

When it comes to insects, people may have preconceived ideas on what and how these living beings are like. For instance, many terrestrial arthropods are popularly considered insects, such as spiders and scorpions (class Arachnida), centipedes (class Chilopoda), millipedes (class Diplopoda), and woodlice (class Crustacea). According to this, with the name "insect," we can refer to any terrestrial or freshwater arthropod other than crustaceans (e.g., crabs, shrimps, and lobsters). However, in the strict sense of the word and according to a zoological point of view, insects are defined as six-legged arthropods with antennae, three pairs of legs, and the triple tagmatization, that is, head, thorax, and abdomen [5]. All these parts are represented in the Figure 2.1.

The number of insects spread across the planet is truly enormous. The latest estimation puts the number of existing species at 5.5 million, many of them still undescribed. Despite this massive figure, there are still many unidentified species, since insects are the most difficult animals to classify taxonomically. This is related to the huge variety and diversity of insects found in tropical regions of the world, especially in rainforests, due to the intervention of different factors, such as the miscellaneous geographic area in the tropics, the greater speciation rates in this part of the world compared to others (caused by the high energy involved), the low climatic variation, the high evolution speed influenced by heat, the increased pressure of pathogens and

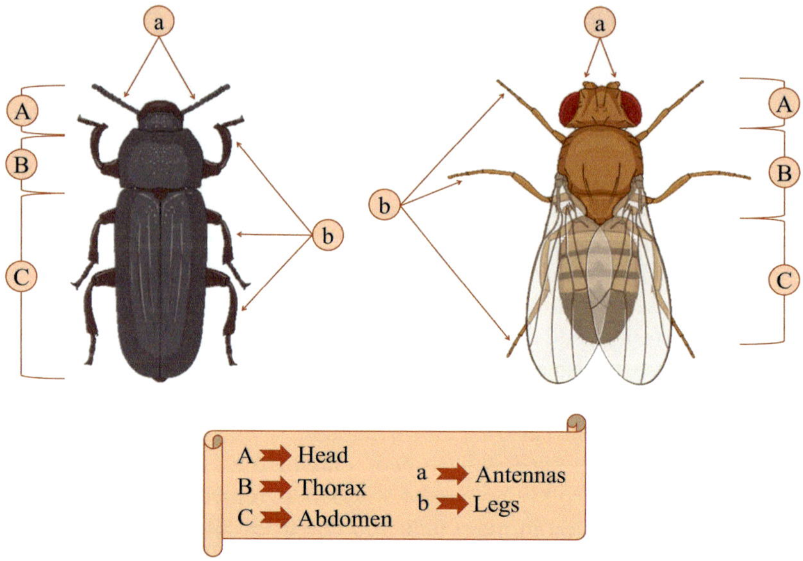

FIGURE 2.1 Different parts of an insect's body.

predators, which promotes higher biotic filtration and speciation, and finally the geometric constraints occurred when the distribution areas of the species overlap [1].

Insects can be grouped according to the ecosystem to which they belong. Thus, they are divided into terrestrial and aquatic organisms. Terrestrial insects have a more developed tracheal system and stronger exoskeletons. In contrast, insects adapted to aquatic environments have developed swimming hairs on their legs and excrete a more diluted substance. For a more in-depth classification, taxonomic studies pay special attention to features, such as general structure, body symmetry, proportion of the head compared to the body, and posture. In addition, the relationship between different body lengths and skeletal structure is especially useful. In this way, insects can be classified into different orders. Some of the largest ones described are as follows: Protura (e.g., coneheads), Collembola (e.g., springtails), Plecoptera (e.g., stoneflies), Orthoptera (e.g., grasshoppers and crickets), Dermaptera (e.g., earwigs), Isoptera (e.g., termites), Ephemeroptera (e.g., mayflies), Hemiptera (e.g., true bugs), Homoptera (e.g., aphids, cicadas, leafhoppers, and scale insect), Odonata (e.g., dragonflies and damselflies), Coleoptera (e.g., beetles), Siphonaptera (e.g., fleas), Diptera (e.g., true flies), Lepidoptera (e.g., moths and butterflies), and Hymenoptera (e.g., ants, wasps, and bees) (Figure 2.2) [6].

Identifying and classifying insects is a hard endeavor for entomologists due to the large number of these existing animals. Traditionally, taxonomists attempt to determine the position of the collected insects in the taxonomic catalog down to the lowest possible level based on their morphological characteristics. However, this approach presents a barrier when identifying undescribed species, which can only be distinguished from other existing insects through an exhaustive analysis of their characteristics. For instance, the use of DNA-based technologies such as barcoding by sequencing certain conservative variable regions of the genome may help confirm new spaces in the case that differentiation by analysis of phenotypic features is not possible. Still, although this method facilitates the

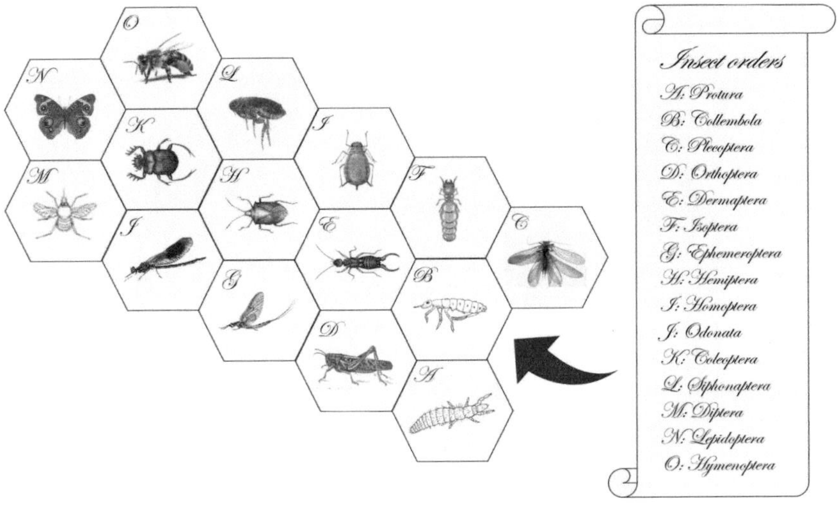

FIGURE 2.2 Different orders of insects and visual examples of each of them. Drawings taken from Cerdá [7].

discovery of new insect species, identification is occurring at a slower pace than required. At this point, artificial intelligence (AI) such as the machine learning tool can be applied to recognize and organize complex data patterns for classification and outlier detection tasks [8]. Similarly, deep learning – an advanced branch of machine learning – improves the detection and accurate identification of insects [9]. Thanks to these alternative approaches, cryptic organisms, which are groups of two or more taxa classified as a single nominal species because they have non-designable morphological characteristics, can be solved [10, 11].

The different distribution of insect species worldwide is represented in Figure 2.3.

2.3 WHAT DOES EDIBLE INSECT MEAN?

Insects have been part of the human diet as a valuable source of nutrients for centuries, a tradition that continues today with more than 2,100 insect species consumed worldwide [13]. These invertebrates are eaten by more than 3,000 different ethnic groups of people, representing approximately 30% of the world's population, in 130 countries around the world, especially in tropical areas of continents as diverse as Africa, Latin America, and Asia, where there is greater diversity and consequently a wider range of possibilities. This fact makes the use of insects in the human diet both a cultural and biogeographic event [4, 14].

In the vast world of insects, only a small amount has been reported to be consumed [4]. Why is this? First of all, it is necessary to understand the meaning of the word "edible." According to the Merriam-Webster dictionary, edible can be defined as "something that is suitable or safe to eat." Hence, the premise of safety seems to be an important point to keep in mind. In fact, the few definitions of "edible insect" currently found in the scientific literature mention the "food safety" as a differentiating element to distinguish between edible

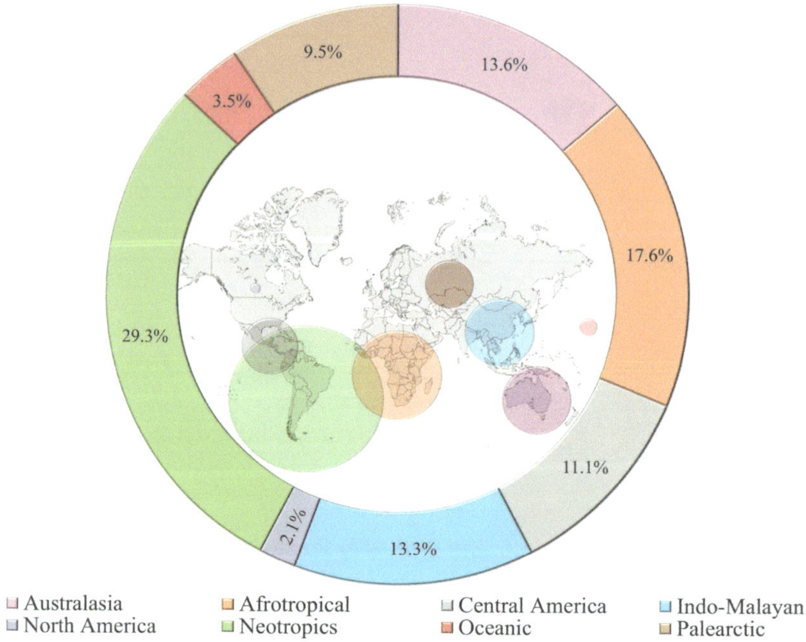

FIGURE 2.3 Estimated percentage distribution of terrestrial insect species across different biogeographic regions. Data from Stork [12].

and inedible. Grabowski et al. [5] defined "edible insect" as one that has been proven to be safe for health throughout history, even if consumed sporadically. According to this definition and considering the estimated number of insects still unknown (around 4.5 million) [1], there is a very large gap of potentially undiscovered edible insects. Itterbeeck and Pelozuelo [15] shared a similar point of view, considering safety as a key condition to stablish differences in the edibility of insects. Thus, it seems clear that an "edible insect" is any insect that can be consumed without harmful consequences for human health. Therefore, the historical tradition of consumption or simple physical observation after eating has led to the selection of a group of insects, which are already an integral part of the diet of many communities worldwide.

2.4 TAXONOMIC CLASSIFICATION OF EDIBLE INSECTS

Up to 2,205 different species of edible insects have been identified, belonging to 25 different orders. Coleoptera, with 32% of total edible insect species counted, is the most prevalent order, followed by Hymenoptera, Lepidoptera, Orthoptera, and Hemiptera/Heteroptera, among others, with 15.5%, 15.2%, 14.1%, and 11.4%, respectively. Each of these orders is represented by several families, which in turn comprise a certain number of

TABLE 2.1 Taxonomic Classification of the Most Representative Edible Insects.

ORDER	*MAIN FAMILIES*	*NUMBER OF SPECIES*	*PERCENTAGE OVER TOTAL*
Coleoptera	Scarabaeidae	258	32%
	Cerambycidae	142	
	Dytiscinae	60	
	Curculionidae	52	
	Passalidae	35	
	Lucanidae	26	
	Buprestidae	23	
Hymenoptera	Apidae	153	15.5%
	Vespidae	105	
	Formicidae	72	
Lepidoptera	Saturniidae	86	15.2%
	Hepialidae	47	
	Sphecidae	31	
	Notodontidae	24	
	Cossidae	22	
Orthoptera	Acrididae	176	14.1%
	Gryllidae	42	
	Tettigoniidae	42	
Hemiptera/ Heteroptera	Cicadidae	73	11.4%
	Pentatomidae	39	
Isoptera	Termitidae	66	3.4%
Odonata	Libellulidae	28	2.4%
	Aeschnidae	11	
	Gomphidae	7	
Diptera	Tipulidae	6	1.8%
	Oestridae	5	
	Ephydridae	5	
Dictyoptera	Mantidae	17	1.7%
	Biattidae	12	

Source: Data from Omuse et al. [4]

species. Within the main order, Coleoptera, the families Scarabaeidae and Cerambycidae are the most abundant, encompassing 258 and 142 edible species, respectively. Other massive orders, such as Hymenoptera and Lepidoptera, are represented by families with around or more than 100 species [4]. The largest orders and families of edible insects, together with the number of species they comprise, are summarized in Table 2.1.

2.4.1 Discrimination Between Edible Insects According to the Main Orders: Morphology, Habitat, and Nutritional Properties

Shared features among insects are used by entomologists to establish sets and create lines of kinship for the purpose of classifying the different types discovered and thus aiding in their identification. Within the taxonomic category of the class Insecta, insects are grouped into orders that represent lineages that, although divergent, are usually recognizable, since invertebrates in this category have a set of homologous characteristics [16].

The following paragraphs describe the characteristics of the most representative orders of edible insects.

2.4.1.1 Coleoptera

The order Coleoptera includes a very distinctive group of insects with 705 edible species described, known by the common name of beetles, which are the most described and diverse insects worldwide [4, 17]. Considering this fact, it is not surprising that it is the order with the most recorded edible species. Despite being such a large and varied group of insects (they represent the 40% of insects and 30% of all animals), there are some common morphological characteristics among these invertebrates. They possess a thick, enduring exoskeleton that covers and protects most of their bodies. A similar function is performed by their front wings, known as elytra, which covers the membranous flight wings and the abdomen, protecting the body from blows and predation. Unlike this shared morphology, the body size of beetles can be extremely variable, ranging from 0.3 to 0.4 mm to more than 17 mm [6, 17].

Bettles is a widely consumed insect group that is represented in different parts of the world. In Asia, the beetles, such as *Sitophilus oryzae* L., *Hydrophilus hastatus* L., *Rhynchophorus ferrugineus* Olivier, *H. palpalis* Brullé, *Cybister tripunctatus* Olivier, *Eretes sticticus* L., and *H. cavisternum* Bedel, can be found in at least five countries. Edible beetles, such as *R. palmarum* L. and *Macrodontia cervicornis* L., also live in South America. Africa is another prominent continent in terms of insect consumption. The order Coleoptera is also present in this region with different beetle species in the menu, including *R. phoenicis* F., *Oryctes boas* Fabricius, and *Goliathus* spp. In contrast, edible beetles and insects in general are not abundant in North America and Oceania. However, on this last continent, the edible beetles, such as *Mallodon costatus* Montrouzier and *R. bilineatus* Montrouzier, can be found [4].

Beetles are notable for their high fiber and protein content. Fat also occupies a significant fraction of the insect's body. On the other hand, values for macronutrients can remarkably vary depending on species. For this reason, it is not weird to find more than 60% of variability for protein, with values ranging from below 20% to around 70% of edible portion. Fat may comprise values between 6.1% and 51.6% as the second largest component in coleopterans. Fiber and total carbohydrates account for 1–11.8% and 3.6–51.6% of total insect, respectively. These figures show a very complete macronutrient profile. In this sense, the contribution of beetles in amino acids is decisive, since they provide essential compounds, such as lysine – a non-synthesized amino acid in the human body – and arginine – pivotal amino acid for growth in children –. However, their content is also highly related to the species. The consumption of edible coleopterans may also favor the inclusion of beneficial fatty acids in the diet, such as oleic (C18:1n–9) and linoleic (C18:2n–6) acids (5.4–66.4% and 7.8–51.2% of total fatty acids, respectively). Both the amount of fat and the fatty acid composition can be affected by sex, diet, species, life stage, and environmental temperature [18].

Regarding micronutrients, the order Coleoptera is a rich source of minerals, such as calcium, magnesium, phosphorous, potassium, and sodium. Positively, 100 g of coleopterans cannot reach the value of 1,500 mg of sodium, which is the maximum daily intake for this mineral. Iron also stands out as an important mineral compound in the coleopteran matrix, ranging from 1 to around 30 mg/100 g. According to the maximum daily intake for iron (7.5–58.5 mg/day), the consumption of 100–150 g of coleopteran insects could meet the requirement of this mineral. Vitamins are also valuable components in the beetles. Several compounds, such as vitamins A (8.5–12.4 mg/100 g), C (1.2–14.9 mg/100 g), E (0.5–18.8 mg/100 g), and some of group B (e.g., B1, B2, and B6), can be found. As with macronutrients, the richness in minerals and vitamins is closely related to the species [18].

The existence of anti–nutrients, such as oxalates, phytates, and tannins, in the insect's body might reduce the digestibility of some nutrients, including minerals and proteins. The order Coleoptera is not an exception to these compounds, and their presence should be examined.

2.4.1.2 Hymenoptera

Up to 341 edible insect species belonging to this order have been described [4]. They are the only ones that have a stinger and are capable of stinging [6]. Among the most recognized insects of this order are bees, wasps, and hornets. Ants are also very representative of this group. The order Hymenoptera comprises a broader diversity of insects, covering a large number of species. Some examples of hymenopterans include paper wasps, potter wasps, yellow jackets, bumble bees, velvet ants, wood wasps, and horntails [19].

Hymenoptera insects are classified into the suborders Symphyta and Apocrita. The first comprises the most primitive hymenopterans and represents almost 5% of the insects found in this order. Symphyta encompasses insects, such as sawflies, wood wasps, and horntails. Most symphytans have a unique caterpillar-like larva that feeds on leaves, shoots, and cones. On the other hand, the suborder Apocrita includes about 96% of all hymenopterans, which in turn are divided into Aculeata (stinging hymenopterans), including ants, social wasps, and bees, and Parasitica, a large and miscellaneous group of commonly small, inconspicuous insects, most of which parasitize other insects and spiders [19].

The enormous diversity of hymenopterans hinders the recognition of some species, since the body morphology (e.g., mandibles) is significantly different. Body color covers the entire visible spectrum with colorful combinations and patterns. Size is also variable, ranging from 0.1 to 75 mm in length and reaching up to 20 cm in females of Megarhyssa species if the extended ovipositor and antennae are included [19].

The order Hymenoptera is widely distributed worldwide. Edible hymenopterans are found in several continents, such as Asia, including the species *Apis dorsata* F., *A. mellifera*, and *Xylocopa* sp., and South America, with species, such as *Atta cephalotes* L., *Trigona clavipes* F., *Cephalotrigona capitata* Smith, *Tetragonisca angustula* Latreille, and *A. sexdens* L. Africa also exhibits an important number of edible hymenopterans, including *A. mellifera* L., *Carebara vidua* Smith, *C. lignata* Westwood, *Hypotrigona gribodoi* Magretti, and *Meliponula bocandei* Spinola. In the Oceania region, the hymenopteran *Oecophylla smaragdina* F. is present in at least two countries [4].

Hymenopterans such as bees were reported to be a good food source. Proximate composition of worker honeybees (*A. mellifera adansonii*) from several ecological areas of Nigeria showed a high-water content (76.4–79.9%) and approximately the same protein and carbohydrate content (7.4–8.7% and 6.7–8.2% wet basis, respectively), representing both together more than 50% of the dry matter. In contrast, the fat content is lower than 5% on a wet basis. The fiber content is also low (around 2.2% wet basis). To meet the estimated daily protein requirement (23–56 g), an individual should consume between 400 and 800 g of honeybees. Minerals, such as sodium, potassium, magnesium, calcium, chlorine, and phosphorous, are present in these insects. Potassium was found in the highest concentration of all the minerals mentioned. This finding is related to the higher presence of potassium and sodium in plants, since honeybees are phytophagous insects, which means that they feed on plants. The nutritive composition of honeybees is supposedly affected by environmental temperature, showing a negative correlation. That is, the higher the temperature, the lower the nutrient content, which causes an increase in the insect activity and the consumption of its stored body metabolites [20].

Finally, the levels of antinutrients for the honeybees studied showed much lower tannin (14–14.2 mg/100 g) and oxalate (2.4–2.7 mg/100 g) than the levels considered safe for health (tannin: 150–200 mg/100 g; oxalate: 105 mg/100 g). However, the amount of phytates was over the recommended limit estimated by the WHO (World Health Organization) (168.6–172.9>22.1 mg/100 g), potentially attributed to the high phytic acid content in pollen [20].

Sheileja et al. [21] studied the chemical composition of the Asian giant hornet *Vespa magnifica*, widely consumed in the Northeast India. The authors found dissimilar values compared to those reported for honeybees. Both protein and fat reached more than 30 and around 20% of the insect's weight, respectively. On the contrary, carbohydrates did not reach 1% and fiber hardly exceeded 4%. The tannin content is much higher than the recommended limit (2,450 > 150–200 mg/100 g).

2.4.1.3 Lepidoptera

The order Lepidoptera is the third largest order of edible insects by number of species (335). Butterflies and moths are very representative insects of this order [4, 6]. An adult Lepidoptera is easily identified by the presence of two pairs of membranous wings covered with flat scales, a peculiarity that gives the name to this order. Some Lepidoptera have a proboscis, which is a trunk-like morphological feature. However, this singularity is not naturally present in some insects of this order, and in others, it has been lost or reduced secondarily [22]. It consists of a pair of C-shaped fibers – the maxillary galeae – and is the device used by insects, such as butterflies and moths, to take food [23]. On the other hand, the head of these insects has well-developed antennae of varied morphology and compound eyes. Lepidoptera have wings anchored to the mesothorax and metathorax while the prothorax is small, representing the most visible surface of these insects and showing characteristic drawings or combinations of colors, typical of this order [22].

Regarding the distribution of edible lepidopterans throughout the world, the species *Bombyx mori* L. can be found in the Asian continent, while *Cirina forda* Westwood, *Bunaea alcinoë* Stoll, *Eumeta cervine* Druce, *Anaphe panda* Boisduval, *C. butyrospermi* Vuillet, *Dactyloceras lucina* Drury, *Imbrasia ertli* Rebel, *Platysphinx stigmatica* Mabille, *Clania moddermanni* Heylaerts, and *I. epimethea* Drury, among others, are typical from the African continent [4].

The contribution of Lepidoptera members to nutrition can be significant. Protein represents more than half of the insect weight in 31 lepidopteran species, which is a remarkable figure compared to other animal protein matrices. Fat may also be a preponderant macronutrient in the order Lepidoptera, but the amount reported in the literature for these insects is highly variable. Thus, the isopteran *Phasus triangularis* can reach 77% of fat and *Gonimbrasia alopia* only 1.9%. In contrast, lepidopterans usually present lower than 10% of fiber, but some exceptions such as *Latebraria amphipyrioides* with 29% are possible [24].

The presence of a wide range of minerals, such as calcium, copper, iron, magnesium, manganese, phosphorous, potassium, sodium, and zinc was reported in several lepidopterans, namely: *Aegiala hesperiasis*, *A. venata*, *B. mori*, *C. forda*, *Clanis bilineata*, *Comadia redtenbacheri*, *Gynanisa maja*, and *Usta terpsichore*. Edible lepidopterans seem not to be a good source of vitamins according to the data extracted from a number of these insects used for nutritive evaluation, with values below the daily minimum requirement suggested by the FAO (Food and Agriculture Organization)/WHO. Vitamins A, B1, B2, B3, B5, B7, B9, C, and E are present in lepidopterans, such as *A. infracta*, *B. mori*, *Galleria mellonella*, and *I. truncata* [24].

2.4.1.4 Orthoptera

Orthoptera is one of the most diverse insect orders with 310 edible insect species described [4]. Apart from presenting significant differences among species, the insects of this order are found in tropical and subtropical regions of the globe, but also in high latitudes and altitudes, from tropical and desert regions to the Arctic. Their populations are often numerous, inhabiting terrestrial areas, although they are also found in aquatic environments. Regarding their morphology, Orthoptera are insects with big back legs to jump, well-developed, compound eyes, a large prothorax with a shield-like pronotum that curves downward laterally, and mandibulate mouthparts for biting and chewing. Except for the subfamily Oecanthinae, all Orthoptera have a hypognathous head, with mouthparts directed downward and head vertical with respect to the body axis. The jaws are large and very different among groups. On the other hand, the presence of wings is frequent. The first pair of wings is related to sound production in some groups of insects, and the second pair of wings is used for battling [25].

The Orthoptera is divided in turn into two suborders: Ensifera, which comprises true crickets (Gryllidae), mole crickets (Gryllotalpidae), camel crickets (Raphidophoridae), bush crickets or katydids (Tettigoniidae), grigs (Prophalangopsidae), and Cooloola monsters (Coolooidae), and the suborder Caelifera, comprising grasshoppers (Acrididae), ground-hoppers (Tetrigidae), and pygmy mole crickets (Tridactylidea). The first suborder mentioned (Ensifera) is characterized by presenting long antennae (frequently longer than the body), sword-like or needle-like ovipositors, and auditory organs. On the other hand, the suborder Caelifera is differentiated by having shorter antennae, ovipositor with only two valvular pairs, and abdominal tympana [25].

The order Orthoptera is consumed in the Asian region, where several species may be part of the diet in different communities. Specifically, orthopterans, including *Schistocerca gregaria* Forsskål, *Gryllotalpa africana* Palisot de Beauvois, and *Tarbinskiellus portentosus* L., are considered adequate for consumption. In Africa, orthopterans, such as *Schistocerca gregaria* Forsskål, *Acanthacris rufcornis* F., *Brachytrupes membranaceus* Drury, *Nomadacris septemfasciata* Audinet–Serville, *Zonocerus variegatus* L., *Phymateus viridipes brunneri* Bolivar, and *Ruspolia diferens* Serville, among others, can be consumed, proving to be a region very rich in this type of insects. Conversely, in Oceania, the presence of orthopterans is scarce (e.g., *Teleogryllus commodus* Walker) [4].

Edible orthopterans seem to follow the nutritional pattern of other edible insects previously described. Fombong et al. [26] assessed three widely consumed orthopterans, such as *Gryllus bimaculatus*, *Locusta migratoria*, and *Schistocerca gregaria*. All these insects showed a high protein content, representing between 10% and 18% of the total insect weight and 50% and 70% on a dry basis. This protein presented all essential amino acids (EEAs), with glutamic acid, alanine, and leucine being the most abundant. Fat was another outstanding compound in these insects' bodies as expected if compared to most edible insects. Values between 3.4% and 5.5% on a wet basis and between 19.1% and 30.5% on a dry basis were found for the three insect species evaluated. Oleic acid (22.9–40.8% of total fatty acids) and palmitic acid (C16:0) (26.1–43% of total fatty acids) were the primary fatty acids in the fat of these three insects. On the other hand, linoleic acid was the main polyunsaturated fatty acid (PUFA), ranging from around 5–30% of total fatty acids. Satisfactorily, the omega-6:omega-3 ratio in the species *L. migratoria* and *S. gregaria* resulted much lower than 5 (1.6 and 0.5, respectively), the recommended limit by the FAO for a healthy diet, but not in *Gryllus bimaculatus* (17.2). Minerals were reported to be important in the three orthopterans studied, standing out phosphorous and potassium. Copper, iron, manganese, and zinc were also present. Vitamin B12 was abundantly found (0.2–1.4 μg/100 g dry weight).

2.4.1.5 Hemiptera

Up to 251 edible insects of the order Hemiptera were counted [4]. They evolved from terrestrial to aquatic habitats and are divided into two infraorders: semi-aquatic (gerromorphs) and aquatic (nepomorphs) hemipterans. The former live primarily on the water's surface, while the latter live submerged. Almost all aquatic hemipterans depend on air for breathing and are capable of walking and flying in the terrestrial environment as adults. Aquatic hemipterans belong to the suborder Heteroptera, a very wide group of insects known as "true bugs." These insects are united by the presence of metathoracic glands in adults, dorsal abdominal glands in immature individuals, and a labium inserted anteriorly on the head [27].

Aquatic hemipterans are hemimetabolous insects that possess a piercing mouthpart with three or four segments located beneath the head. This part of the body is adapted to predate, feeding on digested fluids. The hindwings of adults are membranous as in other insect orders, but the forewings are like a leathery hemelytra plus a thin membranous margin. The body size of hemipterans can reach almost two orders of magnitude. For instance, the largest aquatic hemipteran is about 11 cm in length, whereas adults of some pygmy backswimmers are only 1.5 mm. Legs of hemipterans are well adapted for skating on the water's surface in the case of gerromorphans and for swimming and gripping in the case of nepomorphans. Hydrofuge hairs on the legs and body surface allow gerromorphans to move without breaking the surface tension of water, forming trough-shaped depressions at the surface. Regarding wings, some groups of hemipterans are very skilled flyers capable of traveling long distances, dispersing to new habitats (most insects of the families Belostomatidae, Corixidae, and Notonectidae, and some Gerridae). On the contrary, other groups of hemipterans lack functional wings (some Gerridae) or functional flight muscles (some insects of the families Belostomatidae, Naucoridae, and Notonectidae) [27].

Hemiptera is the second most consumed insect order in Asia, with 128 edible species, while in other continents their presence is scarcer. In North America and Africa, edible hemipterans occupy the fourth place of most consumed insects with 68 and 42 species, respectively. On the other hand, only 16 edible hemipterans are consumed in Oceania. The hemipteran *Nezara viridula* L. is listed as one of the most consumed and distributed insects worldwide [4].

Two edible insects of the order Hemiptera were analyzed, obtaining dissimilar results regarding the amount of lipids, since it was higher than that of protein. The *Aspongopus nepalensis* Westwood 1837 insect presented almost 40% of fat on a dry weight and surprisingly only 10.6% of protein. According to the authors of this work, the reported lipid content is higher than that of most insects evaluated in the literature. Fiber was also higher than the protein content (33.5 > 10.6% dry weight). Monounsaturated fatty acids (MUFAs) accounted for just over 50% of total fatty acids, with oleic acid being the main compound (46.4% of total fatty acids). Palmitic acid was also present in a high proportion (32.3% of total fatty acids) [28]. Another hemipteran, *Coridius* sp., showed similar results by presenting a higher fat content than protein (38.7 > 36.5% dry weight). Fiber levels were practically the same as those of the *Aspongopus nepalensis* species analyzed in the previous work (8.7% dry weight) [29].

Some minerals and vitamins are present in significant amounts for the aforementioned hemipterans. Calcium, magnesium, potassium, and sodium are prominent in *A. nepalensis*, while calcium, iron, and potassium are in *Coridius* sp. Vitamin A is the most abundant in *A. nepalensis* (34.4 μg/100 g) [28, 29].

2.4.1.6 Isoptera

The order Isoptera is composed of a type of insect commonly known as termites, with up to 76 edible species identified worldwide [4]. The meaning of the name Isoptera attempts to describe the two pairs of straight, similar wings of termites as reproductive adults. Isopterans can be confused with ants, which belong to the order Hymenoptera as discussed before in another section. However, termites have a number of morphological distinctive characteristics that allow us to distinguish them from "true ants." Thus, the order Isoptera possesses straight antennae and a broad waist between the thorax and the abdomen. On the contrary, ants have elbowed antennae and a narrow waist. The color of termites can vary between white and black. In fact, sometimes they are called "white ants." [30]

Termites are insects that can be easily distinguished from others, since they have very distinctive morphologies among all types. Families of termites are divided into Mastotermitidae, Kalotermitidae, Termopsidae, Hodotermitidae, Rhinotermitidae, and Serritermitidae, all of them belonging to the group of lower termites, and the family Termitidae, belonging to the group of higher termites. The former has symbiotic intestinal protozoa and bacteria, while the latter only has intestinal bacteria [30].

With respect to the isopteran morphology, these insects have a characteristic head with a dorsal part arranged as in most orthopteroid insects and a labrum located at the front end, followed by a clypeus. The labrum is often small and tongue-shaped in all wings. Some termites possess a pit in the middle-front, called the fontanelle, with a defensive function in soldiers. On the other hand, the lower part of the head is structurally complex, with mouthparts composed of maxillae, labium, and mandibulae. The thorax is segmented in prothorax, mesothorax, and metathorax, as in all insects, with a pair of legs in each of these divisions. The mesothorax and metathorax also have a pair of wings. The legs and wings are attached to plates along the thorax. Despite the presence of wings, termites are not good flyers; they just launch themselves into the air and fly randomly. Termites have several legs with a classical standard pattern for insects. Starting closest to the body, these are coxa, trochanter, femur, tibia, and tarsus. On the other hand, the abdomen is composed of ten segments, divided into upper and lower plates, called tergites and sternites, respectively. Nine of the ten tergites are broad while the tenth is elongated and tapered. The first sternite is small or absent, and the second to the sixth are wider than long. The seventh sternite of the female alate is large and frequently covers sternites eight and nine. The tenth sternite splits in half, giving rise to a pair of plates, one on each side of the body [31].

The consumption of edible isopterans along the world is not so extent as other insect orders. Even so, a significant number of termite species were identified as edible in Africa (15), highlighting individuals of the genus *Macrotermes*. Termites of this genus and subgenus *Bellicositermes* belong to the 11 most ubiquitous edible insects worldwide, especially in Asia, Africa, and South America [4].

Several winged termites of the species *M. bellicosus* inhabiting Africa showed big differences in macronutrients. That is, protein content oscillates between 12.1% and 69.8% of the insect's body and fat is between 1.9% and 52.3%. Fiber levels also present a wide range (0.2–16.4% of the insect's body). Regarding fatty acids, MUFAs are the most abundant fatty acid group and oleic acid the main fatty acid (41.7–54.5% of total fatty acids). The proportion of SFAs is remarkably high, with palmitic acid being the second most abundant fatty acid in these insects (26–38.9%). Linoleic acid is the third most important fatty acid in *M. bellicosus* (5–11.5%). Differences in fatty acids within termites are attributed to species. Vitamins A, B1, B2, B3, B6, C, D, and E are present at different concentrations that depend on the insect species, geographical location, and even processing method. Up to 50% of the recommended daily intake for vitamins A and B1 in infants and adults and 100% for vitamin B2 could be provided by winged termites. Iron and zinc highlight over other minerals, and *M. bellicosus* could provide the recommended daily intake of these compounds in children aged between 4 and 13 years [32].

A group of termites belonging to the species *M. falciger*, *M. herus*, and *M. jeanneli* from Ethiopia displayed a 4:6 ratio between essential and non-EAAs, highlighting leucine, followed by lysine as the most important EAAs. In this regard, it is worth noting that lysine is the limiting amino acid in cereals, especially those based on wheat, rice, and maize, representing a nutritional disadvantage compared to the isopterans of the mentioned species. Regarding the fatty acid profile, the isopterans studied showed a high proportion of MUFAs (46% of total fatty acids), and oleic acid is the main fatty acid. SFAs are the second fatty acid group in importance, with palmitic and stearic acids as the reference fatty acids within this group. This fact was also observed in other species of termites, such as *Odontotermes* sp., *M. subhyalinus*, *M. bellicosus*, *P. militaris*, *P. spiniger*, and *Nasutitermes* sp. The minor fatty acid group in *Macrotermes* sp. is the PUFAs, which contained almost all linoleic acid. These data are consistent with those reported for the *Macrotermes* species analyzed in the previous study. This is also applicable to the mineral content, since iron is one of the most prominent minerals together with calcium, which reached the highest concentration in the *Macrotermes* species studied (379.7 mg/100 g dry basis) [33].

2.4.1.7 Odonata

The insect order Odonata is comprised of hemimetabolous insects that are widespread in most aquatic environments, including several brackish and terrestrial habitats [34]. Up to 54 different species were identified worldwide [4]. Damselflies and dragonflies are typical and very recognized insects belonging to this order. The extant Odonata are divided into three suborders: Anisoptera, Anisozygoptera, and Zygoptera. Anisoptera are characterized by large, strong, and fast fliers, capable of turning quickly in place; their fore- and hindwings are broad and of different shapes. In contrast, Zygoptera are smaller and fly slow in any direction without turning their body. The fore- and hindwings are almost identical to Anisoptera and Anisozygoptera, but the latter have a body shape and anal appendages similar to Anisoptera [35].

Dragonflies belong to the suborder Anisoptera. These insects vary in color, shape, and size, but they possess a general feature: the tendency to have a thick abdomen. Most spread their wings laterally when resting. On the other hand, damselflies belong to the suborder Zygoptera. Unlike dragonflies, these odonatans have svelte abdomens and a tendency to hold their wings over the back when resting. In addition, damselflies have characteristic widely spaced eyes [36].

At least 61 species of odonatans of almost 2000 distributed worldwide are edible. The most consumed dragonflies correspond to the Libellulidae family (33 species), followed by Aeshnidae (nine species) and Gomphidae (seven species) families, whereas the most consumed damselflies correspond to the Coenagrionidae family (five species). Interestingly, dragonflies are nutritionally well considered and appreciated as one of the best edible insect resources [37].

Up to 43 of 61 species of edible dragonflies were observed only in Asia, and nine species of odonatans are highly consumed, these are *Aeschna* sp., *Argia* sp., *Ceriagrion* sp., *Crocothemis servilia*, *Diplacodes trivialis*, *Enallagma* sp., *Ischnura* sp., *Leucorrhina* sp., *Libellula pulchella*, *Pachydiplax* sp., *Pantala flavescens*, and *Sympetrum striolatum*. From all species of edible dragonflies identified, only the *Ischnura* sp. (Zygoptera: Coenagrionidae) is distributed across all continents [37].

Dragonflies are a great source of protein, ranging from around 45–76%. Among some edible species of odonatans, including *Aeschna multicolor*, *Anax parthenope*, *Crocothemes servillia*, *Epophthalmia elegans*, *Gomphus cuneatus*, *Ictinogomphus rapax*, *Orthetrum pruinosum*, *Pantala* sp., *Sinictinogom–phus clavatus*, and *Sympetrum* sp., the richest protein matrix belongs to the latter species, while *Pantala* sp. is the poorest. Dragonfly protein could be a potential substitute for other regularly consumed protein source such as meat, meeting the recommended daily intake of protein for adults (50 g/day). Fat is the second largest macronutrient in dragonflies, as in most insects. Values between 4% and 23% were observed for all the aforementioned dragonfly species. On the other hand, fiber is present between 9% and 14%. SFAs, MUFAs, and PUFAs are pretty much balanced, representing 31.6%, 33.9%, and 27.2% of total fatty acids, respectively. Dragonfly protein contains all the necessary amino acids, with a good balance between some of them. Valine is the most prevalent EAA in this protein, while arginine and glutamic acid are the majority non-EAAs. Concentrations of amino acids present in the dragonfly protein, such as lysine, histidine, and methionine, can be higher than those of beef, pork, and chicken meat. Vitamins and minerals are present in dragonflies and damselflies, standing out vitamin C and several vitamins of group B, such as B1, B2, B3, B6, and B12. Minerals, including calcium, copper, iron, magnesium, manganese, phosphorous, potassium, and zinc, are contained in dragonflies' bodies [37].

2.4.1.8 Diptera

The insect order Diptera comprise the two-winged or the called "true flies." It is one of the most recognized and widespread group of insects and one of the most successful group of living organisms on earth. Dipterans are quite diverse, and their number rise to more than 125,000 species [38], but only 39 species were identified as edible [4]. True flies vary in size from robber flies with 7 cm long to midges of little more than 1 mm. The most evolutionary ancient flies are brittle insects, with delicate wings. In contrast, the more-progressive flies, such as house flies and blow flies, are generally bristly, thick, and strong, and powerful fliers than the latter [38].

Members of the order Diptera have a simple, differentiated structure and physiological heterogeneity compared to other members of this group. Dipteran adults present a wide range of length, from 1 to 12 mm. However, there are others whose size may reach between 25 and 60 mm long. Dipterans have a single pair of membranous wings; hind wings are primitive and nonfunctional for flying. On the other hand, the bodies are long, tube shaped. The mouthparts are either sharp tubes for feeding by penetrating flesh and sucking up liquids (e.g., mosquitoes) or blunt pads designed for absorbing liquids [38].

Three mosquito species abundantly found, namely: *Aedes caspius*, *Anopheles hyrcanus*, and *Culex pipiens* were nutritionally analyzed. All of them exhibited high protein levels (54–62% dry basis; 24–28% wet basis) and a significant amount of fat (16–28% dry basis; 9–13% wet basis). The fatty acid profile revealed the presence of a high concentration of palmitic (23.8–30.4% of total fatty acids) and palmitoleic (C16:1*n*–7) (28.5–37% of total fatty acids) fatty acids. Oleic acid was also present in remarkable amounts (18.3–29.1% of total fatty acids) [39].

The studied wild-harvested mosquitoes proved to be a good source of minerals, highlighting the presence of calcium, copper, iron, magnesium, manganese, phosphorous, potassium, sodium, and zinc, with values even higher than those of other conventionally consumed insects. Phosphorous was the most important mineral in the mosquito species examined (13,700–18,800 mg/kg dry weight), followed by potassium and calcium (9,079–10,305 and 6,626–12,113 mg/kg dry weight, respectively) [39].

2.4.1.9 Dictyoptera

The order Dictyoptera is formed by a group of paurometabolous insects, that is, insects with gradual metamorphosis, with 38 edible species identified, and they are divided into two suborders: Blattodea and Mantodea. Species of Blattodea are known with the colloquial name of cockroaches. They possess an oval, flattened body with dark colors, although some species are colored. Their legs are long and spiny and are well adapted for running. In general, the head of cockroaches is partially covered by a broad and oval pronotum. These insects have very parchment-like forewings and membranous hindwings folded in a fan shape, although there are wingless species or with reduced wings. Cockroaches vary greatly in size. They can measure from 3 to 50 mm or more such as the *Blaberus giganteus* cockroach. The suborder Mantodea comprise those insects commonly known as mantids. They are extraordinary predators with a highly developed hunting strategy thanks to certain physical adaptations, which consist in very strong, elongated, and extensible front legs, with very spiny femurs and tibias forming a kind of pincers. These legs are located in the anterior part of a very elongated prothorax. Mantids have highly developed, frontally positioned compound eyes, giving them a very useful stereoscopic vision when hunting. The outstanding mobility of their heads (270° of rotation) allows mantids to follow the movement of preys remaining undetected. These insects can vary in size from 15 mm such as the genus *Mantoida* to more than 120 mm such as the species *Macromantis hyaline* [4, 7].

China is one of the few countries in the world where it is possible to eat cockroaches. They are introduced into the diet as a snack or as a garnish. Close to China, in Vietnam, cockroaches are part of their local cuisine, which are consumed in a variety of ways. These insects are also consumed in other parts of the globe such as in the American continent. Specifically, the species *Periplaneta americana* Linnaeus, 1758, also called American cockroach, is widely recognized as an edible species. They are present in commercial establishments, including restaurants, grocery stores, bakeries, and other facilities. Curiously, although Mexico is a country with a tradition of entomophagy, the use of cockroaches in gastronomy is limited. Even so, these insects can still be found in some regions of the country where they are sold as a delicacy [40].

Cockroaches have significant nutritional value compared to other traditional protein sources such as meat. Protein content is highly variable with values that may comprise between 8% and 80% of the insect's dry weight or between 9% and 23% on a wet basis. This value is usually higher than the common values found in meat, but with a similar amino acid content, meeting the total human requirement. In this regard, not all cockroach species present the same amino acid profile. Thus, *Blaberus craniifer* Burmeister, 1838 exhibits deficiency in cysteine, tyrosine, and tryptophan compared to the *P. americana* Linnaeus, 1978 and Australian cockroach, *Periplaneta australasiae* Fabricius, 1775. Fat is the second most important component in cockroach's body, as in other insects, potentially achieving between 5% and 17% on a wet basis, which is aligned with the recommended daily fat intake (20–35%). MUFAs and SFAs often represent most of the fat content in cockroaches, with more than 73%. Palmitic and stearic acids are the main SFAs, while oleic acid is the most important MUFA and the highest fatty acid in the insect's fat (almost half of the total fatty acids). The PUFA/SFA ratio showed by the American cockroach is nutritionally healthier than those of the chicken breast, drumstick, or pork. This dictyopteran species is also a valuable source of minerals, capable of providing essential micronutrients to population with specific requirements, such as pregnant and breastfeeding mums. Compounds, including calcium, magnesium, phosphorous, potassium, and sodium, and to a lesser extent iron, manganese, copper, iodine, selenium, and zinc, were found in the insect's matrix. [40]

Figures 2.4 and 2.5 show visual differences among the insect orders discussed in the sections mentioned above.

2.5 CONCLUSION

Edible insects have shown to be a broad and variable group of recognized insects distributed across the globe, especially abundant in the continents of America, Asia, and Africa. Up to nine different insect orders with different features were reported to be consumed

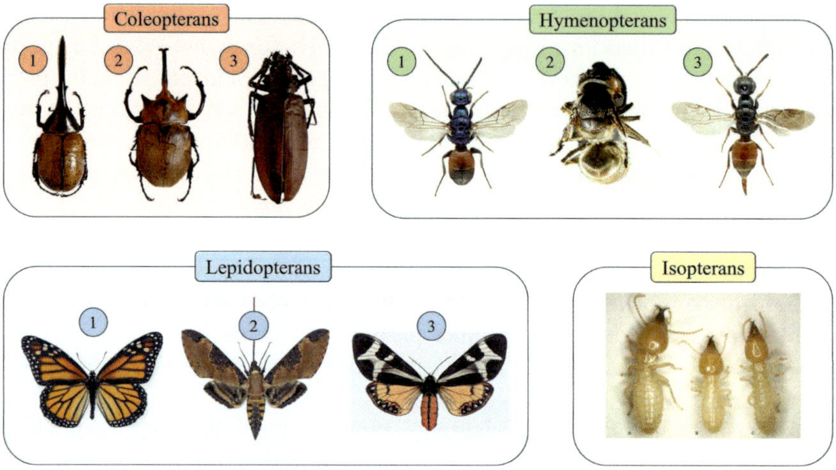

FIGURE 2.4 Example of insects from some of the major orders of edible insects. Coleoptera, 1: *Megasoma elephas*; 2: *Goliathus regius*; 3: *Dynastes hercules*. Hymenoptera, 1: *Cleptes sfenthourakisi*; 2: *Vespa crabro*; 3: *Cleptes christi*. Lepidoptera, 1: *Danaus plexippus*; 2: *Manduca ochus*; 3: *Dysschema howardi*. Isopterans, A: *Coptotermes curvignathus*; B and C: *C. gestroi*. Pictures from Luiselli and Amori [41]; Dalmon et al. [42]; Rosa et al. [43]; Buchmann [44]; Halbritter et al. [45]; Wikantyoso et al. [46].

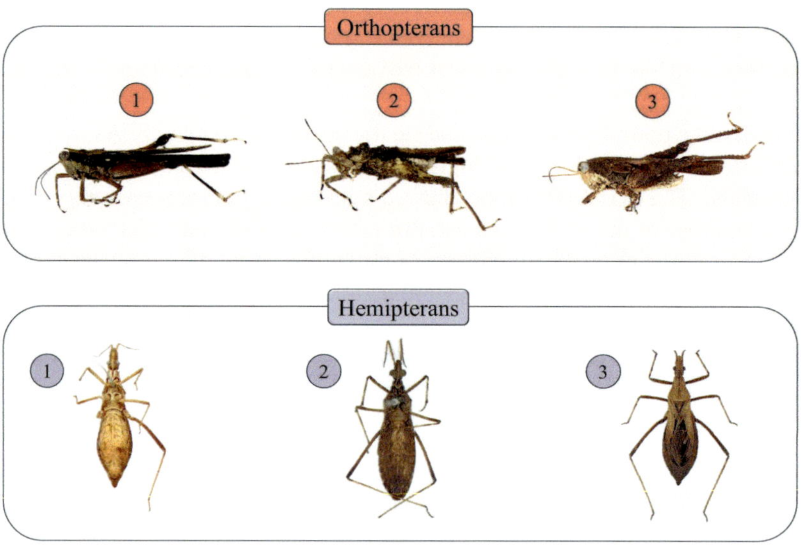

FIGURE 2.5 Example of insects from some of the major orders of edible insects. Orthoptera, 1: *Scaria fasciata*; 2: *Chiriquia serrata*; 3: *Tettigidea lateralis lateralis*. Hemiptera, 1: *Caunus farinator* Reuter; 2: *Bardesanes signatus* Distant; 3: *Argolis signata* Distant. Pictures from Kasalo et al. [47]; Chen et al. [48].

by different human communities in different continents. In reality, more orders of edible insects can be found than those listed in this chapter, but the number of species identified in them is insignificant compared to the total. On the other hand, nutritional properties of edible insects could be largely comparable to other traditional protein sources such as livestock, with protein being the most important nutrient in most of the edible orders.

Therefore, edible insects can be considered ubiquitous animals capable of providing a significant amount of nutrients, potentially contributing to ensuring food security.

2.6 ACKNOWLEDGMENTS

Rubén Agregán and Noemí Echegaray thank GAIN (Axencia Galega de Innovación), Xunta de Galicia, for granting an extension to their postdoctoral scholarships (grant numbers: IN606C-2025/002 and IN606C-2025/001, respectively). These authors and Rubén Domínguez-Valencia, Mirian Pateiro, and José Manuel Lorenzo are members of the Competitive Reference Group "FunFood" (Functional Food), funded by GAIN (grant number IN607A2023/01). This work was (*partially*) supported by Programa Iberoamericano de Ciencia y Tecnología para el Desarrollo (CYTED) (through Red AlProSos 125RT0165).

REFERENCES

[1] Eggleton, P. The State of the World's Insects. *Annu Rev Environ Resour*, 2020, *45* (1), 61–82. https://doi.org/10.1146/annurev-environ-012420-050035.

[2] Olivadese, M.; Dindo, M. L. Edible Insects: A Historical and Cultural Perspective on Entomophagy with a Focus on Western Societies. *Insects*, 2023, *14* (8), 690. https://doi.org/10.3390/insects14080690.

[3] Eswaran, U. G. M.; Karunanithi, S.; Gupta, R. K.; Rout, S.; Srivastav, P. P. Edible Insects as Emerging Food Products–Processing and Product Development Perspective. *J Food Sci Technol*, 2023, *60* (8), 2105–2120. https://doi.org/10.1007/s13197-022-05489-y.

[4] Omuse, E. R.; Tonnang, H. E. Z.; Yusuf, A. A.; Machekano, H.; Egonyu, J. P.; Kimathi, E.; Mohamed, S. F.; Kassie, M.; Subramanian, S.; Onditi, J.; et al. The Global Atlas of Edible Insects: Analysis of Diversity and Commonality Contributing to Food Systems and Sustainability. *Sci Rep*, 2024, *14* (1), 5045. https://doi.org/10.1038/s41598-024-55603-7.

[5] Grabowski, N. T.; Abdulmawjood, A.; Acheuk, F.; Barragán Fonseca, K.; Chhay, T.; Costa Neto, E. M.; Ferri, M.; Franco Olivas, J.; González Aguilar, D. G.; Keo, S.; et al. Review: Insects – a Source of Safe and Sustainable Food? – "Jein" (Yes and No). *Front Sustain Food Syst*, 2022, *5*, 701797. https://doi.org/10.3389/fsufs.2021.701797.

[6] Bektaş, M. *Edible Insects*; Iksad Publishing House: Ankara, 2024.

[7] Cerdá, F. J. Dictiópteros. Museo Del Instituto de Zoología Agrícola. In *Biodiversidad en Venezuela*; Aguilera, M.; Azócar, A.; Jiménez, E. G., Eds.; ExLibris: Caracas, 2003; pp. 340–349.

[8] Badirli, S.; Picard, C. J.; Mohler, G.; Richert, F.; Akata, Z.; Dundar, M. Classifying the Unknown: Insect Identification with Deep Hierarchical Bayesian Learning. *Methods Ecol Evol*, 2023, *14* (6), 1515–1530. https://doi.org/10.1111/2041-210X.14104.

[9] Bjerge, K.; Alison, J.; Dyrmann, M.; Frigaard, C. E.; Mann, H. M. R.; Høye, T. T. Accurate Detection and Identification of Insects from Camera Trap Images with Deep Learning. *PLOS Sustain Transform*, 2023, *2* (3), e0000051. https://doi.org/10.1371/journal.pstr.0000051.

[10] Hending, D. Cryptic Species Conservation: A Review. *Biological Reviews*, 2024, *100*, 258–274. https://doi.org/10.1111/brv.13139.

[11] Heine, H. L. A.; Derkarabetian, S.; Morisawa, R.; Fu, P. A.; Moyes, N. H. W.; Boyer, S. L. Machine Learning Approaches Delimit Cryptic Taxa in a Previously Intractable Species Complex. *Mol Phylogenet Evol*, 2024, *195*, 108061. https://doi.org/10.1016/j.ympev.2024.108061.

[12] Stork, N. E. How Many Species of Insects and Other Terrestrial Arthropods Are There on Earth? *Annu Rev Entomol*, 2018, *63* (1), 31–45. https://doi.org/10.1146/annurev-ento-020117-043348.

[13] Choi, Y.-S.; Lee, J. H.; Kim, T.-K.; Shin, D.-M. Edible Insects in Food. In *Advances in Food and Nutrition Research*; Toldrá, F., Ed.; Elsevier: Amsterdam, 2024; Vol. 108, pp. 223–264. https://doi.org/10.1016/bs.afnr.2023.10.005.

[14] Hunter, G. L. *Edible Insects. A Global History*; Reaktion Books: London, 2021.

[15] Van Itterbeeck, J.; Pelozuelo, L. How Many Edible Insect Species Are There? A Not so Simple Question. *Diversity (Basel)*, 2022, *14* (2), 143. https://doi.org/10.3390/d14020143.

[16] Resh, V. H.; Cardé, R. T. Insecta, Overview. In *Encyclopedia of Insects*; Resh, V. H.; Cardé, R. T., Eds.; Elsevier Inc.: Amsterdam, 2009; pp. 501–502. https://doi.org/10.1016/B978-0-12-374144-8.00142-9.

[17] Bouchard, P.; Smith, A. B. T.; Douglas, H.; Gimmel, M. L.; Brunke, A. J.; Kanda, K. Biodiversity of Coleoptera. In *Insect Biodiversity. Science and Society*; Foottit, R. G.; Adler, P. H., Eds.; Wiley: Hoboken, 2017; pp. 337–417. https://doi.org/10.1002/9781118945568.ch11.

[18] Siddiqui, S. A.; Yüksel, A. N.; Şahin Ercan, S.; Abdul Manap, A. S.; Afzal, S.; Wu, Y. S.; Yudhistira, B.; Ibrahim, S. A. Edible Beetles (Coleoptera) as Human Food – a Comprehensive Review. *J Insects Food Feed*, 2024, 1–59. https://doi.org/10.1163/23524588-00001095.

[19] Huber, J. T. Biodiversity of Hymenoptera. In *Insect Biodiversity*; Foottit, R. G.; Adler, P. H., Eds.; John Wiley & Sons: Hoboken, NJ, 2017; pp. 419–461. https://doi.org/10.1002/9781118945568.ch12.

[20] Bamidele, J. A.; Idowu, A. B.; Ademolu, K.; Osipitan, A. A. Nutritional Composition of *Apis Mellifera Adansonii* L. (Hymenoptera: Apidae) from Three Ecological Zones of Nigeria. *J Apic Res*, 2021, *60* (3), 445–456. https://doi.org/10.1080/00218839.2021.1874709.

[21] Sheileja, T.; Singh, K. M.; Shantibala, T.; Mishra, L. K.; Haldhar, S. M. Nutritional Composition and Safety Aspects of Asian Giant Hornet, *Vespa Magnifica* (Hymenoptera: Vespidae): Commonly Consumed Edible Insect in Manipur, North East India. *J Insects Food Feed*, 2023, *10* (6), 1005–1014.

[22] García-Barros, E.; Romo, H.; Sarto i Monteys, V.; Munguira, M. L.; Baixeras, J.; Moreno, A. V.; García, J. L. Y. Orden Lepidoptera. *IDE@ – SEA*, 2015, *65*, 1–21.

[23] Zhang, C.; Adler, P. H.; Monaenkova, D.; Andrukh, T.; Pometto, S.; Beard, C. E.; Kornev, K. G. Self-Assembly of the Butterfly Proboscis: The Role of Capillary Forces. *J R Soc Interface*, 2018, *15* (144). https://doi.org/10.1098/rsif.2018.0229.

[24] Siddiqui, S. A.; Ngah, N.; Eddy-Doh, A. M.; Ucak, I.; Afreen, M.; Fernando, I.; Singh, S.; Shah, M. A.; Povetkin, S. N.; Castro-Muñoz, R. Edible Lepidoptera as Human Foods – a Comprehensive Review. *J Insects Food Feed*, 2023, *10* (1), 25–49. https://doi.org/10.1163/23524588-20230072.

[25] Sperber, C. F.; Zefa, E.; de Oliveira, E. C.; de Campos, L. D.; Bolfarini, M. P.; Fianco, M.; Lhano, M. G.; Vicente, N.; Szinwelski, N.; de Souza Dias, P. G. B.; et al. Measuring Orthoptera Diversity. In *Measuring Arthropod Biodiversity: A Handbook of Sampling Methods*; Santos, J. C.; Fernandes, G. W., Eds.; Springer International Publishing: Cham, 2020; pp. 257–287. https://doi.org/10.1007/978-3-030-53226-0_11.

[26] Fombong, F. T.; Kinyuru, J.; Ng'ang'a, J.; Ayieko, M.; Tanga, C. M.; Broeck, J. Vanden; Van Der Borght, M. Affordable Processing of Edible Orthopterans Provides a Highly Nutritive Source of Food Ingredients. *Foods*, 2021, *10* (1), 144. https://doi.org/10.3390/foods10010144.

[27] Lytle, D. A. Order Hemiptera. In *Thorp and Covich's Freshwater Invertebrates: Ecology and General Biology: Fourth Edition*; Thorp, J. H.; Rogers, D. C., Eds.; Elsevier Inc.: Amsterdam, 2015; Vol. 1, pp. 951–963. https://doi.org/10.1016/B978-0-12-385026-3.00037-1.
[28] Chakravorty, J.; Ghosh, S.; Meyer-Rochow, V. B. Chemical Composition of *Aspongopus Nepalensis* Westwood 1837 (Hemiptera; Pentatomidae), a Common Food Insect of Tribal People in Arunachal Pradesh (India). *International Journal for Vitamin and Nutrition Research*, 2011, *81* (1), 49–56. https://doi.org/10.1024/0300-9831/a000050.
[29] Sheileja, T.; Singh, K.; Shantibala, T.; Haldhar, S.; Singh Summary, K. Nutritional Aspects of Edible Insect, *Coridius* Sp. (Hemiptera: Dinidoridae) of Manipur. *Journal of Agriculture and Ecology*, 2022, *14*, 158–163. https://doi.org/10.53911/JAE.2022.14223.
[30] Lewis, V. R. Isoptera: (Termites). In *Encyclopedia of Insects*; Resh, V. H.; Cardé, R. T., Eds.; Elsevier Inc.: Amsterdam, 2009; pp. 535–538. https://doi.org/10.1016/B978-0-12-374144-8.00152-1.
[31] Eggleton, P. An Introduction to Termites: Biology, Taxonomy and Functional Morphology. In *Biology of Termites: A Modern Synthesis*; Bignell, D. E.; Roisin, Y.; Lo, N., Eds.; Springer Netherlands: Dordrecht, 2011; pp. 1–26. https://doi.org/10.1007/978-90-481-3977-4_1.
[32] Makore, T. A.; Mvumi, B. M.; Matsungo, T. M. Nutritional Quality of Edible Winged Termites, *Macrotermes* Spp. and Edible Stink Bug, *Encosternum Delegorguei* Spinola Consumed in Sub-Saharan Africa: A Scoping Review. *Discover Food*, 2025, *5* (1), 170. https://doi.org/10.1007/s44187-025-00316-7.
[33] Ghosh, S.; Debelo, D. G.; Lee, W.; Meyer-Rochow, V. B.; Jung, C.; Dekebo, A. Termites in the Human Diet: An Investigation into Their Nutritional Profile. In *African Edible Insects As Alternative Source of Food, Oil, Protein and Bioactive Components*; Mariod, A. A., Ed.; Springer International Publishing: Cham, 2020; pp. 293–306. https://doi.org/10.1007/978-3-030-32952-5_22.
[34] Pessacq, P.; Muzón, J.; Neiss, U. G. Order Odonata. In *Thorp and Covich's Freshwater Invertebrates*; Neusa Hamada, Rogers, D. C.; Thorp, J. H., Eds.; Elsevier: Amsterdam, 2018; pp. 355–366. https://doi.org/10.1016/b978-0-12-804223-6.00014-7.
[35] Wakeling, J. M. Odonatan Wing and Body Morphologies. *Odonatologica*, 1997, *26* (1), 35–52.
[36] Ware, J. L. Odonata. *Current Biology*, 2020, *31*, R51–R66.
[37] Siddiqui, S. A.; Asante, K.; Ngah, N.; Saraswati, Y. R.; Wu, Y. S.; Lahan, M.; Aidoo, O. F.; Fernando, I.; Povetkin, S. N.; Castro-Muñoz, R. Edible Dragonflies and Damselflies (Order Odonata) as Human Food – A Comprehensive Review. *J Insects Food Feed*, 2024, *10* (12), 1947–1972. https://doi.org/10.1163/23524588-20230097.
[38] Sarwar, M. Typical Flies: Natural History, Lifestyle, and Diversity of Diptera. In *Life Cicle and Development of Diptera*; Sarwar, M., Ed.; IntechOpen: London, 2020; pp. 1–50.
[39] Christaki, A.; Zinoviadou, K. G.; Papoti, V. T.; Miaoulis, M.; Chaskopoulou, A. The Nutrient Composition of Three Mosquito (Diptera: Culicidae) Species, *Aedes Caspius*, *Anopheles Hyrcanus*, and *Culex Pipiens*, Harvested from Rice Fields for Their Potential Utilization as Poultry Feed Ingredients. *Sustainability (Switzerland)*, 2022, *14* (21), 13852. https://doi.org/10.3390/su142113852.
[40] Siddiqui, S. A.; Hadus, M. C. I.; Fitriani, A.; Guleria, V.; Kuppusamy, S.; Bhattacharjee, B.; Yudhistira, B.; Maggiolino, A. Edible Cockroaches as Food and Feed-A Systematic Review on Health Benefits, Nutritional Aspects and Consumer Acceptance. *J Insects Food Feed*, 2024. https://doi.org/10.1163/23524588-00001138.
[41] Luiselli, L.; Amori, G. Community Ecology, Macro-Ecological Patterns, and Conservation of Tropical Beetles: An Introduction. *Diversity (Basel)*, 2021, *13* (9), 406. https://doi.org/10.3390/d13090406.
[42] Dalmon, A.; Gayral, P.; Decante, D.; Klopp, C.; Bigot, D.; Thomasson, M.; AHerniou, E.; Alaux, C.; Conte, Y. Le. Viruses in the Invasive Hornet Vespa Velutina. *Viruses*, 2019, *11* (11), 1041. https://doi.org/10.3390/v11111041.

[43] Rosa, P.; Makris, C.; Boustani, M. Description of Three New *Cleptes* Species from Cyprus and Lebanon with Updates and Revisions to the Genus (Hymenoptera, Chrysididae). *Taxonomy*, 2022, *2* (2), 208–226. https://doi.org/10.3390/taxonomy2020016.
[44] Buchmann, S. L. Moths on the Flatbed Scanner: The Art of Joseph Scheer. *Insects*, 2011, *2* (4), 564–583. https://doi.org/10.3390/insects2040564.
[45] Halbritter, D. A.; Gordon, J. M.; Keacher, K. L.; Avery, M. L.; Daniels, J. C. Evaluating an Alleged Mimic of the Monarch Butterfly: *Neophasia* (Lepidoptera: Pieridae) Butterflies Are Palatable to Avian Predators. *Insects*, 2018, *9* (4), 150. https://doi.org/10.3390/insects9040150.
[46] Wikantyoso, B.; Tseng, S. P.; Himmi, S. K.; Yusuf, S.; Yoshimura, T. Morphometric Analysis of *Coptotermes* Spp. Soldier Caste (Blattodea: Rhinotermitidae) in Indonesia and Evidence of *Coptotermes Gestroi* Extreme Head-Capsule Shapes. *Insects*, 2021, *12* (5), 477. https://doi.org/10.3390/insects12050477.
[47] Kasalo, N.; Janzen, D. H.; Hallwachs, W.; Brown, A.; Husemann, M.; Vielsäcker, M.; Domazet-Lošo, T.; Franjević, D.; Subedi, M.; Bogić, D.; et al. Integrative Taxonomy of Costa Rican Tetrigidae (Orthoptera) Reveals Eight New Species. *Diversity (Basel)*, 2025, *17* (3), 190. https://doi.org/10.3390/d17030190.
[48] Chen, Z.; Webb, M. D.; Cai, W. A Revision of the Genus *Argolis* (Hemiptera: Reduviidae: Stenopodainae) from Asia. *Insects*, 2023, *14* (8), 680. https://doi.org/10.3390/insects14080680.

Nutritional Value of Insects: Proteins and Amino Acids

3

Irais Sánchez-Ortega,
José Antonio Rodríguez Ávila,
Rodrigo Jiménez-Pichardo, Rubén Agregán,
Jose Angel Perez-Alvarez, José Manuel Lorenzo,
and Eva María Santos López

3.1 EDIBLE INSECTS' TYPES

The accelerated population growth have resulted in an increased food demand, and it has become a big problem around the world in the recent times. Due to the elevated production costs, ecological impact, lack of spaces, animal wellness, and among other reasons, consumers are looking for alternative protein substitute sources to cover their needs. Plants, fungi, algae, and insects have been identified as potential alternatives to cover the high demand for protein (Halloran *et al.*, 2014; Ishara *et al.*, 2022; Siddiqui *et al.*, 2024).

Among the mentioned sources, edible insects are seen as a sustainable food source that can contribute to food security and address the protein deficiency issue. Some advantages of insect consumption versus traditional animal protein sources are the need of less space, less water and feed consumption, low carbon footprint, nutritious, and be more environmentally friendly (Siddiqui *et al.*, 2024). Insects has been consumed since ancient times, for many cultures around the world, being particularly common in several zones of Asia, Africa, and Latin America, where they are a traditional component of local gastronomy and dietary habits (Halloran *et al.*, 2014).

DOI: 10.1201/9781003488156-3

We can find edible insects in aquatic (12%), terrestrial (88%), and a few species in arboreal environments, where they feed on a variety of plants as their host (Papastavropoulou *et al.*, 2022a). Insects are classified into eight main orders: Blattodea, Coleoptera, Diptera, Hemiptera, Hymenoptera, Lepidoptera, Odonata, and Orthoptera. The most consumed insect types belong to the Coleoptera (32%), Hymenoptera (15.5%), Lepidoptera (15.2%), Orthoptera (14.1%), and Hemiptera (11.4%) orders (Omuse *et al.*, 2024; Ishara *et al.*, 2022; Liceaga, 2022). Nowadays, there are more than 2000 identified edible insect species, and among them, the most commonly consumed ones including beetles, stink bugs, ants, moths, crickets, grasshoppers, worms, and caterpillars (Acuña *et al.*, 2011; Longvah *et al.*, 2011; Melo-Ruiz *et al.*, 2011; Ramos-Elorduy *et al.*, 2011; Rumpold and Schlüter, 2013; van Huis, 2013; Halloran *et al.*, 2014; Lara-Juárez *et al.*, 2015; Payne *et al.*, 2016; Kouřimská and Adámková, 2016; Dobermann *et al.*, 2017; Ribeiro *et al.*, 2018; Köhler *et al.*, 2019; Patel *et al.*, 2019; Raheem *et al.*, 2019; Kim *et al.*, 2019; Patel *et al.*, 2019; Tang *et al.*, 2019; Gahukar, 2020; Weru *et al.*, 2021; Doğan & Çekal, 2022; Ishara *et al.*, 2022; Liceaga, 2022; Papastavropoulou *et al.*, 2022a, 2022b; Aiello *et al.*, 2023; Torres-Castillo and Olazarán-Santibáñez, 2023; Granados-Echegoyen *et al.*, 2024; Siddiqui *et al.*, 2024). The percentage of each order consumed varies by region due to differences in local availability, cultural traditions, and consumer preferences (Ishara *et al.*, 2022; Liceaga, 2022). The rate of insects consumed by humans varies across regions, with estimates suggesting over 2 billion people worldwide include insects as part of their regular diet (van Huis, 2013; Liceaga, 2022; Omuse *et al.*, 2024). Nutritional value of edible insects is due to their high content of protein, essential amino acids, fat, fiber, vitamins, and minerals (Kouřimská and Adámková, 2016; Papastavropoulou *et al.*, 2022a).

Insects have a long history of use as food and medicine by different societies around the world (Ishara *et al.*, 2022), and their consumption is not a new phenomenon (Liceaga, 2022). These insects can be consumed whole, with wings and legs removed, or processed into powders and pastes (Lange and Nakamura, 2021). However, in some countries, the consumption of insects is still taboo or considered unpalatable, and their widespread adoption as a food source continues facing cultural and social barriers (Bresciani *et al.*, 2022; Liceaga, 2022). Some studies have shown that the perception of edible insects can be improved through education and exposure, and that their acceptance can be increased by incorporating them into daily meal recipes or by using them as ingredients in processed products (Liceaga, 2022).

3.2 INSECTS PROTEIN

The use of insects as a food and feed source has gained popularity in recent years as a result of their high nutritional value, environmental sustainability, and potential to address global food security concerns (Ishara *et al.*, 2022). One of the main nutrients that is found in insects is protein, whose content ranging from 40% to 75%, exceeding that of animal protein (12–34.5%) and plant proteins (7–50%) (Omuse *et al.*, 2024). Besides the high content and high-quality proteins, the advantages in the consume of edible insects include good content in essential amino acids, healthy fats, and presence of minerals and vitamins (Melgar-Lalanne *et al.*, 2019; Bresciani *et al.*, 2022; Liceaga, 2022), low environmental impact, low production costs, and the cultural acceptability in certain regions of the world (Hlongwane *et al.*, 2020; Ishara *et al.*, 2022).

Insect protein is highly digestible, with a biological value comparable to animal origin foods like meat, fish, or eggs. However, the nutritional value of insects can vary substantially depending on factors such as the specific insect species, their habitat, the insect's diet, the edible stage of development (egg, larva, nymph, or adult), sex, and the type of processing they undergo, such as being consumed whole (dehydrated, boiled, roasted, fried, pulverized, ground, pickled, smoked) (Meyer-Rochow *et al.*, 2021; Rodríguez-Rodríguez *et al.*, 2022). In addition, the storage of edible insects directly affects the content and availability of nutrients due to potential changes in the physicochemical properties of proteins and lipids (Zhao *et al.*, 2016; Nowakowski *et al.*, 2021; Granados-Echegoyen *et al.*, 2024).

In general, all the insect species are characterized by high protein levels in their composition. Certain insect species, such as crickets, grasshoppers, and mealworms, are particularly rich in protein, containing until 80% protein on a dry weight basis, being comparable to that of conventional dry meat and higher than in soy (Melo-Ruiz *et al.*, 2015; Payne *et al.*, 2016; Zielińska *et al.*, 2015; Churchward-Venne *et al.*, 2017; Dobermann *et al.*, 2017; Ghosh *et al.*, 2017; Akhtar and Isman, 2018; Bbosa *et al.*, 2019; González-Aguilar *et al.*, 2019; Kim *et al.*, 2019; Köhler *et al.*, 2019; Tang *et al.*, 2019; Gahukar, 2020; Bresciani *et al.*, 2022; Granados-Echegoyen *et al.*, 2024). In Hymenoptera insects like ants and wasps, proteins reach levels around 40–70% on a dry weight basis (Melo-Ruiz *et al.*, 2011; Payne *et al.*, 2016; Dobermann *et al.*, 2017; Cruz-Labana *et al.*, 2018; Kim *et al.*, 2019; Tang *et al.*, 2019; Gahukar, 2020; Baigts-Allende *et al.*, 2021; Granados-Echegoyen *et al.*, 2024). Other insect groups like caterpillars, termites, and some aquatic beetles also have a high protein content (40% in average) (Rumpold and Schlüter, 2013; Payne *et al.*, 2016; Dobermann *et al.*, 2017; Akhtar and Isman, 2018; Kim *et al.*, 2019; Melgar-Lalanne *et al.*, 2019; Tang *et al.*, 2019; Gahukar, 2020; Granados-Echegoyen *et al.*, 2024). Lepidoptera insects like caterpillars contain protein concentrations in the range of ranging around 60% on a dry weight basis (Melo-Ruiz *et al.*, 2011; Tang *et al.*, 2019; Weru *et al.*, 2021; Gahukar, 2020; Tang *et al.*, 2019; Kim *et al.*, 2019; Siddiqui *et al.*, 2024). Orthoptera insects represent one of the most consumed and most protein-rich insects, reaching in some cases protein values above 70% (Melo-Ruiz *et al.*, 2015; Payne *et al.*, 2016; Zielińska *et al.*, 2015; Churchward-Venne *et al.*, 2017; Dobermann *et al.*, 2017; Ghosh *et al.*, 2017; Akhtar and Isman, 2018; Bbosa *et al.*, 2019; González-Aguilar *et al.*, 2019; Kim *et al.*, 2019; Köhler *et al.*, 2019; Tang *et al.*, 2019; Gahukar, 2020; Granados-Echegoyen *et al.*, 2024). Blattodea species such as cockroaches, while also containing significant amounts of protein, generally have protein concentrations typically ranging from 50% to 70% on a dry weight basis (Rumpold and Schlüter, 2013; Tang *et al.*, 2019; Weru *et al.*, 2021). Isoptera insects, which include termites, have protein levels lower compared with other insects between 10% and 40% range on a dry weight basis (Rumpold and Schlüter, 2013; Payne *et al.*, 2016; Dobermann *et al.*, 2017; Kim *et al.*, 2019; Tang *et al.*, 2019; Gahukar, 2020; Granados-Echegoyen *et al.*, 2024). Coleoptera (beetles) and Hemiptera (true bugs) are two other major orders of insects that are known to contain significant amounts of protein, often in the 20–65% range on a dry weight basis, though the precise protein content can vary widely by species (Melo-Ruiz *et al.*, 2011; Rumpold and Schlüter, 2013; Payne *et al.*, 2016; Zielińska *et al.*, 2015; Ghosh *et al.*, 2017; Akhtar and Isman, 2018; Kim *et al.*, 2019; Köhler *et al.*, 2019; Tang *et al.*, 2019; Baigts-Allende *et al.*, 2021; Weru *et al.*, 2021; Granados-Echegoyen *et al.*, 2024). Diptera larvae like blackfly and housefly larvae also represent a potentially promising source of protein, containing around 20–60% protein on a dry weight basis (Payne *et al.*, 2016; Dobermann *et al.*, 2017; Bbosa *et al.*, 2019; Tang *et al.*, 2019; Baigts-Allende *et al.*, 2021; Granados-Echegoyen *et al.*, 2024). Weevils and cerambycid beetles contain 20–27% protein on a dry weight basis (Rumpold and Schlüter, 2013; Shah *et al.*, 2022) (Table 3.1).

TABLE 3.1 Several Edible Insects Reported in Literature. They Are Classified into Eight Main Groups, and Some of the Most Common Species under Each One of the Groups Are Mentioned.

ORDER/ INSECT TYPE	*SOME EDIBLE SPECIES*	*REFERENCES*
Blattodea		
Cockroaches	*Blaberus* sp., *Blaptica dubia, Blatta lateralis, Goliathus* spp., *Nauphoeta cinérea, Periplaneta americana, Periplaneta australasiae*	Rumpold and Schlüter, 2013; Raheem *et al.*, 2019; Roncolini *et al.*, 2019; Tang *et al.*, 2019; Liceaga, 2022; Papastavropoulou *et al.*, 2022b
Isoptera		
Termites	*Kalotermes flavicollis, Macrotermes bellicosus, Macrotermes falciger, Macrotermes natalensis, Macrotermes nigeriensis, Macrotermes* spp., *Macrotermes subhyalinus, Nasutitermes* spp., *Odontotermes formosanus, Odontotermes* sp., *Pseudacanthotermes militaris, Pseudacanthotermes spiniger*	Rumpold and Schlüter, 2013; Payne *et al.*, 2016; Dobermann *et al*, 2017; Kim *et al.*, 2019; Patel *et al.*, 2019; Tang *et al.*, 2019; Gahukar, 2020; Weru *et al.*,2021; Ishara *et al.*, 2022; Papastavropoulou *et al.*, 2022a, 2022b; Torres-Castillo *et al.*, 2023
Coleoptera		
Beetles	*Alphitobius diaperinus, Allomyrina dichotoma, Analeptes trifasciata, Aplagiognathus spinosus, Arophalus rusticus, Callipogon barbatus, Chalcophora* sp., *Copris nevinsoni, Cotinis nítida, Cybister flavocicinctus, Eulepida mashona, Gyrinus parcus, Heteroligus meles, Holotrichia parallela, Holotrichia* sp., *Homolepta* sp., *Metamasius spinolae, Oileus rimator, Oryctes boas, Oryctes monoceros, Oryctes owariensis, Orycetes rhinoceros, Pachnoda marginata peregrina, Passalus af. Punctiger, Phyllophaga* sp., *Prostephanus truncates, Protaetia brevitarsis, Rhantus atricolor, Rhynchophorus ferrugineus, Rhynchophorus phoenicis, Scyphophorus acupunctatus, Tenebrio molitor, Tesseratoma papillosa, Trichoderes pini, Zophobas morio*	Rumpold and Schlüter, 2013; Payne *et al.*, 2016; Zielińska *et al.*, 2015; Kouřimská and Adámková, 2016; Dobbermann *et al.*, 2017; Ghosh *et al.*, 2017; Ribeiro *et al.*, 2018; Kim *et al.*, 2019; Köhler *et al.*, 2019; Tang *et al.*, 2019; Gahukar, 2020; Pino-Moreno and Blasquez, 2021; Liceaga, 2022; Papastavropoulou *et al.*, 2022b; Torres-Castillo *et al.*, 2023; Granados-Echegoyen *et al.*, 2024

(*Continued*)

TABLE 3.1 (Continued)

ORDER/ INSECT TYPE	*SOME EDIBLE SPECIES*	*REFERENCES*
Diptera		
Flies	*Copestylum haggi, Copestylum anna, Drosophila melanogaster, Ephydra hians, Eristalis* sp., *Hermetia aurata, Hermetia illucens, Musca domestica, Piophila casei*	Rumpold and Schlüter, 2013; Payne *et al.*, 2016; Dobbermann *et al.*, 2017; Patel *et al.*, 2019; Baigths-Allende *et al.*, 2021; Pino-Moreno and Blasquez, 2021; Tang *et al.*, 2019; Papastavropoulou *et al.*, 2022b; Torres-Castillo *et al.*, 2023; Granados-Echegoyen *et al.*, 2024
Hemiptera		
Cicadas	*Cicadoidea, Hoplophorion monogramma, Krizousacorixa azteca, Meimuna opalifera Walke, Proarna hilaris*	Melo-Ruiz *et al.*, 2011; Kim *et al.*, 2019; Tang *et al.*, 2019; Weru *et al.*, 2021
Stink bugs	*Abedus* sp., *Acantocephala declivi, Agonoscelis pubescens, Apongopus chinensis, Ascra cordifera, Aspongubus viduatus, Atizies taxcoensis, Axayacatl, Belostomatidae, Corizus hyoscyami, Edessa conspersa, Edessa montezumae, Edessa petersii, Edessa* sp., *Encosternum* spp., *Euchistus zopilotensis, Euschistus egglestoni, Euschistus* sp., *Euschistus strennus, Hoplophorion monograma, Hyalopterus pruni, Krizousacorixa azteca, Meimuna opalifera, Neortholomus* sp., *Nezara viridula, Pachilis gigas, Proarna* sp., *Udonga montana, Umbonia reclinata*	Rumpold and Schlüter, 2013; Payne *et al.*, 2016; Kim *et al.*, 2019; Patel *et al., 2019*; Tang *et al.*, 2019; Baigths-Allende *et al.*, 2021
Hymenoptera		
Bees	*Apis cerana, Apis mellifera, Apis mellifera adansonii, Apis mellifera mellifera,* Bee brood, *Lestrimelitta chamelensis, Melipona beeckeii, Melipona fasciata, Nannotrigona testaceicornis, Plebeia mexica, Plebeia* sp., *Scaptotrigona hellwegeri, Scaptotrigona mexicana, Trigona jaty, Trigona angustula, Trigona clavipes, Trigona trinidadensis, Vespula lewisii, Vespula squamosa, Vespula* spp.	Acuña *et al.*, 2011; Rumpold and Schlüter, 2013; Payne *et al.*, 2016; Kim *et al.*, 2019; Tang *et al.*, 2019; Weru *et al.*, 2021; Ishara *et al.*, 2022; Papastavropoulou *et al.*, 2022a; Granados-Echegoyen *et al.*, 2024

(*Continued*)

TABLE 3.1 (Continued)

ORDER/ INSECT TYPE	*SOME EDIBLE SPECIES*	*REFERENCES*
Ants	*Atta cephalotes, Atta laevigata, Atta mexicana, Atta sexdens, Atta texana, Camponotus* spp., *Carebara lignata, Carebara vidua, Liometopum apiculatum, Liometopum occidentale, Liometopum occidentale var. Luctuosum, Melophorus bagoti, Myrmecosystus melliger, Myrmiicinae carebara, Oecophylla smaragdina, Oecophylla* spp., *Pogonomyrmex barbatus, Pogonomyrmex* sp., *Polyrhachis dives, Polyrhachis vicina*	Acuña *et al.*, 2011; Melo-Ruiz *et al.*, 2011; Rumpold and Schlüter, 2013; van Huis, 2013; Lara-Juárez *et al.*, 2015; Payne *et al.*, 2016; Kouřimská and Adámková, 2016; Dobermann *et al.*, 2017; Raheem *et al.*, 2019; Kim *et al.*, 2019; Patel *et al.*, 2019; Tang *et al.*, 2019; Gahukar, 2020; Weru *et al.*, 2021; Liceaga, 2022; Papastavropoulou *et al.*, 2022a; Torres-Castillo *et al.*, 2023; Granados-Echegoyen *et al.*, 2024;
Wasps	*Brachygastra azteca, Brachygastra mellifica, Brachistraga* sp., *Dolichovespula* spp., *Mischocyttarus basimacula, Mischocyttarus cubensis mexicanus, Mischocyttarus pallidipectus, Mischocyttarus* sp., *Parachartegus apicalis, Polistes canadensis, Polistes instabilis, Polistes major, Polybia ignobilis, Polybia occidentalis bohemani, Polybia occidentalis nigratella, Polybia parvulina, Vespa* sp.	Acuña *et al.*, 2011; Rumpold and Schlüter, 2013; Raheem *et al.*, 2019; Kim *et al.*, 2019; Patel *et al.*, 2019; Tang *et al.*, 2019; Baigths-Allende *et al.*, 2021; Granados-Echegoyen *et al.*, 2024
Lepidoptera		
Butterflies	*Catasticta teutila, Eucheira socialis sociales, Eurema salome Jamapa, Phasus triangularis, Phasus* sp., *Phoebis agarithe agarithe, Phoebis philea philea, Phoebis sennae marcellina, Pterourus multicaudata multicaudata*	Ramos-Elourdy *et al.*, 2011; Rumpold and Schlüter, 2013; Weru *et al.*, 2021
Moths	*Aegiale esperiaris, Agrotis infusa, Anaphe infracta, Anaphe panda, Anaphe recticulata, Anaphe venata, Arsenura armada, Ascalapha odorata, Cirina forda, Coloradia pandora, Comadia redtenbacheri, Euxoa auxiliaries, Galleria mellonella, Gonimbrasia belina, Gonometa postica, Heliothis zea, Hylesia frígida, Imbrasia epimethea, Imbrasia oyemensis, Imbrasia opimethea, Imbrasia truncata, Laniifera cyclades, Latebraria amphipyroides, Nudaurelia oyemensis, Omphisa fuscidentalis, Paradirphia fumosa, Samia cynthia ricinii, Spodoptera* spp., *Usta terpsichore, Xyleutes leucomochla, Xyleutes redtembacheri*	Longvah *et al.*, 2011; Acuña *et al.*, 2011; Melo-Ruiz *et al.*, 2011; Rumpold and Schlüter, 2013; Payne *et al.*, 2016; Dobbermann *et al.*, 2017; Ribeiro *et al.*, 2018; Raheem *et al.*, 2019; Kim *et al.*, 2019; Patel *et al.*, 2019; Tang *et al.*, 2019; Gahukar, 2020; Liceaga, 2022; Torres-Castillo *et al.*, 2023;

(*Continued*)

TABLE 3.1 (Continued)

ORDER/ INSECT TYPE	*SOME EDIBLE SPECIES*	*REFERENCES*
Odonata		
Dragonflies	*Aeschna multicolor, Anax parthenope, Anax* sp., *Crocothemes servillia, Epophthalmia elegans, Gomphus cuneatus, Ictinogomphus rapax, Orthetrum pruinosum, Pantala* sp., *Sinictinogomphus clavatus, Sympetrum* sp.	Rumpold and Schlüter, 2013; Kim *et al.*, 2019; Siddiqui *et al.*, 2024
Damselflies	*Lestes praemorsus*	Siddiqui *et al.*, 2024
Orthoptera		
Crickets	*Acheta domesticus, Allonemobius fasciatus, Anurogryllus arboreus, Arphia fallax, Brachytrupes membranaceus, Brachytrupes portentosus, Brachytrypes* spp., *Gryllodes sigillatus, Gryllus assimilis, Gryllus bimaculatus, Gryllus campestris, Gryllus locorojo, Henicus whellani, Teleogryllus emma*	Rumpold and Schlüter, 2013; Zielińska *et al.*, 2015; Kouřimská and Adámková, 2016; Churchward-Venne *et al.*, 2017; Dobbermann *et al.*, 2017; Ghosh *et al.*, 2017; Ribeiro *et al.*, 2018; Raheem *et al.*, 2019; Kim *et al.*, 2019; Köhler *et al.*, 2019; Patel *et al.*, 2019; Tang *et al.*, 2019; Weru *et al.*, 2021; Ishara *et al.*, 2022; Papastavropoulou *et al.*, 2022a, 2022b; Liceaga, 2022; Aiello *et al.*, 2023; Torres-Castillo *et al.*, 2023
Grasshoppers	*Acanthacris ruficornis, Acrida cinérea, Acrida exaltata, Acoeypha picta, Acorypha glaucopsis, Acrida bicolor, Aidemona azteca, Boopedon flaviventris, Conocephalus triops, Cyrtacenthecuis tartarica, Encoptolophus herbaceus, Hieroglyphus banian, Lamercleance cucullata, Mecopoda elongate, Melanoplus differentialis, Melanoplus femurrubrum, Melanoplus mexicanus, Orphulella orizabae, Orphulella tolteca, Orphulella quiroga, Orphulella* sp., *Oxya fuscovittata, Oxya japónica, Oxya* spp., *Oxya velox, Oxya yezoensis, Pterophylla beltrani, Romalea colorata, Ruspolia differens, Ruspolia nitidula, Schistocerca gregaria, Schistocerca piceiforns, Spharagemon equale, Sphenarium histrio, Sphenarium mexicanum, Sphenarium purpurascens, Sphenarium* sp., *Taeniopoda eques, Tropidacris latreillei, Zonocerus variegatus*	Melo-Ruiz *et al.*, 2011; Rumpold and Schlüter, 2013; van Huis, 2013; Melo-Ruiz *et al.*, 2015; Payne *et al.*, 2016; Zielińska *et al.*, 2015; Kouřimská and Adámková, 2016; Churchward-Venne *et al.*, 2017; Dobbermann *et al.*, 2017; Ribeiro *et al.*, 2018; Raheem *et al.*, 2019; Kim *et al.*, 2019; Mishyna *et al.*, 2019; Patel *et al.*, 2019; Tang *et al.*, 2019; Gahukar, 2020; Pino-Moreno and Blasquez, 2021; Liceaga, 2022; Papastavropoulou *et al.*, 2022a, 2022b; Torres-Castillo *et al.*, 2023; Granados-Echegoyen *et al.*, 2024

(Continued)

TABLE 3.1 (Continued)

ORDER/ INSECT TYPE	*SOME EDIBLE SPECIES*	*REFERENCES*
Locusts	*Brachytrupes membranaceus, Chortoicetes terminifera, Cytacanthacris aeruginosus, Idiarthron subquadratum, Locusta migratoria, Patanga succincta, Schistocerca gregaria*	Oonincx *et al.*, 2010; Rumpold and Schlüter, 2013; Kouřimská and Adámková, 2016; Ribeiro *et al.*, 2018; Akhtar *et al.*, 2018; Kim *et al.*, 2019; Köhler *et al.*, 2019; Ishara *et al.*, 2022; Liceaga, 2022; Papastavropoulou *et al.*, 2022b
Other		
Worms	*Alphitobius diaperinus, Hypopta agavis, Laniifera cyclades, Rhynchophorus phoenicis, Zophobas atratus*	Payne *et al.*, 2016; Ribeiro *et al.*, 2018; Patel *et al.*, 2019; Papastavropoulou *et al.*, 2022a, 2022b
Silkworms/ Caterpillars	*Anaphe infracta, Anaphe panda, Antheraea pernyi, Antheraea* spp., *Bombyx japanensis, Bombyx mori, Bunaea alcinoe, Cirina forda, Gonimbrasia belina, Imbrasia alopia, Imbrasia anthina, Imbrasia belina, Imbrasia ebilis, Imbrasia epimethea, Imbrasia melanops, Imbrasis obsura, Imbrasia oyemensis, Imbrasia petiveri, Imbrasia swahlbergii, Omphisa fuscidentalis, Spodoptera littoralis*	van Huis, 2013; Payne *et al.*, 2016; Kouřimská and Adámková, 2016; Ribeiro *et al.*, 2018; Raheem *et al.*, 2019; Kim *et al.*, 2019; Köhler *et al.*, 2019; Patel *et al.*, 2019; Gahukar, 2020; Liceaga, 2022; Papastavropoulou *et al.*, 2022a, 2022b; Aiello *et al.*, 2023
Weevils	*Rhynchophorus ferrugineus, Rhynchophorus palmarum, Rhynchophorus phoenicis, Rhynchoporus* spp., *Scyphophorus acupunctatus, Sitophilus* sp.	van Huis, 2013; Akhtar & Isman, 2018; Raheem *et al.*, 2019; Kim *et al.*, 2019; Patel *et al.*, 2019; Gahukar, 2020; Pino-Moreno and Blasquez, 2021; Gorbunova and Zakharov, 2021; Papastavropoulou *et al.*, 2022b

3.3 QUALITY AND BIOLOGICAL VALUE OF INSECT PROTEINS

Many scientific researchers have demonstrated that edible insects are rich in nutritional value, containing all the main biomolecules as proteins, fat, carbohydrates, vitamins, and minerals at levels recommended to human requirements (Kouřimská and Adámková, 2016; Zielińska *et al.*, 2015; Siddiqui *et al.*, 2024). Insect proteins are not only contained in high amounts but also they are highly digestible and of good biological quality, with amino acid profiles comparable to those from meat, eggs, and dairy products (Manditsera *et al.*, 2019; Meyer-Rochow *et al.*, 2021). The biological value of insect proteins depends on the digestibility of the protein, and this depends largely on the degree of protein hydrolysis after digestion and the amino acid absorption intestine capacity. For insects' proteins, this parameter results favorably compared to conventional protein sources (Liceaga, 2022). Digestibility of insect proteins is generally high, ranging from 77% to 98% depending on the specific insect species and processing methods used. Digestibility of insects' proteins is on par with that of typical animal-derived proteins. Biological value of proteins in insects is due to the diversity of essential amino acids they contain, including lysine, leucine, isoleucine, threonine, and valine, which are critical for human growth and development (Köhler *et al.*, 2019; Oonincx *et al.*, 2021).

Amino acids composition of insects can vary depending on factors like the insect species, life stage, feed, and processing methods. Amino acid content on insects provide a complete protein source that meets human essential amino acid requirements (Rumpold and Schlüter, 2013; Kouřimská and Adámková, 2016; Meyer-Rochow *et al.*, 2021; Rodríguez-Rodríguez *et al.*, 2022). Amino acids in different insects' species have been determined, showing rich amounts of essential amino acids like histidine, isoleucine, leucine, lysine, methionine, phenylalanine, threonine, tryptophan, and valine (Tables 3.2 and 3.3) (Kouřimská and Adámková, 2016). Compared with those amino acids from another animal species, amino acids from insects are high quality and highly digestible (Rumpold and Schlüter, 2013; Meyer-Rochow *et al.*, 2021; Shah *et al.*, 2022).

TABLE 3.2 Protein Content Reported in Some Species of the Most Consumed Edible Insects Worldwide.

ORDER/INSECT TYPE	*EXAMPLES (SPECIES OR PRODUCTS)*	*PROTEIN CONTENT RANGE (%)*	*REFERENCES*
Blattodea			
Cockroaches	*Blaberus* sp.	57.30	Rumpold and Schlüter, 2013
	Blaptica dubia	64.40	Weru *et al.*, 2021
	Blatta lateralis	76.00	Tang *et al.*, 2019
	Periplaneta americana	65.60–66.00	Rumpold and Schlüter, 2013; Tang *et al.*, 2019
	Periplaneta australasiae	62.4–63.00	Rumpold and Schlüter, 2013; Tang *et al.*, 2019

(Continued)

TABLE 3.2 (Continued)

ORDER/INSECT TYPE	*EXAMPLES (SPECIES OR PRODUCTS)*	*PROTEIN CONTENT RANGE (%)*	*REFERENCES*
Isoptera			
Termites	*Macrotermes bellicosus*	10.06–35.30	Dobermann *et al.*, 2017; Kim *et al.*, 2019; Tang *et al.*, 2019; Gahukar, 2020
	Macrotermes falciger	43.26	Kim *et al.*, 2019
	Macrotermes notalensis	22.10	Kim *et al.*, 2019
	Macrotermes spp.	3.90–18.70	Payne *et al.*, 2016;
	Microtermes bellicosus	40.70	Granados-Echegoyen *et al.*, 2024
	Microtermes nigerensis	37.50	Granados-Echegoyen *et al.*, 2024
	Odontotermes sp.	33.67	Gahukar, 2020;
Coleoptera			
Beetles	*Analeptes trifasciata*	29.62	Rumpold and Schlüter, 2013
	Aplagiognathus spinosus	26.00	Rumpold and Schlüter, 2013
	Arophalus rusticus	20.10	Rumpold and Schlüter, 2013
	Callipogon barbatus	5.00–41.00	Akhthar and Isman, 2018; Tang *et al.*, 2019
	Copris nevinsoni	54.43	Rumpold and Schlüter, 2013
	Cybister flavocicinctus	69.01	Rumpold and Schlüter, 2013
	Heteroligus meles	38.10	Kim *et al.*, 2019
	Holotrichia sp.	28.92–51.74	Rumpold and Schlüter, 2013; Köhler *et al.*, 2019
	Homolepta sp.	54.00	Rumpold and Schlüter, 2013
	Oileus rimator	21.30	Weru *et al.*, 2021
	Oryctes boas	26.00	Kim *et al.*, 2019
	Oryctes rhinoceros	52.00	Granados-Echegoyen *et al.*, 2024
	Protaetia bravitarsis	44.20	Granados-Echegoyen *et al.*, 2024
	Rhynchophorus palmarum	5.00–35.00	Akhthar and Isman, 2018
	Rhynchophorus phoenicis	5.20–50.01	Payne *et al.*, 2016; Kim *et al.*, 2019 Tang *et al.*, 2019
	Tenebrio molitor	10.00– 53.22	Payne *et al.*, 2016; Zielińska *et al.*, 2015; Ghosh *et al.*, 2017; Akhtar and Isman, 2018; Tang *et al.*, 2019;

(*Continued*)

TABLE 3.2 (Continued)

ORDER/INSECT TYPE	*EXAMPLES (SPECIES OR PRODUCTS)*	*PROTEIN CONTENT RANGE (%)*	*REFERENCES*
	Zophobas morio	45.00–46.00	Tang *et al.*, 2019; Granados-Echegoyen *et al.*, 2024
Diptera			
Flies	*Caliphora vomitoria*	64.90	Bbosa *et al.*, 2019; Granados-Echegoyen *et al.*, 2024
	Copestylum anna	37.00	Tang *et al.*, 2019
	Ephydra hians	42.00	Tang *et al.*, 2019
	Hermetia illucens	17.50–49.00	Payne *et al.*, 2016; Tang *et al.*, 2019; Baigths-Allende *et al.*, 2021; Granados-Echegoyen *et al.*, 2024
	Musca domestica	48.00–49.48	Dobbermann *et al.*, 2017; Tang *et al.*, 2019
Hemiptera			
Cicadas	*Hoplophorion monogramma*	62.00	Tang *et al.*, 2019
	Krizousacorixa azteca	53.00– 64.00	Melo-Ruiz *et al.*, 2011; Tang *et al.*, 2019
Stink bugs	*Ascra cordifera*	32.03	Baigths-Allende *et al.*, 2021
	Aspongopus nepalensis	10.60	Granados-Echegoyen *et al.*, 2024
	Aspongupus viduatus	27.00	Tang *et al.*, 2019
	Euschistus egglestoni	46.00	Tang *et al.*, 2019
	Pachilis gigas	65.00	Tang *et al.*, 2019
	Umbonia reclinata	29.00	Tang *et al.*, 2019
Hymenoptera			
Bees	*Apis mellifera*	7.01–51.00	Payne *et al.*, 2016; Kim *et al.*, 2019; Tang *et al.*, 2019
	Bee brood	40.00	Tang *et al.*, 2019
Ants	*Atta cephalotes*	43.00	Granados-Echegoyen *et al.*, 2024
	Atta mexicana	46–66.00	Melo-Ruiz *et al.*, 2011; Tang *et al.*, 2019; Granados-Echegoyen *et al.*, 2024
	Carebara vidua	40.83	Gahukar, 2020
	Liometopum apiculatum	40.90–93.92	Melo-Ruiz *et al.*, 2011; Cruz-Labana *et al.*, 2018
	Oecophylla smaragdina	53.46–55.30	Dobbermann *et al.*, 2017; Gahukar, 2020;

(*Continued*)

TABLE 3.2 (Continued)

ORDER/INSECT TYPE	*EXAMPLES (SPECIES OR PRODUCTS)*	*PROTEIN CONTENT RANGE (%)*	*REFERENCES*
Wasps	*Brachygastra mellifica*	38.24	Baigths-Allende *et al.*, 2021
	Brachygastra azteca	70.00	Tang *et al.*, 2019
	Parachartegus apicalis	50.00	Tang *et al.*, 2019
Lepidoptera			
Butterflies			
Moths	*Aegiale hesperiaris*	30.88	Melo-Ruiz et al., 2011
	Anaphe venata	58.00	Tang *et al.*, 2019
	Ascalapha odorata	56.00	Weru *et al.*, 2021
	Cirina forda	31.4	Gahukar, 2020;
	Imbrasia epimethea	62.00	Tang *et al.*, 2019
	Imbrasia truncata	65.00	Tang *et al.*, 2019
	Usta terpsichore	76.00	Tang *et al.*, 2019
Odonata			
Dragonflies	*Aeschna multicolor*	54.24	Kim *et al.*, 2019
	Anax parthenope	65.76	Siddiqui *et al.*, 2024
	Anax sp.	26.22	Kim *et al.*, 2019
	Crocothemes servillia	70.48	Siddiqui *et al.*, 2024
	Epophthalmia elegans	65.23	Siddiqui *et al.*, 2024
	Gomphus cuneatus	64.64	Siddiqui *et al.*, 2024
	Ictinogomphus rapax	62.37	Siddiqui *et al.*, 2024
	Orthetrum pruinosum	71.53	Siddiqui *et al.*, 2024
	Pantala sp	45.14	Siddiqui *et al.*, 2024
	Sinictinogomphus clavatus	63.34	Siddiqui *et al.*, 2024
	Sympetrum sp	76.75	Siddiqui *et al.*, 2024
Damselflies	*Lestes praemorsus*	46.37	Siddiqui *et al.*, 2024
Orthoptera			
Crickets	*Acheta domesticus*	23.70–65.04	Payne *et al.*, 2016; Dobbermann *et al.*, 2017; Köhler *et al.*, 2019; Tang *et al.*, 2019;
	Brachytrypes spp.	6.00–6.25	Kim *et al.*, 2019
	Brachytrypes orientalis	65.70	Gahukar, 2020; Granados-Echegoyen *et al.*, 2024
	Gryllodes silligatus	56.00–71.70	Zielińska *et al.*, 2015; Tang *et al.*, 2019
	Gryllus assimilis	56.00	Tang *et al.*, 2019
	Gryllus bimaculatus	58.32	Ghosh *et al.*, 2017 Kim *et al.*, 2019

(Continued)

TABLE 3.2 (Continued)

ORDER/INSECT TYPE	*EXAMPLES (SPECIES OR PRODUCTS)*	*PROTEIN CONTENT RANGE (%)*	*REFERENCES*
	Schistocerca sp.	61.05–80.3	Dobbermann *et al.*, 2017; Granados-Echegoyen *et al.*, 2024
	Schistocerca gregaria	75.10–76.90	Zielińska *et al.*, 2015;
	Teleogryllus emma	55.65	Ghosh *et al.*, 2017; Kim *et al.*, 2019
Grasshoppers	*Chondacris rosea*	68.88	Gahukar, 2020;
	Melanoplus femurrubrum	74.00–75.40	Melo-Ruiz *et al.*, 2015
	Melanoplus mexicanus	77.00	Churchward-Venne *et al.*, 2017
	Ruspolia differens	44.30–44.59	Dobbermann *et al.*, 2017; Köhler *et al.*, 2019
	Ruspolia nitidula	40.80	Bbosa *et al.*, 2019; Gahukar, 2020; Granados-Echegoyen *et al.*, 2024
	Schistocerca spp.	61.9–63.1	Melo-Ruiz *et al.*, 2015
	Sphenarium histrio	76.55–77.95	Melo-Ruiz *et al.*, 2015
	Sphenarium purpurascens	35.00–76.47	Melo-Ruiz *et al.*, 2011; Melo-Ruiz *et al.*, 2015; Dobbermann *et al.*, 2017; Akhtar, 2018; González-Aguilar *et al.*, 2019
	Taeniopoda eques		Melo-Ruiz *et al.*, 2015
	Zonocerus variegatus	26.80	Köhler *et al.*, 2019
Locusts	*Patanga succincta*	36.31	Köhler *et al.*, 2019
Others			
Worms	*Galleria mellonella*	38.01	Dobbermann *et al.*, 2017
	Mealworms fresh	10.20–57.80	Liceaga, 2022; Papastavropoulou *et al.*, 2022b
Silkworms/ caterpillars	*Anaphe pande*	50.89	Gahukar, 2020
	Bombyx mori	10.10–61.80	Payne *et al.*, 2016; Dobbermann *et al.*, 2017; Köhler *et al.*, 2019
	Cirina forda	47.48	Dobbermann *et al.*, 2017
	Elaphrodes lacteal	53.55	Gahukar, 2020;
	Imbrasia belina	55.41	Gahukar, 2020;
	Imbrasia hecate	51.40	Gahukar, 2020;
	Imbrasia epimethea	57.84	Gahukar, 2020;

(Continued)

TABLE 3.2 (Continued)

ORDER/INSECT TYPE	EXAMPLES (SPECIES OR PRODUCTS)	PROTEIN CONTENT RANGE (%)	REFERENCES
	Imbrasia oyemensis	51.68	Gahukar, 2020;
	Imbrasia petiveri	57.92	Gahukar, 2020;
	Imbrasia truncate	54.49	Gahukar, 2020;
	Samia cynthia ricinii	54.70	Dobbermann *et al.*, 2017
Weevil	*Scyphophorus acupunctatus*	36.80	Ramos-Elorduy *et al.*, 1997
Salmon		22.20 (fresh)	Liceaga, 2022
Chicken		22.2 (fresh), 54.70 (dry basis)	Liceaga, 2022; Ghosh *et al.*, 2017
Beef		22.50 (fresh), 40.50 (dry basis)	Liceaga, 2022; Ghosh *et al.*, 2017
Pork		21.00 (fresh), 27.70 (dry basis)	Liceaga, 2022; Ghosh *et al.*, 2017

There are several studies focused on to determine the *in vitro* digestibility of insects' proteins. These studies vary into the insects probed, the way to obtain them, the hydrolysis method used, the hydrolysis phases, the enzymes used in each phase, and the methods to corroborate protein hydrolysis. In most of the results obtained, the method of digestion, the method used to assess digestibility and the species, or the developmental stage of the studied insect were the factors that significatively affected to the insects' protein digestibility. Nevertheless, in spite of these differences, the protein digestibility of insects can be considered to be high (80–90%) being similar to fish and soy meals (Rodríguez-Rodríguez *et al.*, 2022).

Some investigations have also considered the influence of the processing steps or modifications in insects prior the hydrolysis, such as defatting, freeze-drying, vacuum-packaging, drying, boiling, frying, blanching, freezing, roasting, extruding, and others. Among the positive effects of processing methods in protein digestibility, we can mention for instance, degutting, which eliminates the fiber contained in the intestinal tract of those species that consuming fiber plants; blanching, which inhibits the enzymatic browning process and could cause protein denaturation; extrusion, that destroys antinutritional factors and undesirable enzymes; protein isolation, helps to the elimination of chitin in the protein fraction; heat processing at 60 °C in a drying oven that helps to avoid polymerization of the protein particles (Manditsera *et al.*, 2019; Rodríguez-Rodríguez *et al.*, 2022; Hammer *et al.*, 2023).

3.4 CHARACTERISTICS OF INSECT PROTEINS

Characteristics of insect proteins are similar to those of animal proteins in terms of amino acid profiles, solubility, digestibility, pH sensitivity, water-holding capacity (WHC), oil-holding capacity (OHC), gelling ability, emulsifying, foamability, rheological, and texture

TABLE 3.3 Amino Acids Composition and Content Found in Some Edible Insects' Species Compared with the Frequently Consumed Meats.

INSECTS TESTED	*AMINO ACID*	*CONCENTRATION RANGE (G/100G)*	*CONCENTRATION IN MEAT, BEEF/PORK/CHICKEN (G/100 G)*	*REFERENCES*
Acheta domesticus, Aeschna multicolor, American cockroach, *Anaphe venata, Anax parthenope, Anax* sp., *Apis ceranu, Apis dorsata, Apis floren, Apis mellifera, Ascra cordifera, Aspongubus viduatus, Atta mexicana,* Bee brood, *Blatta lateralis,* Bombay locust, *Bombus ignites, Bombus terrestris, Bombyx mori, Boopedon flaviventris, Brachygastra azteca, Brachygastra mellifica, Callipogon barbatus, Copestylum anna,* Cricket flour, *Crocothemes servillia, Ephydra hians, Epophthalmia elegans, Euschistus egglestoni, Glyptotendipes testaceus, Gomphus cuneatus, Gonimbrasia belina, Gryllodes sigillatus, Hermetia illucens, Holotrichia* sp., *Hoplophorion monogramma, Ictinogomphus rapax, Imbrasia epimethea, Imbrasia truncata, Krizousacorixa azteca, Lestes praemorsus, Liometopum apiculatum, Macrotermes bellicosus, Melanoplus ferrumbum, Musca domestica, Orthetrum pruinosum, Pachilis gigas, Pantala* sp., *Parachartegus apicalis, Patanga succincta, Periplaneta americana, Periplaneta australasiae, Pogonomyrmex occidentalis, Polistes sagittarius, Polistes sulcatus, Polybia nigratella, Polybia occidentalis, Polybia parvulina, Rhynchophorus phoenicis, Samia ricinii, Sarcophaga bullata, Schistocerca* spp., *Schistocerca gregaria, Sinictinogomphus clavatus, Sphenarium histrio, Sphenarium purpurascens, Sympetrum* sp., *Taeniopoda eques, Tenebrio molitor, Umbonia reclinata, Usta terpsichore, Vespa aurara, Vespa basalis, Vespa mandarinia, Vespa tropica duenlis, Vespa velutina, Zophobas morio*	Alanine**	0.97–10.95	1.32/1.25/1.15	Longvah *et al.*, 2011; Melo-Ruiz *et al.*, 2011; Belluco *et al.*, 2013; Melo-Ruiz *et al.*, 2015; Payne *et al.*, 2016; Zielińska *et al.*, 2015; Churchward-Venne *et al.*, 2017; Akhtar and Isman, 2018; Cruz-Labana *et al.*, 2018; Köhler *et al.*, 2019; Tang *et al.*, 2019; Roncolini *et al.*, 2019; Baigts-Allende *et al.*, 2021; Mishyna *et al.*, 2021; Ojha *et al.*, 2021; Oonincx *et al.*, 2021; Papastavropoulou *et al.*, 2022b; Yang *et al.*, 2023; Granados-Echegoyen *et al.*, 2024; Siddiqui *et al.*, 2024
	Arginine**	0.69–7.88	1.30/1.33/1.34	
	Aspartic acid**	1.54–13.97	2.07/1.13/1.91	
	Valine*	0.84–9.79	1.02/1.21/1.04	
	Glutamic acid**	2.03–16.75	3.61/3.26/2.85	
	Glycine**	0.78–7.82	1.08/1.02/0.90	
	Threonine*	0.75–4.42	0.87/0.94/0.95	
	Isoleucine*	0.69–6.91	0.92/1.14/0.98	
	Leucine*	1.04–10.00	1.82/1.74/1.64	
	Histidine*	0.42–5.09	0.82/0.82/0.69	
	Cysteine/Cystine**	0.08– 3.60	0.23/0.29/0.25	
	Lysine*	1.03–8.64	1.94/1.80/1.79	
	Methionine*	0.36–4.50	0.61/0.59/0.62	
	Proline**	1.02–9.15	0.89/0.85/0.73	
	Serine**	0.82–5.94	0.86/0.90/0.88	
	Tryptophan*	0.41– 1.63	0.24/0.22/0.28	
	Tyrosine**	1.88–9.57	2.24/0.90/2.10	
	Phenylalanine*	0.50–14.40	0.86/0.83/0.82	

*Essential; **Non essential

ability. However, the protein profile changes even among insects in the same species depending on life stages. They can generate a particular group of proteins showing different behavior. Considering these variations, it is important to know about physicochemical characteristics of insect proteins and how the variation of factors such as heat, pH, and NaCl concentration, among others can modify their intrinsic properties (Gravel and Doyen, 2019; Sales-Queiroz *et al.*, 2023).

3.4.1 Functional Characteristics

The processing of insect proteins is essential for enhancing their functionality and making them suitable for incorporation into food products. The key processing techniques include thermal processing (drying, roasting, and boiling), high-pressure processing, and enzymatic hydrolysis, each of which has specific effects on the nutritional and functional properties of the resulting insect protein ingredients (Ojha *et al.*, 2021; Pan *et al.*, 2022). Thermal processing methods are commonly used to improve the digestibility of insect proteins by deactivating or denaturing certain antinutritional factors. These methods can also reduce allergenicity, making insect proteins safer for consumption. However, thermal processing can also affect the sensory characteristics of insect proteins, such as their flavor and color, which need to be carefully controlled (Sales-Queiroz *et al.*, 2023; Ojha *et al.*, 2021). High-pressure processing is another technique used to enhance the functional properties of insect proteins. It can improve the solubility, emulsification, and gelling abilities of insect proteins, making them more versatile in food formulations. This nonthermal processing method preserves the nutritional quality of proteins while enhancing their techno-functional properties, which are essential for food applications (Sales-Queiroz *et al.*, 2023). Enzymatic hydrolysis involves breaking down insect proteins into smaller peptides and amino acids using proteases. This process enhances the techno-functional and bioactive properties of insect proteins, making them more suitable for a range of food applications. Hydrolyzed insect proteins often exhibit improved solubility, emulsification, and antioxidant properties, which can be leveraged in various food products (Purschke *et al.*, 2018; Leni *et al.*, 2020).

3.4.1.1 Amino Acid Profile

Most insect species studied contain significant amounts of essential amino acids like leucine, isoleucine, valine, lysine, and methionine. Insects like *Tenebrio molitor* (mealworm) larvae and various cricket species also exhibit high levels of essential amino acids, making them a viable alternative protein source. In some cases, insects can meet or exceed the daily amino acid requirements of human adults and children (Akhtar and Isman, 2018; Baigts-Allende *et al.*, 2021; Papastavropoulou *et al.*, 2022b). Methionine and cysteine, both sulfur-containing amino acids, are critical for human nutrition. However, cysteine is often found in lower concentrations or is absent in certain insect species like black soldier fly larvae and jumil (Baigths-Allende *et al.*, 2021). Despite this, methionine content in these species is usually sufficient, helping to meet the sulfur amino acid requirements. Besides, the amino acid profile of insects can be significantly influenced by their diet and environmental conditions. For example, insects reared in controlled environments with specific diets can have higher protein and amino acid contents compared to wild-caught insects (Oonincx and Finke, 2021). Thus, insects provide a rich source of amino acids, particularly essential amino acids;

however, variability in amino acid content across different species and environmental conditions suggests that further research and optimization are necessary to use the nutritional potential of insect proteins (Baigts-Allende *et al.*, 2021).

3.4.1.2 Solubility

Solubility of insect proteins and derived ingredients is an important functional property that depends on protein structure, charge, size, amino acid composition, structural conformation, and pH (Mintah *et al.*, 2020; Mishyna *et al.*, 2021; Sales-Queiroz *et al.*, 2023). High protein solubility is often a good indicator that protein is highly digestible, which is a desirable characteristic for food applications (Nikolić and Tepavčević, 2024). Apart from intrinsic and extrinsic variations due to insect species, extraction method, and processing conditions including ionic strength, pH changes, and temperature, its functionality is related on how the protein interacts with the solvent used (Gravel and Doyen, 2019; Ma *et al.*, 2023; Sales-Queiroz *et al.*, 2023). Solubility of insect proteins can range from around 50–90%, which is comparable to that of soy protein, faba beans, and yellow peas (Mishyna *et al.*, 2021; Ma *et al.*, 2023). Proteins with intact structures are more susceptible to interact with water, which enhances their solubility as hydrophilic amino acid side groups are predominantly at the surface (Mintah *et al.*, 2020¸ Sales-Queiroz *et al.*, 2023). Additionally, smaller protein sizes contribute to a better solubility. One of the most significant factors affecting the solubility of insect proteins is the pH level. Generally, insect proteins show the lowest solubility near their isoelectric point as it happens in other proteins, where the net charge of the protein is neutral, resulting in reduced interaction with water molecules (Sales-Queiroz *et al.*, 2023). However, insect proteins display significantly higher solubility under alkaline conditions. Mishyna *et al.* (2021) reported that protein solubility exceeds 90% at pH 10–11 when alkaline extraction and isoelectric precipitation methods are employed. This increased solubility under alkaline conditions is due to the repulsion between highly charged protein molecules, allowing them to better interact with water. In addition, the highest solubility is generally observed at extreme pH values because more hydrophobic groups are exposed as a result of protein unfolding (Ma *et al.*, 2023).

3.4.1.3 pH

The pH of insect proteins generally falls within the acidic range, between pH 4 and 6, making them suitable for food formulations and being lower than animal-derived proteins. Moreover, the functional properties of insect proteins, including solubility, emulsification, and foaming, are significantly influenced by pH. For example, the solubility of *Locusta migratoria* proteins is higher at pH 9, emulsifying activity peaks at pH 5, while foaming ability improves at pH 3, especially with the addition of 3% NaCl (Purschke *et al.*, 2018). The isoelectric point (pI) is the pH at which a protein has no net electrical charge. At this pH, protein solubility is at its lowest because proteins tend to aggregate and precipitate due to a lack of charge repulsion. For many edible insects, including yellow mealworms (*Tenebrio molitor*) and black soldier flies (*Hermetia illucens*), the pI typically falls in the acidic pH range of 4–5. This means that around these pH levels, insect proteins are less soluble, making it challenging to use them in aqueous food systems without further processing or modification (Mishyna *et al.*, 2021). For example, yellow mealworm proteins have been found to have their lowest solubility at around pH 4.5. At this point, the proteins aggregate, limiting their interaction with water and reducing their functional applicability in food formulations. This behavior is similar to that of other proteins, such as those from legumes, which also exhibit minimal solubility near their isoelectric points (Zielińska *et al.*, 2018).

3.4.1.4 Digestibility

Digestibility refers to how efficiently proteins are broken down into their constituent amino acids during digestion. Different insect species have different protein compositions, which can affect how easily their proteins are digested (Oonincx and Finke, 2021). The digestibility of some insect proteins can be comparable to some common food proteins such as whole beef (89%), pork (90%), turkey (78%), and salmon (85%) (Zielińska *et al.*, 2018; Ma *et al.*, 2023). Digestibility of insect proteins can be comparable to or even exceed that of plant-based proteins like legumes, which often contain antinutritional factors that inhibit protein absorption. Insects, in contrast, generally lack such inhibitors, which enhances their digestibility. For instance, the protein digestibility of crickets and mealworms has been reported to be around 77–85%, which is on par with or higher than many commonly consumed plant proteins (Zielińska *et al.*, 2018).

Several factors such as thermal unfolding, aggregation, carbonylamine reactions, and interactions with other compounds such as carbohydrates or polyphenols have an impact on protein digestibility. Furthermore, the polypeptide chain's accessibility and flexibility also affect protein digestibility (Manditsera *et al.*, 2019).

Processing methods also play a significant role in the digestibility of insect proteins. For instance, heat treatments such as roasting, drying, and boiling can affect protein structure (Liceaga, 2022). While moderate heat processing can improve digestibility by denaturing proteins and making them more accessible to digestive enzymes, excessive heat treatment can lead to the formation of insoluble aggregates that reduce digestibility. Mishyna *et al.* (2021) noted that thermal processing can cause protein denaturation, which alters protein conformation and can either improve or hinder digestibility depending on the extent of the treatment.

Enzymatic hydrolysis and fermentation are effective ways to improve the digestibility of insect proteins, by breaking down it into smaller peptides and amino acids, which increases the availability of these proteins for absorption (Rodríguez-Rodríguez *et al.*, 2022; Ma *et al.*, 2023). Studies have demonstrated that enzymatic hydrolysis significantly enhances the digestibility of insect proteins, particularly those from crickets, mealworms, and black soldier flies (Villaseñor *et al.*, 2021).

3.4.2 Structural and Colloidal Properties

The structural features of insect proteins, such as surface hydrophobicity and the presence of free sulfhydryl groups, play a crucial role in determining their functionality. For example, heating can cause proteins to unfold, increasing surface hydrophobicity by exposing hydrophobic regions, which can either improve or reduce functional properties like solubility and emulsification (Borges *et al.*, 2022). Moreover, the WHC and OHC of insect proteins are key functional attributes that influence in their potential applications in various food systems (Ma *et al.*, 2023).

3.4.2.1 Hydrophobicity and Presence of Sulfhydryl Groups

In edible insect proteins, sulfhydryl groups are critical for understanding how insect proteins behave during processing and how they can be optimized for food applications. During thermal processing or other treatments, the sulfhydryl groups can be altered (Zhang *et al.*, 2021; Ma *et al.*, 2023). For instance, when insect proteins are exposed to heat, proteins may unfold, breaking disulfide bonds and exposing free sulfhydryl groups. This exposure can lead to changes in protein functionality, such as increased solubility or altered emulsifying properties (Borges *et al.*, 2022). The presence of free sulfhydryl groups can also increase the reactivity of the protein, enabling it to participate in further chemical reactions, which may affect texture, flavor, and stability in food products. The availability of free sulfhydryl groups is also critical in protein aggregation and gel formation. During the gelation process, such as in heat-induced gels, free sulfhydryl groups can form new disulfide bonds, which contribute to the formation of a stable protein network. This is particularly relevant in edible insect protein applications where gelation is desired, such as in the development of meat analogs or protein-based snacks. Moreover, the WHC and OHC of insect proteins can be affected by the reactivity of sulfhydryl groups (Ma *et al.*, 2023).

3.4.2.2 WHC and OHC

WHC refers to the ability of proteins to retain water when subjected to external forces such as heat or pressure, and it is especially important in the development of food products. The WHC of edible insect proteins is influenced by several factors, including the protein structure, processing methods, and environmental conditions such as pH and ionic strength. Insect proteins typically have a high proportion of hydrophobic amino acids which may impact their WHC (Ma *et al.*, 2023). Processing methods, such as drying, heat treatment, and enzymatic hydrolysis, can also affect the WHC of insect proteins. Heat treatments, can cause protein denaturation, leading to changes in protein structure that may either enhance or reduce the WHC depending on the extent of the treatment (Belluco *et al.*, 2013). Enzymatic hydrolysis breaks down large protein molecules into smaller peptides, and hydrolysis exposes more functional groups that can interact with water, thus improving the protein's ability to hold moisture (Villaseñor *et al.*, 2021).

Different insect species exhibit varying levels of WHC depending on their protein composition and processing. For example, proteins from yellow mealworms (*Tenebrio molitor*) and crickets (*Acheta domesticus*) have been reported to have favorable WHC, making them suitable for use in high-moisture food products (Zielińska, 2022). A study by Ma *et al.* (2023) found that mealworm proteins displayed a relatively high WHC, which makes them promising candidates for meat substitutes and other moisture-retentive food applications. Similarly, black soldier fly larvae (*Hermetia illucens*) proteins have been studied for their WHC properties. The use of insect proteins as ingredients in emulsified products, such as sausages and processed meat alternatives, relies heavily on their WHC to maintain the desired texture and moisture content during processing and cooking (Zielińska *et al.*, 2018).

On the other hand, OHC is the facility of the protein to absorb and retain oil that is influenced by their hydrophobicity and molecular structure. Proteins with a higher content of hydrophobic amino acids are better able to interact with and bind lipids, thus enhancing their OHC (Zhang *et al.*, 2021b). This characteristic is particularly important in food formulations that require fat retention, such as meat analogs, baked goods, and emulsified products (Ma *et al.*, 2023). For instance, research on *Tenebrio molitor* (yellow mealworm) protein found that it had an OHC of 2.74 g of oil per gram of protein, which is relatively lower than that of some other edible insects like *Gryllodes sigillatus* (crickets) and *Schistocerca gregaria* (desert locusts), which exhibit OHC values of 3.33 and 3.22 g of oil per g of protein, respectively (Zielińska *et al.*, 2018).

Processing methods such as drying, grinding, and protein extraction can influence the OHC of insect proteins. Heat treatments, which cause protein denaturation, can either enhance or diminish OHC depending on the extent of the structural changes in the protein. (Zhang *et al.*, 2021b). Hydrolysis of insect proteins, a process that breaks down proteins into smaller peptides, has been found to increase OHC by exposing more hydrophobic groups, thus allowing for greater oil absorption. Insect proteins with high OHC are valuable in the production of meat substitutes and other fat-rich products, as they can retain oils during cooking, resulting in a juicier and more tender final product (Zielińska *et al.*, 2018). For example, *Tenebrio molitor* protein, despite having a lower OHC than some other insect proteins, has been successfully used in meat analogs where its oil retention properties help create a desirable texture (van Huis *et al.*, 2013).

3.4.3 Gelling and Foaming Properties

3.4.3.1 Gelling

The gelation properties of proteins depend on their amino acid composition, molecular structure, and interaction with environmental factors such as temperature and pH. For insect proteins, their high content of hydrophobic amino acids and cysteine residues plays a crucial role in the gelation process by promoting the formation of disulfide bonds and hydrophobic interactions (Yi *et al.*, 2013). Studies have shown that edible insect proteins, such as those derived from crickets, mealworms, and locusts, possess gelling properties similar to those of more traditional protein sources like soy or egg whites (Pan *et al.*, 2022). For example, cricket (*Acheta domesticus*) proteins have demonstrated the ability to form firm gels, particularly at neutral to alkaline pH levels. Yi *et al.* (2013) investigated the gelation of insect proteins and found that *A. domesticus* formed a firm gel at pH 7, while other edible insect proteins required higher concentrations or pH adjustments to achieve similar gelation results. In terms of gel strength and texture, *Tenebrio molitor* (mealworm) proteins have also been reported to form gels under specific conditions, particularly when subjected to heat (Yi *et al.*, 2013). The gelation behavior of mealworm proteins is influenced by the concentration of protein in the solution and the heating method applied. Like plant proteins, insect proteins benefit from heat-induced denaturation, where unfolding proteins expose hydrophobic regions, facilitating protein–protein interactions that lead to gel formation (Zielińska *et al.*, 2018). Among factors affecting gelation, we can mention: (1) pH which when is near to the isoelectric point, proteins have minimal solubility and tend to aggregate, causing firmer gels. However, gels formed at extreme pH levels tend to have a softer texture due to the increased repulsion between charged protein molecules (Yi *et al.*, 2013; Borges *et al.*, 2022). (2) Protein concentration also plays a role because higher protein concentrations generally result in stronger gels, as there is a greater number of protein molecules available to form a stable network (Borges *et al.*, 2022).

3.4.3.2 Foaming

Foaming capacity refers to the ability of a protein to form a foam when air is incorporated, while foam stability is the ability of the foam to resist collapse over time. These are closely linked to protein amphiphilic nature. Proteins are composed of both hydrophilic (water-attracting) and hydrophobic (water-repelling) regions. During foam formation, proteins migrate to the air–water interface, where they reduce surface tension and stabilize air bubbles. The hydrophobic regions of the protein interact with air, while the hydrophilic regions interact with the surrounding liquid, allowing stable foam to form (Zielińska *et al.*, 2018). For edible insect proteins, this amphiphilic property allows them to exhibit similar foaming capabilities to egg whites and other animal proteins. Studies have shown that the foaming capacity of insect proteins is comparable to that of conventional proteins used in food products, depending on factors such as protein concentration, pH, and processing conditions (Belluco *et al.*, 2013).

For edible insect proteins, factors such as protein concentration, pH, and the extent of protein denaturation significantly affect these properties. For instance, proteins extracted from *Acheta domesticus* (house crickets) and *Tenebrio molitor* (mealworms) have been shown to have good foaming properties at neutral pH levels (Yi *et al.*, 2013). These proteins can produce stable foams, which are essential for their use in aerated food products like mousses or protein shakes. In a study on the functional properties of insect proteins (Zhao *et al.*, 2016), *Tenebrio molitor* protein was found to exhibit foam stability comparable to traditional foaming agents like whey protein and egg white, particularly after heat treatment. Heating causes protein denaturation, which improves foam stability by allowing more protein molecules to unfold and form stronger intermolecular interactions at the air–water interface (Zhao *et al.*, 2016; Santiago *et al.*, 2021).

When compared to traditional proteins like egg whites or soy proteins, edible insect proteins have shown promising foaming properties. For example, the foaming capacity of *Acheta domesticus* proteins has been found to be on par with egg white proteins, which are commonly used as foaming agents in the food industry (Rumpold and Schlüter, 2013). In contrast, the low foaming capacity of some insect proteins, such as those from *B. mori*, is attributed to their conformational characteristics, which limit their air bubble encapsulation ability (Ma *et al.*, 2023). In many food products, insect proteins can act as natural foaming agents to help achieve the desired volume and texture in cakes or bread. Additionally, in beverages or desserts where foam stability is important, such as protein shakes, mousses, and whipped toppings, insect proteins can provide a sustainable and functional alternative to dairy or egg-based proteins. (van Huis, 2022).

3.4.3.3 Emulsification

The emulsifying properties of insect proteins are influenced by their amphiphilic nature, by adsorbing at the oil–water interface, reducing surface tension, and stabilizing the dispersed oil droplets in water or vice versa. When insect proteins are introduced into an oil–water system, their hydrophobic regions interact with the oil phase, while their hydrophilic regions interact with the water phase (Baigts-Allende *et al.*, 2021; Gravel and Doyen, 2019). This property allows the proteins to stabilize the oil droplets in an emulsion, preventing coalescence and phase separation (Yi *et al.*, 2013). Studies have shown that insect proteins can create stable emulsions, similar to traditional protein emulsifiers like soy, egg, and whey proteins (Hall *et al.*, 2016; Zielińska *et al.*, 2018). Emulsifying capacity and stability of different insect proteins can reach up to 100%, with defatting and extraction processes influencing these values (Mishyna *et al.*, 2020). The emulsifying stability of *S. gregaria* protein, particularly when combined with sonication, shows significant promise for industrial applications (Mishyna *et al.*, 2020). *T. molitor* protein has been shown to have a higher emulsifying stability (51.3%) compared to milk protein concentrates (33.50%), highlighting its potential as an alternative emulsifier in food systems (Zhao *et al.*, 2016; Ma *et al.*, 2023). Furthermore, it has been observed that the use of enzymatic hydrolysis improves the emulsion capacity, and it could be explained by unfolding the protein and exposing more hydrophobic regions, previously inaccessible amino acids, enhancing hydrophobic interactions, and emulsion formation (Rumpold and Schlüter, 2013; Villaseñor *et al.*, 2021; Sales-Queiroz *et al.*, 2023; Belluco *et al.*, 2013). One factor influencing emulsion stability is the molecular weight of the protein. Proteins with smaller molecular weights, such as those obtained through enzymatic hydrolysis, heating, or changing the pH tend to form more stable emulsions because they can migrate more easily to the oil–water interface and form a cohesive layer around the oil droplets (Van Huis *et al.*, 2013; Yi *et al.*, 2013).

3.5 INSECT PROTEIN EXTRACTION METHODS

The development of new technologies for processing insect biomass is essential for improving the technological and functional properties of insect proteins. These innovations focus on optimizing key attributes such as solubility, water and oil retention capacities, emulsification, foaming ability, and gelling properties, while maintaining product safety and nutritional quality (van-Huis *et al.*, 2021; Granados-Echegoyen *et al.*, 2024). The efficient extraction of these proteins is essential to maximize their usability and functionality. Various extraction methods have been developed and refined, ranging from traditional techniques to advanced, green technologies offering potential applications in the food industry (Belluco *et al.*, 2013; Rumpold & Schlüter, 2013; Zielińska *et al.*, 2018; Ma *et al.*, 2023).

Conventional methods for extracting insect proteins typically involve the use of solvents, pH adjustment, and mechanical processes. These methods are widely used due to their simplicity and effectiveness in separating proteins from other components like fat and chitin. Among them, we can mention:

- **Aqueous Extraction.** In this method, insect biomass is homogenized in water, followed by centrifugation to separate the soluble proteins from the insoluble materials such as chitin and fat. The solubilized proteins are then collected and concentrated by removing water through evaporation or freeze-drying. Studies have shown that aqueous extraction can yield up to 28.5% of protein from insects such as crickets and mealworms. However, the efficiency of this method is generally lower than other extraction techniques, as it may not efficiently isolate all proteins due to their hydrophobic nature and association with lipids (Belluco *et al.*, 2013).
- **Alkaline Extraction.** Proteins are solubilized at pH 8–11, which enhances solubility by ionizing acidic and neutral amino acids. The solubilized proteins

are then precipitated by adjusting the pH to their isoelectric point (pH 4–5), resulting in a concentrated protein fraction. This method not only yields up to 70% crude protein but also helps remove undesirable components like fat and the exoskeleton (Rahman *et al.*, 2023). This method is particularly effective in increasing protein yield from edible insects by promoting the unfolding of proteins and their dissociation from other biomolecules. Alkaline extraction offers higher protein recovery rates than aqueous, with some studies reporting up to 79.7% yield from insect species such as *Acheta domesticus* (Rumpold and Schlüter, 2013). Moreover, proteins extracted via this method often exhibit superior functional properties due to the solubilization and refolding of protein structures (Hall *et al.*, 2016).

- **Salt-Assisted Extraction.** This method takes advantage of salts to enhance the solubilization of proteins by modifying the ionic strength of the medium. The protein extraction step involves the use of a salt solution (NaCl, $MgSO_4$), which helps dissolve the proteins into the aqueous phase. The proteins are then separated by centrifugation, followed by precipitation (Yi *et al.*, 2013; Zhao *et al.*, 2016). A study on *Tenebrio molitor* demonstrated that extraction using NaCl increased protein recovery compared to traditional aqueous extraction methods, and the resulting protein isolates exhibited enhanced solubility (Zielińska *et al.*, 2018; Jiang *et al.*, 2021). Similarly, in *Hermetia illucens*, salt-assisted extraction was shown to improve the emulsifying properties of the proteins, making them suitable for use in food formulations that require stable emulsions. However, the method also has some limitations, mainly the need to remove excess salt from the final protein product, which can affect the taste and functionality of the protein in food formulations (Hall *et al.*, 2016).
- **Organic Solvent Defatting.** Before protein extraction, defatting is often performed using organic solvents such as n-hexane, petroleum ether, methanol, or ethanol, which efficiently dissolve fats while leaving behind the protein and other nonlipid components from insect powders. While effective, these solvents can lead to protein loss due to the affinity of some proteins for the solvents, potentially reducing the overall protein yield (Ma *et al.*, 2023). The insect is ground into a powder to increase the surface area for extraction, then solvent is added at specific ratios (commonly 3:1 or 5:1 solvent:sample). The sample is stirred at room or slightly elevated temperatures during 1–4 hours to dissolve the fats (Mishyna *et al.*, 2020). Finally, protein extraction is done once the fat has been removed, solubilizing the proteins in water or using alkaline extraction methods, followed by protein precipitation (Gravel and Doyen, 2019). Studies have shown that defatting insect samples with hexane can reduce the fat content to as low as 1–2%, making it easier to extract a purer protein fraction with enhanced functionality (Belluco *et al.*, 2013). The removal of lipids exposes more active protein sites, allowing for better interaction with water, oil, or other components in food systems (Zielińska *et al.*, 2018).

In response to the limitations of conventional methods, nonconventional or advanced green extraction techniques have been developed. The purpose of these methods is to improve protein yield, maintain protein functionality, and reduce the environmental impact of the extraction process. Some of these methods include:

- **Enzymatic Hydrolysis.** Common proteases used in the hydrolysis of insect proteins include alcalase, flavourzyme, and papain, each of which operates under specific temperatures and pH conditions to maximize protein breakdown (Yi *et al.*, 2013). While this method is effective in producing bioactive protein hydrolysates, it is rarely used for large-scale protein extraction due to its high cost, slow processing time, and challenges in scaling up (Pan *et al.*, 2022). This process begins with the grinding and defatting of insect biomass, followed by the introduction of enzymes under controlled conditions (Zielińska *et al.*, 2018). The advantages of this method are the improvement in protein solubility, production of bioactive peptides, enhance emulsification and gelation properties, reduce potential allergenicity, and increase the digestibility (Belluco *et al.*, 2013; Rumpold and Schlüter, 2013; Van Huis *et al.*, 2013; Zhao *et al.*, 2016; Hall *et al.*, 2016). On the other hand, it presents several challenges such as controlling the degree of hydrolysis, the cost of enzymes, and the optimization because the diverse range of insect species means that different enzymes and processing conditions may be required (Zhao *et al.*, 2016).
- **Membrane Filtration.** This technique is employed to concentrate and purify proteins after extraction based on size and molecular weight. These methods provide a gentle and effective way to extract high-quality proteins while minimizing protein denaturation and maintaining functional properties (Rumpold and Schlüter, 2013). The advantages of the technique include mild processing conditions being a nonthermal technique, which reduces the risk of denaturation (Belluco *et al.*, 2013) and allows for high protein recovery with minimal contamination from fats and carbohydrates (Zhao *et al.*, 2016). The process can be easily scaled up, making them suitable for large-scale industrial applications (Van Huis *et al.*, 2013). On the other hand, some disadvantages are present like fouling where particles accumulate on the membrane surface, reducing its efficiency. In insect protein extraction, fats and polysaccharides present in the insect biomass can obstruct the membrane pores, leading to reduced protein yield and higher operational costs (Zielińska *et al.*, 2018).
- **Ultrasound-Assisted Extraction.** It is a green alternative that uses high-frequency sound waves to disrupt cell walls and facilitate the release of intracellular proteins, which enhances the extraction process. This method is advantageous due to its simplicity, time efficiency, and reduced solvent usage and is often combined with solvent or alkaline extraction to improve the overall efficiency of protein recovery (Mishyna *et al.*, 2021). Ultrasonication enhances the solubility and yield of insect proteins by increasing the surface area for extraction and reducing the time required for the process, improving protein yield by up to 15–20% compared to traditional extraction methods (Van Huis *et al.*, 2013; Mishyna *et al.*, 2021).
- **Microwave-Assisted Extraction.** This interaction generates heat, causing rapid temperature increases and facilitating the breakdown of insect tissues, leading to the release of proteins and other bioactive compounds. The main advantage is its ability to shorten extraction time while maintaining protein integrity and reducing solvent consumption (Wu *et al.*, 2023). The technique has been shown to enhance the solubility and bioavailability of insect proteins (Aguilar-Toalá

et al., 2022). The efficiency of this method depends on several factors, including microwave power, extraction time, solvent type, and temperature. Research has shown that optimal conditions for extracting edible insect proteins involve moderate microwave power, typically around 500–800 watts, combined with short extraction times (10–15 min). Using water or a mixture of water and ethanol as solvents often yields the highest protein extraction rates while maintaining functional properties such as emulsification and solubility (Aguilar-Toalá *et al.*, 2022; Zhang *et al.*, 2021b).

- **Pulsed Electric Field Extraction.** This method is an emerging nonthermal method that has gained significant attention for its ability to enhance the extraction of bioactive compounds. When insect cells are subjected to this treatment, the electric field disrupts the cell membrane, leading to enhanced protein solubilization and extraction efficiency (Van Huis *et al.*, 2013). In addition, it preserves the functional and nutritional properties of the extracted proteins. However, pulsed electric field extraction is less commonly used due to its lower efficiency compared to other advanced methods (Rumpold and Schlüter, 2013).
- **Dry Fractionation Techniques.** It is a physical separation process that involves the partitioning of proteins and other components from insect biomass without the use of solvents or chemicals, avoiding the generation of wastewater and reduces energy consumption. This method typically involves fine milling and air classification to separate proteins from other components. While dry fractionation is less widely used, it is effective in producing protein-rich fractions with reduced chitin content (Ma *et al.*, 2023). Additionally, dry fractionation helps maintain the native functionality of insect proteins, which is often compromised during chemical or thermal extraction processes (Boye and Barbana, 2012). Despite its advantages, the complexity of insect composition, particularly the presence of chitin and fats, makes it difficult to achieve high purity protein fractions. The efficiency of dry fractionation depends heavily on the particle size and the effectiveness of the separation process, which can vary depending on the insect species and processing conditions (Zielińska *et al.*, 2018).

As a result of the different extraction treatments, insects offer various forms of protein ingredients, including concentrates, isolates, and hydrolysates, each with distinct characteristics, benefits, and disadvantages. These products, derived through mechanical, chemical, and enzymatic processes, are highly versatile, lending themselves to diverse applications in food (Rumpold and Schlüter, 2013; Kim *et al.*, 2019).

Protein concentrates are one of the simplest forms of insect protein products, typically obtained through mechanical processes like grinding, sieving, or wet fractionation. These methods retain some nonprotein components such as fats, fibers, and carbohydrates (Rahman *et al.*, 2023). Due to the relatively simple processing involved, insect protein concentrates are easier and cheaper to produce compared to isolates and hydrolysates. This makes them suitable for a broader range of applications, including animal feed and human food consumption (Kim *et al.*, 2019). Among the pros of protein concentrates is their potential to overcome the consumer's resistance to eat whole insects because when these insects are transformed into a more familiar powder form, such as a protein concentrate, they are more accepted. In this form, insect protein concentrates can be incorporated into a wide range of food products, such as baked goods, protein bars, and supplements. However, protein concentrates have limitations. Since they retain some nonprotein components, their purity is lower than that of isolates, which may limit their applications in products that require high protein content or purity (Kim *et al.*, 2019).

Protein isolates are more refined than concentrates, undergoing additional processing to remove a higher proportion of nonprotein components like carbohydrates and fats. The isolation process typically involves defatting the insect biomass followed by techniques such as acid or alkaline extraction, enzymatic treatment, and isoelectric precipitation. This results in a highly purified protein product, often with a protein content exceeding 90%, suitable for use in food formulations, supplements, and high-protein applications like energy bars and meat substitutes (Rumpold and Schlüter, 2013). Crickets and grasshoppers are particularly good candidates for protein isolation due to their high protein content, which can reach up to 70% of their dry weight. The functional properties of insect protein isolates, such as solubility, gelation, and emulsification, make them highly attractive to the food industry, particularly in products like protein shakes and meat substitutes (Rumpold and Schlüter, 2013). The cons of protein isolates are that they are more expensive and difficult to produce than concentrates. Additionally, some bioactive compounds and micronutrients present in whole insects may be lost during the extensive processing required to produce isolates, potentially reducing their nutritional value.

Protein hydrolysates are derived from the enzymatic or chemical breakdown of proteins into smaller peptides and amino acids. This process enhances the bioavailability and functional properties of the resulting product. Insect protein hydrolysates are typically produced using proteolytic enzymes such as alcalase, flavourzyme, and papain, which break down the peptide bonds, giving products with improved digestibility, solubility, emulsification, and foaming properties (Yi *et al.*, 2013). Insects like *Tenebrio molitor* (mealworms), *Acheta domesticus* (crickets), and *Hermetia illuce*ns (black soldier fly larvae) are commonly used to produce protein hydrolysates due to their high protein content (Zielińska *et al.*, 2018). Moreover, the bioactive peptides derived from insect protein hydrolysates have been shown to offer various health benefits. For example, peptides from *Tenebrio molitor* can inhibit angiotensin-converting enzyme, suggesting their potential use in cardiovascular health supplements (Belluco *et al.*, 2013). Another advantage of protein hydrolysates is their antimicrobial activity. Certain peptides derived from insect proteins can inhibit the growth of harmful microorganisms, offering potential applications in food preservation, and their antioxidant properties make them suitable for health foods aimed at reducing oxidative stress (Zielińska *et al.*, 2017). However, producing protein hydrolysates requires complex enzymatic treatments, which can be costly. Additionally, hydrolysates may have a bitter taste due to the presence of smaller peptides and free amino acids, which could limit their use in certain food products.

3.6 INSECT PROTEINS IN FOOD PRODUCTS

The unique properties of insect proteins, such as their balanced amino acid, make them a valuable ingredient in enhancing the nutritional as well as the functional properties of many food products. Insect proteins can be incorporated into a wide variety of food products, including snacks, bakery goods, meat products, and dairy alternatives (Lamsal *et al.*, 2019; Borges *et al.*, 2022; Bresciani *et al.*, 2022; Ishara *et al.*, 2022; Zeng *et al.*, 2022).

The formulation of products with whole insects as snacks or candies is an alternative to develop high protein foods. However, whole insects are still not well accepted by consumers; therefore, when insects are incorporated into foods as powdered ingredients, consumers tend to have better acceptance (Caparros-Megido *et al.*, 2016; Lamsal *et al.*, 2019).

3.6.1 Flours and Bakery Products

Insect proteins have been successfully used in the obtention of flours and formulation of pasta and bakery products, such as cookies, crackers, cupcakes, muffins, brownies, pie, cakes, bread, pita chips, and biscuits (González *et al.*, 2019; Roncolini *et al.*, 2019; Kowalski *et al.*, 2022; Amoah *et al.*, 2023). In most cases, the addition of insect protein powder to these products can increase their protein content, improve their amino acid profile, and enhance their texture and sensory properties. The incorporation of 10% of cricket powder to bread results in a volume and a crumb hardness similar to that elaborated with only wheat flour, and in pasta, the best cooking characteristics were obtained at 5% of added powder (Bresciani *et al.*, 2022). Similar results were obtained when added to wheat flour of cockroach (*Nauphoeta cinerea*), mealworm (*Tenebrio molitor*), *Hermetia illucens*, or *Acheta domestica* powders (5 or 10%) to elaborate bread, leading to improve the protein and amino acid contents, contributing to bread volume and softness. Besides, the products showed high acceptability from consumers (Bresciani *et al.*, 2022; de Oliveira *et al.*, 2017; Roncolini *et al.*, 2019; González *et al.*, 2019; Amoah *et al.*, 2023). On the other hand, depending on the type and amount of insect protein used, darker colors and stronger flavors can be produced (de Oliveira *et al.*, 2017; Roncolini *et al.*, 2019). Inclusion of insects' flours also may modify rheological properties as water absorption and stability of doughs (González *et al.*, 2019).

3.6.2 Meat Products

By their nutritional value, insect proteins have been studied for their inclusion in meat products to improve their protein content and functional properties (Kim *et al.*, 2022). Products, like sausages, burgers, jerky analogs, and meat analogs, have been obtained (Caparros-Megido *et al.*, 2016; Kurek *et al.*, 2022; Zeng *et al.*, 2022; Krawczyk *et al.*, 2024). On the positive side, studies have shown that adding up to 10% insect protein to products like sausages can improve water and fat stabilization without compromising texture or sensory quality (Kim *et al.*, 2019). Additionally, insect protein can significantly enhance the protein and mineral content of meat products, such as emulsified sausages and pork meat batters, when mealworms or silkworms are used (Park *et al.*, 2017). In some cases, the incorporation of insect protein can even help maintain optimal meat-like textures, as demonstrated by Smetana *et al.* (2018), who achieved this with 40% insect protein in a soy-based meat analog using high-temperature extrusion. Furthermore, insect proteins combined with textured vegetable proteins can produce tender, high-nutritional-value jerky analogs (Kim *et al.*, 2022). As disadvantages, high levels of insect protein substitution can lead to undesirable changes in texture and flavor, hardness of products, or lower protein content (Kim *et al.*, 2022). When insects are used in plant-based meat analogs, there are further challenges, such as changes in color and fatty acid composition. However, sensory analyses have shown that insect protein inclusion (at levels around 5%) does not significantly affect the overall physicochemical attributes of products, making it an acceptable addition to meat alternatives (Bessa *et al.*, 2020; Krawczyk *et al.*, 2024).

3.6.3 Dairy Products

Several studies have focused on the application of insect protein in dairy products. For instance, insect-based milk substitutes have been developed from *Tenebrio molitor* larvae, offering a product with a lower environmental impact than conventional milk but presenting challenges such as enzymatic browning and weak fractionation (Tello *et al.*, 2021). In refrigerated storage trials using insect flour as a substitute for milk powder, the nutritional composition of the dairy products remained stable, with 4.7% protein and 0.95% fat content, showing that insects can maintain the product's nutritional integrity (Neves *et al.*, 2024). While these trials faced some challenges such as enzymatic browning, further research into processing techniques, like pulse electric field technology, could enhance the nutritional quality and sensory attributes of these products (Tello *et al.*, 2021). One of the major challenges in using insect proteins in dairy products is the sensory impact. In studies where insect flours, such as black soldier fly flour, were used in fermented dairy products, consumers noted negative perceptions related to the product's taste, aroma, and appearance (Neves *et al.*, 2024). Black soldier fly flour, in particular, contributed to a darker color and a distinct taste. Additionally, incomplete dissolution of the insect flour in some products led to changes in texture, further negatively affecting consumer acceptance (Neves *et al.*, 2024). Another significant challenge involves the technical and functional properties of insect proteins when incorporated into dairy products. For instance, in ice cream, the addition of soldier fly larvae protein reduced the amount of air incorporated, leading to a denser texture and altered adhesiveness (Zeng *et al.*, 2022; Diaz-Bustamante *et al.*, 2023). These changes in texture are not always desirable in dairy products, which tend to rely on specific textural qualities, such as creaminess and smoothness, for consumer satisfaction. Additionally, some insect flours contribute to syneresis in fermented dairy products, further complicating their functional use (Neves *et al.*, 2024).

3.7 INSECT IN MEXICAN FOOD

Entomophagy is a common practice mainly in subtropical and tropical regions of the world because diversity and abundance are higher. However, China, Japan, and Mexico are the only countries where entomophagy is widespread and are located entirely or partially in warm zones (Jongema, 2017; Tang *et al.*, 2019; Papastavropoulou *et al.*, 2022a).

In Mexico, insects have been eaten since ancient times being an important part of their culture and tradition. Currently, a great variety of species continue to be consumed (del Mastro, 2021), and nowadays, Mexico has the highest number of edible insect species (around 450) in North America and in the world (Omuse *et al.*, 2024; Granados-Echegoyen

FIGURE 3.1 Preparation of different insects' types: (a) Grasshoppers found in a local market in Oaxaca, prepared as snack fried with garlic and salt; corn tortilla chips with grasshoppers' sauce. (b) Roasted *chinicuiles* (*Comadia redtenbacheri*), prepared into a *Tlayuda* (crispy big tortilla) whit *chicatana* ants and vegetables; *chinicuiles* sauce, and chinicuiles mixed with chili powder on a lemon slice. (c) Fried *escamoles* (*Liometopum apiculatum*) with onion, chili, and epazote (*Dysphania ambrosioides*); corn "*gorditas*" made with *chicatanas* ants (*Atta mexicana*) and *escamoles*.

et al., 2024). States which have the highest insects' consumption are Oaxaca, Hidalgo, Guerrero, Puebla, Mexico State, Chiapas, Michoacan, Tlaxcala, Veracruz, and Yucatan (Ramos-Elorduy, 1997).

Several edible insects can be consumed raw or alive since they have a pleasant flavor. However, it is recommended to cook them to enhance their flavors, and they can be better appreciated. They are prepared toasted, roasted, in tacos, in sauces, as an appetizer, and as additives to various dishes (pizzas, salads omelets, or desserts). Edible insects in Mexico are produced and consumed depending on the season of availability. In big cities, insects are offered, at high prices, both in restaurants as well as in streets and local markets. Insects consumed in Mexico include a variety of species, such as "chapulines" (grasshoppers), "jumiles" (stink bugs), "xamues" (giant bug), "cuchamás" (green caterpillars), "gusanos" (maguey worms), dragonfly larvaes, "ahuautles" (mosquito eggs), "escamoles" (ant larvae), "chinicuiles" (agave red worm), and "chicatanas" (flying ants), which are consumed alive, toasted, fried, or as an ingredient in preparations like sauces and snacks (Figure 3.1) (Ramos-Elorduy, 1997; Acuña *et al*., 2011; Lara-Juárez *et al*., 2015). Among the most common edible insects in Mexico, we can mention:

- Grasshoppers ("Chapulines"). Obtained from *milpa* (fields planted with maize and alfalfa, among other crops) from late spring to early winter. They are prepared cleaned and toasted in a little oil, with some garlic, lemon, salt, and chili, and their consumption could be as a condiment, snack food, or main dish (Cohen *et al.*, 2009; Youssef and Spence, 2021).

- *Edessa* spp. ("Jumiles"). These insects are collected in forest areas under the dry leaves of *Agave salmiana* or under the base of the trunk of some trees (*Quercus* sp.) between November and March. They are consumed in an adult stage, typically in a sauce made by grinding the bugs with garlic, chili, and other spices, and then using the mixture as a topping for tortillas (Silva-García *et al.*, 2018).
- *Liometopum apiculatum* ("Escamoles"). They are harvested only during the period from April to June, prior to the rainy season. The way to cook them is fried with butter, chili, and onion, and sometimes, they are added with some herbs, such as epazote (*Dysphania ambrosioides*) (Lara-Juárez *et al.*, 2015; Ríos, 2024).
- *Comadia redtenbacheri* ("Chinicuiles"). Their collection is done during the rainy season from the maguey plant. They are consumed in central Mexico, mainly in the Mezquital Valley, in their larvae stage. Their preparations could be as main dishes or as garnishes, prepared as snacks, powders, or condiments, fried or grilled to be eaten with tortillas, or incorporated into sauces. They are also included in bottles of mezcal or as the main ingredient of "chinicuil" salt, which is a common complement to this distilled alcoholic beverage (Molina-Vega *et al.*, 2021).
- *Atta mexicana* and *Atta cephalotes* ("Chicatana" ant). They are collected after an intense rain, starting between May to July months. Their forms of consumption are diverse. The most common are into sauce, fried, roasted, and raw chicatanas (Escamilla-Prado *et al.*, 2012).

Besides the traditional insects' intake, recently, many restaurants have increased their gastronomic offers by incorporating insects in their preparations as a way to revalue Mexican ingredients, rescue the ancestral flavors, and to enhance the texture and mouthfeel of traditional Mexican dishes such as sauces, dips, and marinades. The utilization of insect proteins in Mexican cuisine represents an opportunity to improve the nutritional quality and functional properties of traditional food products while also promoting the integration of sustainable and culturally relevant protein sources into the Mexican diet (Abdullahi *et al.*, 2022).

3.8 INSECT PROTEIN ALLERGENS

As we mentioned widely, insects' protein consumption represents a suitable alternative to substitute another kind of animal protein for human intake. Nevertheless, despite their advantages, insects' consumption presents some health risks, including chemical, microbiological, parasitic, physical, and toxicological hazards (Pan *et al.*, 2022; Papastavropoulou *et al.*, 2022b). One of the most worries about these risks is the potential for allergenic reactions. Symptoms caused by allergic reactions to insect proteins may be slight such as rashes and itching, or severe, including anaphylaxis (Kausar, 2018). These reactions can occur not only through ingestion but also via inhalation or skin contact like in jobs where workers handle insects regularly (De Gier and Verhoeckx, 2018; Papastavropoulou *et al.*, 2022a). Common allergens identified in insect species include tropomyosin, arginine kinase, chitinase-like proteins, and glycoproteins (Liu *et al.*, 2009; Kamemura *et al.*, 2019; Lamberti *et al.*, 2021). People who are already sensitized to certain allergens, such as those found in crustaceans and house dust mites, may experience cross-reactivity upon consuming insect proteins due to the structural similarities of allergens found in insect species, such as tropomyosin and arginine kinase, which are structurally similar to those found in crustaceans (Ribeiro *et al.*, 2018; de Marchi *et al.*, 2021; Liceaga, 2022; Yang *et al.*, 2023).

The cross-reactivity between insect proteins and crustacean allergens poses a significant risk for individuals with pre-existing allergies. Tropomyosin and arginine kinase, often referred to as pan-allergens, are found in both insects and crustaceans, leading to co-sensitization between these organisms. Persons who are allergic to crustaceans may also react to insect proteins, as observed in studies involving *Tenebrio molitor* proteins, which share IgE-binding cross-reactivity with shrimp proteins (Meyer-Rochow *et al.*, 2021). Consequently, consumers allergic to shellfish should be cautioned about the potential for allergic reactions when consuming insect-based products, and labeling should reflect this risk (Ribeiro *et al.*, 2018).

It has been observed that some thermal processing treatments, such as boiling, frying, and baking, may alter the allergenic properties of insect proteins. Heat treatments can reduce the allergenic activity of proteins such as tropomyosin and arginine kinase, but their effectiveness varies depending on the insect species and specific proteins involved (Lamberti *et al.*, 2021). Enzymatic hydrolysis is another processing technique explored to reduce allergenicity. This process breaks down proteins into smaller peptides, which can reduce their potential to activate allergic responses (Ahmed *et al.*, 2021). Recent studies have indicated that microwave or high hydrostatic pressure-assisted hydrolysis can significantly reduce the allergenicity of tropomyosin by altering its secondary structure and reducing its IgE-binding capacity (Hall and Liceaga, 2020). Finally, proper labeling of insect-based food products is another crucial aspect of ensuring consumer safety. Regulatory bodies must establish clear guidelines and thresholds for allergenic proteins in insect-based foods, ensuring that consumers are informed of potential risks. Labeling should explicitly warn individuals with shellfish allergies of the potential for cross-reactivity with insect proteins (de Marchi *et al.*, 2021).

The promotion of insect consumption is hampered by a lack of comprehensive knowledge regarding the risks associated with their proteins, particularly in developing countries. Additionally, inconsistent or nonexistent regulations complicate the safe production and marketing of insect-based foods. As the consumption of insects becomes more widespread, there is a pressing need for clear regulations and guidelines to manage allergenic risks effectively (Papastavropoulou *et al.*, 2022b).

3.9 ACKNOWLEDGMENTS

We thank to Andrés García, Antonio Luis Guzmán, Irene Martínez, Miriam Medina, Elia Morales, and Maribel Zamora for their support in providing us the pictures of insect-based dishes presented in this chapter. Authors are members of the AlProSos network (sustainable regional protein-rich foods) funded by CYTED (ref. 125RT0165).

REFERENCES

Abdullahi, N., Igwe, E. C., Dandago, M. A., Rilwan, A., Jibril, H., & Iliyasu, R. (2022). Utilization of edible-insects as protein substitute in food and effects of processing on their nutrient contents and protein functionalities. *Food ScienTech Journal*, *4*(2), 83. https://doi.org/10.33512/fsj.v4i2.14527

Acuña, A. M., Caso, L., Aliphat, M. M., & Vergara, C. H. (2011). Edible insects as part of the traditional food system of the popoloca town of Los Reyes Metzontla, Mexico. *Journal of Ethnobiology*, *31*(1), 150–169. https://doi.org/10.2993/0278-0771-31.1.150

Aguilar-Toalá, J. E., Cruz-Monterrosa, R. G., & Liceaga, A. M. (2022). Beyond human nutrition of edible insects: Health benefits and safety aspects. *Insects*, *13*(11). MDPI. https://doi.org/10.3390/insects13111007

Ahmed, I., Chen, H., Li, J., Wang, B., Li, Z. and Huang, G. (2021). Enzymatic crosslinking and food allergenicity: A comprehensive review. *Comprehensive Reviews in Food Science and Food Safety Ahmed*, 1–24 https://doi.org/10.1111/1541-4337.12855.

Aiello, D., Barbera, M., Bongiorno, D., Cammarata, M., Censi, V., Indelicato, S., Mazzotti, F., Napoli, A., Piazzese, D., & Saiano, F. (2023). Edible insects an alternative nutritional source of bioactive compounds: A review. *Molecules*, *28*(2). MDPI. https://doi.org/10.3390/molecules28020699

Akhtar, Y., & Isman, M. B. (2018). Insects as an alternative protein source. In *Proteins in Food Processing* (2nd ed., pp. 263–288). Elsevier. https://doi.org/10.1016/B978-0-08-100722-8.00011-5

Amoah, I., Cobbinah, J. C., Yeboah, J. A., Essiam, F. A., Lim, J. J., Tandoh, M. A., & Rush, E. (2023). Edible insect powder for enrichment of bakery products – A review of nutritional, physical characteristics and acceptability of bakery products to consumers. *Future Foods*, *8*. https://doi.org/10.1016/j.fufo.2023.100251

Baigts-Allende, D., Doost, A. S., Ramírez-Rodrigues, M., Dewettinck, K., van der Meeren, P., de Meulenaer, B., & Tzompa-Sosa, D. (2021). Insect protein concentrates from Mexican edible insects: Structural and functional characterization. *LWT*, *152*. https://doi.org/10.1016/j.lwt.2021.112267

Bbosa, T., Tamale Ndagire, C., Muzira Mukisa, I., Fiaboe, K. K. M., & Nakimbugwe, D. (2019). Nutritional characteristics of selected insects in Uganda for use as alternative protein sources in food and feed. *Journal of Insect Science (Online)*, *19*(6). https://doi.org/10.1093/jisesa/iez124

Belluco, S., Losasso, C., Maggioletti, M., Alonzi, C. C., Paoletti, M. G., & Ricci, A. (2013). Edible insects in a food safety and nutritional perspective: A critical review. *Comprehensive Reviews in Food Science and Food Safety*, *12*(3), 296–313. https://doi.org/10.1111/1541-4337.12014

Bessa, L. W., Pieterse, E., Sigge, G., & Hoffman, L. C. (2020). Insects as human food; from farm to fork. *Journal of the Science of Food and Agriculture*, *100*(14), 5017–5022. John Wiley and Sons Ltd. https://doi.org/10.1002/jsfa.8860

Borges, M. M., da Costa, D. V., Trombete, F. M., & Câmara, A. K. F. I. (2022). Edible insects as a sustainable alternative to food products: an insight into quality aspects of reformulated bakery and meat products. *Current Opinion in Food Science*, *46*. Elsevier Ltd. https://doi.org/10.1016/j.cofs.2022.100864

Boye, J., & Barbana, C. (2012). Protein processing in food and bioproduct manufacturing and techniques for analysis. In N. T. Dunford (Ed.), *Food and Industrial Bioproducts and Bioprocessing* (pp. 85–113). https://doi.org/10.1002/9781119946083.ch3.

Bresciani, A., Cardone, G., Jucker, C., Savoldelli, S., & Marti, A. (2022). Technological performance of cricket powder (*Acheta domesticus* L.) in wheat-based formulations. *Insects*, *13*(6). https://doi.org/10.3390/insects13060546

Caparros-Megido, R., Gierts, C., Blecker, C., Brostaux, Y., Haubruge, É., Alabi, T., & Francis, F. (2016). Consumer acceptance of insect-based alternative meat products in Western countries. *Food Quality and Preference*, *52*, 237–243. https://doi.org/10.1016/j.foodqual.2016.05.004

Churchward-Venne, T. A., Pinckaers, P. J. M., van Loon, J. J. A., & van Loon, L. J. C. (2017). Consideration of insects as a source of dietary protein for human consumption. *Nutrition Reviews*, *75*(12), 1035–1045. https://doi.org/10.1093/nutrit/nux057

Cohen, J. H., Sánchez, N. D. M., & Montiel-Ishino, F. (2009). Chapulines and food choices in rural Oaxaca. *Gastronomica*, *9*(1), 61–65. https://doi.org/10.1525/gfc.2009.9.1.61

Cruz-Labana, J. D., Crosby-Galván, M. M., Delgado-Alvarado, A., Alcántara-Carbajal, J. L., Cuca-García, J. M., & Tarango-Arámbula, L. A. (2018). Nutritional content of Liometopum apiculatum Mayr larvae ("escamoles") by vegetation type in north-central Mexico. *Journal of Asia-Pacific Entomology*, *21*(4), 1239–1245. https://doi.org/10.1016/j.aspen.2018.09.008

De Gier, S., & Verhoeckx, K. (2018). Insect (food) allergy and allergens. *Molecular Immunology*, *100*, 82–106. https://doi.org/10.1016/j.molimm.2018.03.015

de Marchi, L., Wangorsch, A., & Zoccatelli, G. (2021). Allergens from edible insects: Cross-reactivity and effects of processing. https://doi.org/10.1007/s11882-021-01012-z/Published

de Oliveira, L. M., da Silva Lucas, A. J., Lopes CAdaval, A., & Sallas Mellado, M. (2017) Bread enriched with flour from cinereous cockroach (*Nauphoeta cinerea*). *Innovative Food Science & Emerging Technologies*, *44*, 30–35. https://doi.org/10.1016/j.ifset.2017.08.015

Del Mastro, N. L. (2021). Evolution of the interest on edible insects. *American Journal of Biological and Environmental Statistics*, *7*(2), 52. https://doi.org/10.11648/j.ajbes.20210702.13

Diaz-Bustamante, M. L., Keppler, J. K., Reyes, L. H., & Alvarez Solano, O. A. (2023). Trends and prospects in dairy protein replacement in yogurt and cheese. *Heliyon*, *9*(6). https://doi.org/10.1016/j.heliyon.2023.e16974

Dobermann, D., Swift, J. A., & Field, L. M. (2017). Opportunities and hurdles of edible insects for food and feed. Nutrition Bulletin, 42(4), 293–308. Blackwell Publishing Ltd. https://doi.org/10.1111/nbu.12291

Doğan, E., & Çekal, N. (2022). Sustainable food alternative in gastronomy: Edible insects (Entomophagy). *International Journal of Agriculture Environment and Food Sciences*, *6*(2), 246–253. https://doi.org/10.31015/jaefs.2022.2.7

Escamilla-Prado, Y., Escamilla-Femat, E., Gómez-Utrilla, S., Miguel, J., Andrade, T., Ramos-Elorduy, P., Pino-Moreno, J., & Manuel, J. (2012). *Tropical and Subtropical Agroecosystems*, *15*(2), 101–109. http://www.redalyc.org/articulo.oa?id=93924626013

Gahukar, R. T. (2020). Edible insects collected from forests for family livelihood and wellness of rural communities: A review. Global Food Security, 25, 100348. Elsevier B.V. https://doi.org/10.1016/j.gfs.2020.100348

Ghosh, S., Lee, S. M., Jung, C., & Meyer-Rochow, V. B. (2017). Nutritional composition of five commercial edible insects in South Korea. *Journal of Asia-Pacific Entomology*, *20*(2), 686–694. https://doi.org/10.1016/j.aspen.2017.04.003

González, C. M., Garzón, R., & Rosell, C. M. (2019). Insects as ingredients for bakery goods. A comparison study of H. illucens, A. domestica and T. molitor flours. *Innovative Food Science and Emerging Technologies*, *51*, 205–210. https://doi.org/10.1016/j.ifset.2018.03.021

González-Aguilar, D., Galván-Lozano, D., Pacheco-Gallardo, C., & Cabrera-Díaz, E. (2019). Determination of protein of edible insects. *ECORFAN Journal Republic of Nicaragua*, 12–16. https://doi.org/10.35429/ejrn.2019.9.5.12.16

Gorbunova, N. A. & Zakharov, A. N. (2021). Edible insects as a source of alternative protein. A review. *Theory and Practice of Meat Processing*, *6*(1). https://doi.org/10.21323/2414–438X2021-6–1–23–32

Granados-Echegoyen, C., Vásquez-López, A., Calderón-Cortés, N., Gallego-Ocampo, H. L., Gómez-Rodríguez, C. H., Rodríguez-Vélez, J. M., Sarmiento-Cordero, M. A., Salamanca-Canizales, L. J., Rodríguez-Vélez, B., Arroyo-Balán, F., & Andrade-Hoyos, P. (2024). Brief overview of edible insects: Exploring consumption and promising sustainable uses in Latin America. Frontiers in Sustainable Food Systems, 8. Frontiers Media SA. https://doi.org/10.3389/fsufs.2024.1385081

Gravel, A., & Doyen, A. (2019). The use of edible insect proteins in food: Challenges and issues related to their functional properties. Innovative Food Science & Emerging Technologies, 59, 102272. Elsevier Ltd. https://doi.org/10.1016/j.ifset.2019.102272

Hall, F., & Liceaga, A. (2020). Effect of microwave-assisted enzymatic hydrolysis of cricket (*Gryllodes sigillatus*) protein on ACE and DPP-IV inhibition and tropomyosin-IgG binding. Journal of Functional Foods, 64, 103634. https://doi.org/10.1016/j.jff.2019.103634

Hall, F. G., Jones, O. G., O'Haire, M. E., & Liceaga, A. M. (2016). Functional properties of tropical banded cricket (Gryllodes sigillatus) protein hydrolysates. Food Chemistry. https://doi.org/10.1016/j.foodchem.2016.11.138

Halloran, A., Muenke, C., Vantomme, P., & van Huis, A. (2014). Insects in the human food chain: Global status and opportunities. Food Chain, 4(2). http://dx.doi.org/10.3362/2046-1887.2014.011 ISSN: 2046–1879 (print) 2046–1887 (online)

Hammer, L., Moretti, D., Abbühl-Eng, L., Kandiah, P., Hilaj, N., Portmann, R., & Egger, L. (2023). Mealworm larvae (*Tenebrio molitor*) and crickets (*Acheta domesticus*) show high total protein in vitro digestibility and can provide good-to-excellent protein quality as determined by in vitro DIAAS. *Frontiers in Nutrition, 10*. https://doi.org/10.3389/fnut.2023.1150581

Hlongwane, Z., Slotow, R., & Munyai, T. C. (2020). Nutritional composition of edible insects consumed in Africa: A systematic review. Nutrients, 12, 2786. https://doi.org/10.3390/nu12092786

Ishara, J., Ayagirwe, R., Karume, K., Mushagalusa, G. N., Bugeme, D., Niassy, S., Udomkun, P., & Kinyuru, J. (2022). Inventory reveals wide biodiversity of edible insects in the Eastern Democratic Republic of Congo. *Scientific Reports*, *12*(1). https://doi.org/10.1038/s41598-022-05607-y

Jiang, Y., Zhu, Y., Zheng, Y., Liu, Z., Zhong, Y., Deng, Y., & Zhao, Y. (2021). Effects of salting-in/out-assisted extractions on structural, physicochemical and functional properties of Tenebrio molitor larvae protein isolates. *Food Chemistry*, *338*, 128158. https://doi.org/10.1016/j.foodchem.2020.128158

Jongema, Y. (2017). *Worldwide list of recorded edible insects*. Department of Entomology, Wageningen University & Research.

Kamemura, N., Sugimoto, M., Tamehiro, N., Adachi, R., Tomonari, S., Watanabe, T., et al. (2019). Cross-allergenicity of crustacean and the edible insect Gryllus bimaculatus in patients with shrimp allergy. Molecular Immunology, 106, 127–134. https://doi.org/10.1016/j.molimm.2018.12.015

Kausar, M. A. (2018). A review on respiratory allergy caused by insects. Bioinformation, 14(9), 540–553.

Kim, T. K., Cha, J. Y., Yong, H. I., Jang, H. W., Jung, S., & Choi, Y. S. (2022). Application of edible insects as novel protein sources and strategies for improving their processing. Food Science of Animal Resources, 42(3), 372–388. https://doi.org/10.5851/kosfa.2022.e10

Kim, T. K., Yong, H. I., Kim, Y. B., Kim, H. W., & Choi, Y. S. (2019). Edible insects as a protein source: A review of public perception, processing technology, and research trends. Food Science of Animal Resources, 39(4), 521–540. https://doi.org/10.5851/kosfa.2019.e53

Köhler, R., Kariuki, L., Lambert, C., & Biesalski, H. K. (2019). Protein, amino acid and mineral composition of some edible insects from Thailand. Journal of Asia-Pacific Entomology, 22(1), 372–378. https://doi.org/10.1016/j.aspen.2019.02.002

Kouřimská, L., & Adámková, A. (2016). Nutritional and sensory quality of edible insects. NFS Journal, 4, 22–26. https://doi.org/10.1016/j.nfs.2016.07.001

Kowalski, S., Mikulec, A., Mickowska, B., Skotnicka, M., & Mazurek, A. (2022). Wheat bread supplementation with various edible insect flours: Influence of chemical composition on nutritional and technological aspects. LWT, 159, 113220. https://doi.org/10.1016/j.lwt.2022.113220

Krawczyk, A., Fernández-López, J., & Zimoch-Korzycka, A. (2024). Insect protein as a component of meat analogue burger. Foods, 13(12), 1806. https://doi.org/10.3390/foods13121806

Kurek, M. A., Onopiuk, A., Pogorzelska-Nowicka, E., Szpicer, A., Zalewska, M., & Półtorak, A. (2022). Novel protein sources for applications in meat-alternative products: Insight and challenges. Foods, 11(7), 957. https://doi.org/10.3390/foods11070957

Lamberti, C., Nebbia, S., Cirrincione, S., Brussino, L., Giorgis, V., Romito, A., Marchese, C., Manfredi, M., Marengo, E., Giuffrida, M. G., et al. (2021). Thermal processing of insect allergens and IgE cross-recognition in Italian patients allergic to shrimp, house dust mite, and mealworm. Food Research International, 148, 110567.

Lamsal, B., Wang, H., Pinsirodom, P., & Dossey, A. T. (2019). Applications of insect-derived protein ingredients in food and feed industry. Journal of the American Oil Chemists' Society, 96(2), 105–123. https://doi.org/10.1002/aocs.12180

Lange, K. W., & Nakamura, Y. (2021). Edible insects as future food: Chances and challenges. Journal of Future Foods, 1(1), 38–46. https://doi.org/10.1016/j.jfutfo.2021.10.001

Lara-Juárez, P., Aguirre-Rivera, J. R., Castillo-Lara, P., & Reyes-Agüero, J. A. (2015). Biología y aprovechamiento de la hormiga de escamoles, Liometopum apiculatum Mayr (Hymenoptera: Formicidae). Acta Zoológica Mexicana (n.s.), 31(2).

Leni, G., Tedeschi, T., Faccini, A., Pratesi, F., Folli, C., Puxeddu, I., Migliorini, P., Gianotten, N., Jacobs, J., Depraetere, S., et al. (2020). Shotgun proteomics, in-silico evaluation and immunoblotting assays for allergenicity assessment of lesser mealworm, black soldier fly and their protein hydrolysates. Scientific Reports, *10,* 1228. https://doi.org/10.1038/s41598-020-57863-5

Liceaga, A. M. (2022). Edible insects, a valuable protein source from ancient to modern times. In Advances in Food and Nutrition Research (Vol. 101, **pp.** 129–152). Academic Press. https://doi.org/10.1016/bs.afnr.2022.04.002

Liu, Z., Xia, L., Wu, Y., Xia, Q., Chen, J., & Roux, K. H. (2009). Identification and characterization of an arginine kinase as a major allergen from silkworm (Bombyx mori) larvae. International Archives of Allergy and Immunology, 150, 8–14. https://doi.org/10.1159/000210375

Longvah, T., Mangthya, K., & Ramulu, P. (2011). Nutrient composition and protein quality evaluation of eri silkworm (Samia ricinii) prepupae and pupae. Food Chemistry, 128(2), 400–403. https://doi.org/10.1016/j.foodchem.2011.03.041

Ma, Z., Mondor, M., Goycoolea Valencia, F., & Hernández-Álvarez, A. J. (2023). Current state of insect proteins: Extraction technologies, bioactive peptides and allergenicity of edible insect proteins. Food & Function, 14(18), 8129–8156. https://doi.org/10.1039/d3fo02865h

Manditsera, F. A., Luning, P. A., Fogliano, V., & Lakemond, C. M. M. (2019). Effect of domestic cooking methods on protein digestibility and mineral bioaccessibility of wild harvested adult edible insects. *Food Research International, 121*, 404–411. https://doi.org/10.1016/j.foodres.2019.03.052

Melgar-Lalanne, G., Hernández-Álvarez, A. J., & Salinas-Castro, A. (2019). Edible insects processing: Traditional and innovative technologies. *Comprehensive Reviews in Food Science and Food Safety*, *18*(4), 1166–1191. https://doi.org/10.1111/1541-4337.12463

Melo-Ruiz, V., García, M., Sandoval, H., Jiménez, H. D., & Calvo, C. (2011). Quality proteins from edible indigenous insect food of Latin America and Asia. *Emirates Journal of Food and Agriculture*, *23*, 283–289.

Melo-Ruiz, V., Sandoval-Trujillo, H., Quirino-Barreda, T., Sánchez-Herrera, K., Díaz-García, R., & Calvo-Carrillo, C. (2015). Chemical composition and amino acids content of five species of edible Grasshoppers from Mexico. *Emirates Journal of Food and Agriculture*, *27*(8), 654–658. https://doi.org/10.9755/ejfa.2015.04.093

Meyer-Rochow, V. B., Gahukar, R. T., Ghosh, S., & Jung, C. (2021). Chemical composition, nutrient quality and acceptability of edible insects are affected by species, developmental stage, gender, diet, and processing method. Foods, 10(5). MDPI AG. https://doi.org/10.3390/foods10051036

Mintah, B. K., He, R., Agyekum, A. A., Dabbour, M., Golly, M. K., & Ma, H. (2020). Edible insect protein for food applications: Extraction, composition, and functional properties. *Journal of Food Process Engineering*, *43*(4). https://doi.org/10.1111/jfpe.13362

Mishyna, M., Chen, J., & Benjamin, O. (2020). Sensory attributes of edible insects and insect-based foods – Future outlooks for enhancing consumer appeal. Trends in Food Science & Technology, 95, 141–148. Elsevier Ltd. https://doi.org/10.1016/j.tifs.2019.11.016

Mishyna, M., Keppler, J. K., & Chen, J. (2021). Techno-functional properties of edible insect proteins and effects of processing. Current Opinion in Colloid & Interface Science, 56. Elsevier Ltd. https://doi.org/10.1016/j.cocis.2021.101508

Molina-Vega, A., Hernández-Domínguez, E. M., Villa-García, M., & Álvarez-Cervantes, J. (2021). Comadia redtenbacheri (Lepidoptera: Cossidae) and Aegiale hesperiaris (Lepidoptera: Hesperiidae), two important edible insects of Agave salmiana (Asparagales: Asparagaceae): A review. International Journal of Tropical Insect Science, 41(3), 1977–1988. Springer Science and Business Media Deutschland GmbH. https://doi.org/10.1007/s42690-020-00396-1

Neves, V., Campos, L., Ribeiro, N., Costa, R., Correia, P., Gonçalves, J., & Henriques, M. (2024). Insect flour as milk protein substitute in fermented dairy products. *Food Bioscience, 60*. https://doi.org/10.1016/j.fbio.2024.104379

Nikolić, K., & Tepavčević, J. (2024). Functional properties of insect proteins. *AIDASCO Reviews, 2*(1), 26–31. https://doi.org/10.59783/aire.2024.38

Nowakowski, A. C., Miller, A. C., Miller, M. E., Xiao, H., & Wu, X. (2021). Potential health benefits of edible insects. Critical Reviews in Food Science and Nutrition, 62(13), 3499–3508. Taylor and Francis Ltd. https://doi.org/10.1080/10408398.2020.1867053

Ojha, S., Bekhit, A. E. D., Grune, T., & Schlüter, O. K. (2021). Bioavailability of nutrients from edible insects. Current Opinion in Food Science, 41, 240–248. Elsevier Ltd. https://doi.org/10.1016/j.cofs.2021.08.003

Omuse, E. R., Tonnang, H. E. Z., Yusuf, A. A., Machekano, H., Egonyu, J. P., Kimathi, E., Mohamed, S. F., Kassie, M., Subramanian, S., Onditi, J., Mwangi, S., Ekesi, S., & Niassy, S. (2024). The global atlas of edible insects: analysis of diversity and commonality contributing to food systems and sustainability. *Scientific Reports, 14*(1). https://doi.org/10.1038/s41598-024-55603-7

Oonincx, D. G. A. B., & Finke, M. D. (2021). Nutritional value of insects and ways to manipulate their composition. *Journal of Insects as Food and Feed, 7*(5), 639–659. https://doi.org/10.3920/JIFF2020.0050

Oonincx, D. G. A. B., van Itterbeeck, J., Heetkamp, M. J. W., van den Brand, H., van Loon, J. J. A., & van Huis, A. (2010). An exploration on greenhouse gas and ammonia production by insect species suitable for animal or human consumption. *PLoS ONE, 5*(12), e14445. https://doi.org/10.1371/journal.pone.0014445

Pan, J., Xu, H., Cheng, Y., Mintah, B. K., Dabbour, M., Yang, F., Chen, W., Zhang, Z., Dai, C., He, R., & Ma, H. (2022). Recent insight on edible insect protein: Extraction, functional properties, allergenicity, bioactivity, and applications. *Foods, 11*(19). MDPI. https://doi.org/10.3390/foods11192931

Papastavropoulou, K., Koupa, A., Kritikou, E., Kostakis, M., & Proestos, C. (2022a). Edible insects: Benefits and potential risk for consumers and the food industry. *Biointerface Research in Applied Chemistry, 12*(4), 5131–5149. https://doi.org/10.33263/BRIAC124.51315149

Papastavropoulou, K., Xiao, J., & Proestos, C. (2022b). Edible insects: Tendency or necessity (a review). eFood, 4(1), e58. John Wiley and Sons Inc. https://doi.org/10.1002/efd2.58

Park, Y. S., Choi, Y. S., Hwang, K. E., Kim, T. K., Lee, C. W., Shin, D. M., & Han, S. G. (2017). Physicochemical properties of meat batter added with edible silkworm pupae (Bombyx mori) and transglutaminase. Korean Journal of Food Science of Animal Resources, 37(3), 351–359. https://doi.org/10.5851/kosfa.2017.37.3.351.

Patel, S., Suleria, H. A. R., & Rauf, A. (2019). Edible insects as innovative foods: Nutritional and functional assessments. Trends in Food Science & Technology, 86, 352–359. https://doi.org/10.1016/j.tifs.2019.02.033

Payne, C. L. R., Scarborough, P., Rayner, M., & Nonaka, K. (2016). A systematic review of nutrient composition data available for twelve commercially available edible insects, and comparison with reference values. Trends in Food Science & Technology, 47, 69–77. https://doi.org/10.1016/j.tifs.2015.10.012

Pino-Moreno, J. M., & Blasquez, J. R.-E. (2021). Taxonomic analysis of some edible insects from the state of Michoacán, Mexico. *Frontiers in Veterinary Science, 8*, 629194. https://doi.org/10.3389/fvets.2021.629194

Purschke, B., Meinlschmidt, P., Horn, C., Rieder, O., & Jäger, H. (2018). Improvement of techno-functional properties of edible insect protein from migratory locust by enzymatic hydrolysis. European Food Research and Technology, 244(6), 999–1013. https://doi.org/10.1007/s00217-017-3017-9

Raheem, D., Carrascosa, C., Oluwole, O. B., Nieuwland, M., Saraiva, A., Millán, R., & Raposo, A. (2019). Traditional consumption of and rearing edible insects in Africa, Asia and Europe. Critical Reviews in Food Science and Nutrition, 59(14), 2169–2188. https://doi.org/10.1080/10408398.2018.1440191

Rahman, M. M., Byanju, B., & Lamsal, B. P. (2023). Protein, lipid, and chitin fractions from insects: Method of extraction, functional properties, and potential applications. Critical Reviews in Food Science and Nutrition, 64(18), 6415–6431. https://doi.org/10.1080/10408398.2023.2168620

Ramos-Elorduy, J. (1997). Insects: A sustainable source of food? Ecology of Food and Nutrition, *36*(2–4), 247–276. https://doi.org/10.1080/03670244.1997.9991519

Ramos-Elorduy, J., Manuel, J., Moreno, P., Prado, E. E., Alvarado Perez, M., Otero, J. L., & Ladron De Guevara, O. (1997). Nutritional value of edible insects from the State of Oaxaca, Mexico. Journal of Food Composition and Analysis, 10, 142–157.

Ramos-Elorduy, J., Moreno, J. M. P., Vázquez, A. I., Landero, I., Oliva-Rivera, H., & Camacho, V. H. M. (2011). Edible Lepidoptera in Mexico: Geographic distribution, ethnicity, economic and nutritional importance for rural people. Journal of Ethnobiology and Ethnomedicine, 7, 2. https://doi.org/10.1186/1746-4269-7-2

Ribeiro, J. C., Cunha, L. M., Sousa-Pinto, B., & Fonseca, J. (2018). Allergic risks of consuming edible insects: A systematic review. Molecular Nutrition & Food Research, 62(1), e1700030. https://doi.org/10.1002/mnfr.201700030

Ríos, R. E. (2024). Escamoles, gastronomic heritage and tourist attraction in Mineral de Pozos (Guanajuato, Mexico). Turismo y Sociedad, 35, 107–128. https://doi.org/10.18601/01207555.n35.05

Rodríguez-Rodríguez, M., Barroso, F. G., Fabrikov, D., & Sánchez-Muros, M. J. (2022). In vitro crude protein digestibility of insects: A review. Insects, 13(8), 682. https://doi.org/10.3390/insects13080682

Roncolini, A., Milanović, V., Cardinali, F., Osimani, A., Garofalo, C., Sabbatini, R., Clementi, F., Pasquini, M., Mozzon, M., Foligni, R., Raffaelli, N., Zamporlini, F., Minazzato, G., Trombetta, M. F., van Buitenen, A., van Campenhout, L., & Aquilanti, L. (2019). Protein fortification with mealworm (Tenebrio molitor L.) powder: Effect on textural, microbiological, nutritional and sensory features of bread. PLoS One, 14(2), e0211747. https://doi.org/10.1371/journal.pone.0211747

Rumpold, B. A., & Schlüter, O. K. (2013). Nutritional composition and safety aspects of edible insects. Molecular Nutrition & Food Research, 57(5), 802–823. https://doi.org/10.1002/mnfr.201200735

Sales-Queiroz, L., Nogueira Silva, N. F., Jessen, F., Mohammadifar, M. A., Stephani, R., Fernandes de Carvalho, A., Perrone, Í. T., & Casanova, F. (2023). Edible insect as an alternative protein source: A review on the chemistry and functionalities of proteins under different processing methods. Heliyon, 9(4), e14831. Elsevier Ltd. https://doi.org/10.1016/j.heliyon.2023.e14831

Santiago, L. A., Fadel, O. M., & Tavares, G. M. (2021). How does the thermal-aggregation behavior of black cricket protein isolate affect its foaming and gelling properties? Food Hydrocolloids, 110, 106169. https://doi.org/10.1016/j.foodhyd.2020.106169

Shah, A. A., Totakul, P., Matra, M., Cherdthong, A., Hanboonsong, Y., & Wanapat, M. (2022). Nutritional composition of various insects and potential uses as alternative protein sources in animal diets. Animal Bioscience, 35(2), 317–331. https://doi.org/10.5713/ab.21.0447.

Siddiqui, S. A., Fernando, I., Povetkin, S. N., Asante, K., Ngah, N., Castro-Muñoz, R., Saraswati, Y. R., Wu, Y. S., Lahan, M., & Aidoo, O. F. (2024). Edible dragonflies and damselflies (order Odonata) as human food – A comprehensive review. Journal of Insects as Food and Feed, 16, 1–26. Brill Wageningen Academic. https://doi.org/10.1163/23524588-20230097

Silva-García, F., Figueroa-Castro, P., López-Martínez, V., & Pérez-Villalba, E. (2018). Primera cita de los jumiles comestibles Edessa bifida y Edessa championi (Hemiptera: Pentatomidae) y su importancia en la cultura entomofágica en la comunidad rural de "El Zompantle", Taxco de Alarcón, Guerrero, México. *Acta Zoológica Mexicana (N.S.)*, *34*, 1–3. https://doi.org/10.21829/azm.2018.3411193

Smetana, S., Larki, N. A., Pernutz, C., Franke, K., Bindrich, U., Toepfl, S., et al. (2018) Structure design of insect-based meat analogs with high-moisture extrusion. Journal of Food Engineering, 229, 83–85. https://doi.org/10.1016/j.jfoodeng.2017.06.035

Tang, C., Yang, D., Liao, H., Sun, H., Liu, C., Wei, L., & Li, F. (2019). Edible insects as a food source: A review. Food Production, Processing and Nutrition, 1(1). https://doi.org/10.1186/s43014-019-0008-1

Tello, A., Aganovic, K., Parniakov, O., Carter, A., Heinz, V., & Smetana, S. (2021). Product development and environmental impact of an insect-based milk alternative. Future Foods, 4, 100080. https://doi.org/10.1016/j.fufo.2021.100080

Torres-Castillo, J. A., & Olazarán-Santibáñez, F. E. (2023). Insects as source of phenolic and antioxidant entomochemicals in the food industry. Frontiers in Nutrition, 10, 1133342. https://doi.org/10.3389/fnut.2023.1133342

Van Huis, A. (2013). Potential of insects as food and feed in assuring food security. Annual Review of Entomology, 58, 563–583. https://doi.org/10.1146/annurev-ento-120811-153704

Van Huis, A. (2022). Edible insects: Challenges and prospects. Entomological Research, 52, 161–177. https://doi.org/10.1111/1748-5967.12582

van Huis, A., Rumpold, b., Maya, c., & Roos, N. (2021). Nutritional qualities and enhancement of edible insects. *Annual Review of Nutrition*, *41*, 551–576. https://doi.org/10.1146/annurev-nutr-041520-010856

Van Huis, A., Van Itterbeeck, J., Klunder, H., Mertens, E., Halloran, A., Muir, G., & Vantomme, P. (2013). Edible insects: Future prospects for food and feed security. FAO. ISBN 978-92-5-107595-1, ISBN 978-92-5-107596-8 (PDF)

Villaseñor, V. M., Enriquez-Vara, J. N., Urías-Silva, J. E., & Mojica, L. (2021). Edible insects: Techno-functional properties, food and feed applications, and biological potential. Food Reviews International, 38(S1), 866–892. https://doi.org/10.1080/87559129.2021.1890116

Weru, J., Chege, P., & Kinyuru, J. (2021). Nutritional potential of edible insects: A systematic review of published data. International Journal of Tropical Insect Science, 41(3), 2015–2037. https://doi.org/10.1007/s42690-021-00464-0

Wu, X., Zhao, W., Wang, X., Bai, Z., & Ma, L. (2023). A novel variable power microwave (VPM) drying technology for lowering energy consumption and improving the in vitro protein digestibility of black soldier fly larvae. Innovative Food Science & Emerging Technologies, 89, 103470. https://doi.org/10.1016/j.ifset.2023.103470

Yang, J., Zhou, S., Kuang, H., Tang, C., & Song, J. (2023). Edible insects as ingredients in food products: Nutrition, functional properties, allergenicity of insect proteins, and processing modifications. Critical Reviews in Food Science and Nutrition. https://doi.org/10.1080/10408398.2023.2223644

Yi, L., Lakemond, C. M. M., Sagis, L. M. C., Eisner-Schadler, V., van Huis, A., & van Boekel, M. A. J. S. (2013). Extraction and characterisation of protein fractions from five insect species. Food Chemistry, 141(4), 3341–3348. https://doi.org/10.1016/j.foodchem.2013.05.115

Youssef, J., & Spence, C. (2021). Introducing diners to the range of experiences in creative Mexican cuisine, including the consumption of insects. International Journal of Gastronomy and Food Science, 25, 100371. https://doi.org/10.1016/j.ijgfs.2021.100371

Zeng, Y., Chen, E., Zhang, X., Li, D., Wang, Q., & Sun, Y. (2022). Nutritional value and physicochemical characteristics of alternative protein for meat and dairy – A review. Foods, 11(21), 3326. https://doi.org/10.3390/foods11213326

Zhang, Y., Dong, L., Zhang, J., Shi, J., Wang, Y., & Wang, S. (2021a). Adverse effects of thermal food processing on the structural, nutritional and biological properties of proteins. Annual Review of Food Science and Technology, 12, 259–286. https://doi.org/10.1146/annurev-food-062320-012215

Zhang, Y., Sharan, S., Rinnan, A., & Orlien, V. (2021b). Survey on methods for investigating protein functionality and related molecular characteristics. Foods, 10(11), 2848. https://doi.org/10.3390/foods10112848

Zhao, X., Vázquez-Gutiérrez, J. L., Johansson, D. P., Landberg, R., & Langton, M. (2016) Yellow mealworm protein for food purposes – Extraction and functional properties. PLoS One, 11(2), e0147791. https://doi.org/10.1371/journal.pone.0147791

Zielińska, E. (2022) Evaluating the functional characteristics of certain insect flours (non-defatted/defatted flour) and their protein preparations. Molecules, 27(19), 6339. https://doi.org/10.3390/molecules27196339

Zielińska, E., Baraniak, B., & Karaś, M. (2017). Antioxidant and anti-inflammatory activities of hydrolysates and peptide fractions obtained by enzymatic hydrolysis of selected heat-treated edible insects. Nutrients, 9, 970. https://doi.org/10.3390/nu9090970

Zielińska, E., Baraniak, B., Karaś, M., Rybczyńska, K., & Jakubczyk, A. (2015). Selected species of edible insects as a source of nutrient composition. *Food Research International*, *77*, 460–466. https://doi.org/10.1016/j.foodres.2015.09.008

Zielińska, E., Karaś, M., & Baraniak, B. (2018). Comparison of functional properties of edible insects and protein preparations thereof. *LWT*, *91*, 168–174. https://doi.org/10.1016/j.lwt.2018.01.058

Nutritional Value of Insects: Lipids and Fatty Acids

4

Diana Martin, David Villanueva-Bermejo, Tiziana Fornari, Mónica R. García-Risco, and Luis Vázquez

4.1 INTRODUCTION

Proteins are the major and most popular nutritional component of edible insects. However, it is important to consider that lipids are the second major macronutrient that is found in most insect species. In fact, the lipid fraction contributes not only to the energy content of insects but also to their overall health benefits and even culinary versatility. Due to these reasons, mainly the energy contribution of lipids, the consumption of insects has been traditionally part of the diet of those cultures that lacks enough access to food and resources, the insect being a food with a high energy density due to the sum of proteins and lipids, which makes these animals as relevant contributors to food security (Van Huis, 2015).

The lipid fraction can be consumed as part of the insect, but the oils and fats from insects are being explored as additional products to be valued besides the protein insect meals (Antunes et al., 2024; Kolobe et al., 2023). This is because, for those high lipid content insect species, this lipid fraction is being technologically removed, since it may interfere with the production of protein-rich products. Additionally, in the case of animal feeding, the defatting process is also necessary for insect species, since fat may hinder the processing of the feeds, affect the palatability and impact on the productive yields and lipid composition of the fed animals (Alfiko et al., 2022). Therefore, the independent value of the lipid co-product from edible insects is being established for its specific commercial use, especially when interesting nutritional values are present in such oils.

DOI: 10.1201/9781003488156-4

As in any other animals, lipids serve various physiological functions in insects, including energy storage, cell membrane structure and signalling pathways. Specially, the use of lipids as a primary metabolic substrate in insects makes this component a large reservoir of energy, which is used during periods of prolonged energy demand, suffering important variations, both quantitatively and qualitatively (Downer and Matthews, 1976). Anatomically, the main tissue that is remarked to contain the greatest concentration of lipids in insects is the fat body, together with the haemolymph and the cuticle. The fat body is an extensive and diffused aggregate of cells that can be considered analogous to the adipose tissue and liver of the vertebrate animals. This is because this tissue is the storage of proteins, carbohydrates and lipids but also serves as an organ of intermediary metabolism and lipid synthesis (Downer and Matthews, 1976). Concerning the haemolymph, this is the primary tissue by which molecules are transported for absorption or synthesis in insects, including lipids. Therefore, the lipid content and composition of haemolymph vary extremely with the physiological state of the animal (Toprak et al., 2020). Finally, the cuticle is the hard structure that externally cover the insects having an important function to prevent from water loss. Thus, lipids, as hydrophobic molecules, are a basic constituent of the cuticle and represent and important fraction of the total lipid content of insects (Fernández et al., 2016). Therefore, it sounds easy to understand that due to the metabolic, reservoir and structural functions of the lipid tissues of insects, the specific content and composition in lipid species and fatty acids vary among the different lipid tissues. Thus, glycerides as triglycerides (TAGs) and diglycerides (DAGs) are the major lipids in fat body and haemolymph, while fatty acids and wax esters are the main lipids in cuticle (Ganina et al., 2022). Additionally, the complex pattern of insect development, especially those that suffer metamorphosis, result in intense and rapid physiological and biochemical transformation of the lipid tissues, including the modification of the lipid components, during the life history of an insect. Furthermore, it is important to remark that, together with these general endogenous factors related to the physiology and development stage of insects, these animals exhibit a wide range of lipid compositions, depending on other wide range of factors, both endogenous, such as the own insect species or the sex of the animals, and exogenous, such as the provided substrate for feeding and the subsequent technological processing during their industrial transformation (Meyer-Rochow et al., 2021; Riekkinen et al., 2022).

This chapter aims to provide a comprehensive overview of the lipid fraction of edible insects, examining their nutritional significance for potential applications in food systems. Thus, the lipid content of edible insects, the fatty acid profile, as well as the lipid classes, and minor lipid compounds, considering their implications for human health, will be reviewed. Furthermore, the influence of the major factors that can be intervened by the human practices to modulate the lipid composition of insects will be described, mainly the diet of insects and the subsequent technological processing. Specifically, with close to 2000 species of edible insects identified, this chapter will be focused on those that are currently approved for food and feed in Europe and that are also those most popular worldwide. These species include *Tenebrio molitor*, *Acheta domesticus*, *Locusta migratoria*, *Alphitobius diaperinus*, *Hermetia illucens*, *Musca domestica*, *Gryllodes sigillatus*, *Gryllus assimilis* and *Bombyx mori*.

4.2 LIPIDS AND FATTY ACIDS OF *TENEBRIO MOLITOR*

T. molitor (yellow mealworm) belongs to the family Tenebrionidae (order Coleoptera), and it is one of the most studied edible insect species. Its popularity is also due to the fact that it was the first authorized insect species for food uses in Europe. Specifically, dried preparations of *T. molitor* larvae were approved as a novel food in Europe in 2021 (European Commission, 2021a). Thereafter, frozen and powder forms of the larvae from these insects were also approved to be used in foods for the general population (European Commission, 2022a). Therefore, this insect is typically produced and commercialized under larvae form or subsequent transformations of these larvae.

The lipid content of *T. molitor* larvae can be considered high, but variable, presenting values in the range of 15–35% (Table 4.1).

The lipid fraction is rich in unsaturated fatty acids (UFAs) (Table 4.2), especially monounsaturated fatty acids (MUFA), being oleic acid the main fatty acid (40–50% of total fatty acids). This composition is remarkable because *T. molitor* larvae can be considered the richest insect species in this fatty acid. Therefore, due to the well-known health value of oleic acid, this is one of the main reasons that makes the *T. molitor* lipid fraction of great health interest.

The amount of saturated (SFA) and polyunsaturated (PUFA) fatty acids are quite similar (about 25%, respectively). The most prevalent SFA and PUFA are palmitic (15–20%) and linoleic acids (20–30%), respectively (Table 4.2). Thus, oleic, linoleic and palmitic acids comprise more than 85% of total fatty acids.

The main n-3 fatty acid of *T. molitor* fat is linolenic acid. However, the amount of this fatty acid is usually below 2% and far from other edible insect species, such as *Locusta migratoria* (Section 4) or *Bombyx mori* (Section 10). In contrast to other edible insects, such as *Hermetia illucens* (Section 6), no short-chain fatty acids are generally present, and the concentration of lauric acid (main medium-chain fatty acid (MCFA)) in *T. molitor* fat is less than 1%.

Therefore, considering the global fatty acid profile and despite its low n-3 fatty acid content, the abundance of UFAs (>70% of total fatty acids) and a PUFA/SFA value close to 1 in *T. molitor* results in a fat with a good nutritional quality in terms of fatty acid profile. Furthermore, hypocholesterolemic/hypercholesterolemic ratios (HH) in the range of

TABLE 4.1 Lipid Content of *T. molitor* Larvae According to References from the Scientific Literature.

LIPID CONTENT (G/100 G DM)	*REFERENCE*	*LIPID CONTENT (G/100 G DM)*	*REFERENCE*
29.5	Khanal et al. (2023)	32.3	Son et al. (2020)
28.8	Lawal et al. (2022)	34.5	Ghosh et al. (2017)
26.6	Orkusz et al. (2024)	32.0	Paul et al. (2017)
13.9	Mattioli et al. (2024)	24.7	Zielinska et al. (2015)

TABLE 4.2 Fatty Acid Profile (g/100 g of fatty acids) of *Tenebrio molitor* Larvae, According to Different References Found in the Scientific Literature.

C12:0	–	0.23	0.34	0.42	0.21
C14:0	2.97	4.20	4.5	4.07	2.63
C15:0	–	0.10	–	0.11	0.16
C16:0	18.99	15.20	17.8	17.75	18.0
C17:0	0.19	0.17	0.11	0.10	0.19
C18:0	3.87	2.74	2.6	2.49	3.84
C20:0	0.41	0.08	0.11	0.12	0.17
C16:1	1.46	2.37	2.0	1.95	2.07
C18:1	40.20	49.10	50.1	48.68	40.86
C20:1	–	0.08	–	0.13	0.16
C18:2 n-6	25.41	25.40	21.93	23.09	29.68
C20:2 n-6	–	–	0.11	0.08	–
C18:3α, n-3	0.68	0.43	0.45	0.96	1.61
SFA	27.06	22.70	25.4	25.06	25.32
MUFA	46.13	51.55	52.1	50.76	43.27
PUFA	26.76	25.83	22.5	24.13	31.37
PUFA/SFA	0.99	1.14	0.88	0.96	1.24
n-3	0.75	0.43	0.45	0.96	1.61
n-6	25.41	25.4	22.1	23.17	29.68
n-6/n-3	33.88	59.07	49.0	24.14	18.44
References	Orkusz et al. (2024)	Khanal et al. (2023)	Son et al. (2020)	Lee et al. (2024)	Zielinska et al. (2015)

3.0–3.7 and atherogenicity (AI) and thrombogenicity (TI) indices between 0.4–0.7 and 0.6–0.7, respectively, have been reported (Orkusz et al., 2023; Kotsou, Chatzimitakos, Athanasiadis, Bozinou, Rumbos et al., 2023; Lawal et al., 2022; Otero et al., 2020). Nevertheless, the high n-6/n-3 ratio of *T. molitor*, in the range of 20–60, should be remarked, since a desirable health n-6/n-3 ratio is generally considered to be in the range of 2:1–4:1 to reduce inflammation and associated health risks. Furthermore, *T. molitor* is one of the edible insects with the highest n-6/n-3 ratios. Nevertheless, the high n-6/n-3 ratio of this insect does not make this food of worse quality compared to other conventional sources that also typically contain similar high n-6/n-3 ratios, as it is the case of many vegetable sources. Therefore, considering all the global positive and negative considerations of the fatty acid profile of *T. molitor* larvae, it is relevant to finally remark that despite being an animal source, it can be placed in an intermediate position between the typical fatty acid profile of conventional animal and vegetable sources (Otero et al., 2020).

Besides the global lipid content and fatty acid profile described, it is important to highlight that *T. molitor* is able to be fed on many different food substrates and the amount of lipids, and its composition can be modified based on the insect diet. Numerous studies have been carried out, mainly using agrifood by-products, to evaluate the impact of the diet on the nutritional composition of mealworms, including the lipid fraction (Jankauskiene et al., 2024; Romero-Lorente et al., 2022; Lawal et al., 2021; Mattioli et al., 2021; Boukid et al., 2021). For example, Khanal et al. (2023) applied an oat-based and wheat-based by-products diet. In this case, the lipid amount of the larvae was higher when the oat-based was supplied. Dragojlovic et al. (2022) investigated different times of rearing, and the effect of diets composed of wheat bran along with carrot, cabbage or a mix of carrot, cabbage and flaxseed. The addition of carrots led to a reduction in the lipid content in comparison to the other diets. In addition, the incorporation of flaxseed resulted in a strong increment in the concentration of linolenic acid n-3 and hence a reduction of the n-6/n-3 ratio. Kotsou, Chatzimitakos, Athanasiadis, Bozinou, Athanassiou et al. (2023) studied the inclusion of spent coffee grounds at different ratios (0–25%) on a wheat bran-based feeding. Even though the protein content of larvae increased, no significant difference was observed on the amount of lipids with the addition of the coffee by-product after 8 weeks of feeding. Nevertheless, a significant reduction in the amount of linoleic acid (24% less) and a proportional increase in the concentration of oleic acid occurred with the inclusion of 25% of spent coffee grounds in comparison to the fatty acid profile from the control diet. In another study, Kotsou, Chatzimitakos, Athanasiadis, Bozinou, Rumbos et al. (2023) studied the inclusion of *Moringa oleifera* leaves at different proportions. The addition of 50% of this plant to wheat bran brought about a 35% reduction in the lipid content in comparison to the control diet. Inclusion rates below 50% did not result in significant modifications. Concerning the fatty acid composition, the inclusion of *Moringa oleifera* gave rise to a slight reduction in linoleic and linolenic acids and an increment in oleic acid. Therefore, these examples clearly demonstrate how human intervention by using feeding strategies can be an effective tool to modify both the fatty acid profile and the total lipid content of *T. molitor*, enhancing the nutritional quality of these insects to achieve a healthier composition.

Concerning the lipid classes, TAGs are the major lipid class of *T. molitor.* Costa et al. (2020) reported a lipid fraction consisting of TAG (56.9%), free fatty acids (FFAs, 30.2%), DAGs (2.0%) and monoglycerides (MAG, 1.3%). With respect to polar lipid classes, Tognocchi et al. (2023) found a concentration of phospholipids (PL) of 6.90% in the total lipids, with sphingomyelin (4.30%) being the main PL. Costa et al. (2020) indicated a similar PL concentration (5.4%).

With respect to other minor lipids, it should be noticed that, like most animals, edible insects also typically contain cholesterol. For example, Tognocchi et al. (2023) described values of sterols of 0.64%, with cholesterol (0.34%) being the most abundant sterol in the lipid fraction. Similarly, Otero et al. (2020) described values of cholesterol of 0.47 g/100 g of lipids. However, cholesterol was found in higher quantities (4.2%) in the study of Costa et al. (2020). Additionally, phytosterols, such as brassicasterol, campesterol, δ-avenasterol and β-sitosterol, have also been described in the lipid fraction of edible insects, including *T. molitor*, but their quantity being much lower than that of cholesterol. Tocopherols are other minor lipid compounds analysed in mealworm lipids. Son et al. (2020) found these compounds in low concentration (0.014% of total lipids) and γ-tocopherol accounted for 85.6% of total tocopherols. The presence of tocopherols was also reported by Mattioli et al. (2024). In this case, the main tocopherol found was α-tocopherol. In addition, γ- and α-tocotrienol were also detected. Squalene was also present in the *T. molitor* lipids but at concentrations below 0.01% of total lipids (Son et al., 2020). As described for the fatty acid profile and total lipid content, insect diets may also have a strong influence on the composition in minor lipids profile. For example, Kotsou, Chatzimitakos, Athanasiadis,

Bozinou, Rumbos et al. (2023) reported a β-carotene and vitamin A concentration of 1.05 μg/g and 3.14 μg retinol activity equivalents (RAE)/100 g in *T. molitor* larvae fed on wheat bran, respectively. The concentration of both compounds increased up to 6.22 μg/g and 18.56 μg RAE/100 g when wheat brand was supplemented with 50% of *Moringa oleifera* leaves. The impact of the diet was also described by Mattioli et al. (2021), who found variable concentrations of retinol, zeaxanthin and tocopherols in larvae as a result of the different diets.

Another factor of great influence on the lipid composition of edible insects is the technological processing to which these animals are subjected after their rearing in farms. Typically, insects should be slaughtered, followed by drying for microbial stabilisation, and finally, they can be defatted for obtaining the two major fractions, as the protein meal and the oils and fats as co-product. Martins da Silva et al. (2024) studied the effect of the two most popular slaughter methods typically used for edible insects, such as blanching and freezing. Slightly lower levels of oleic acid occurred by freezing that were attributed to oxidation. Kröncke et al. (2019) investigated the effect of different drying methods (oven, vacuum or freeze-drying). The method applied had an impact on the color and volatile profile but not on the fatty acid composition. Vlahova-Vangelova et al. (2024) observed that microwave drying generated a higher acid value and the appearance of secondary products of lipid oxidation compared to freeze-drying and hot air drying. Otero et al. (2020) studied the effect of two extraction processes, ultrasound-assisted extraction and pressurized liquid extraction, on the lipid composition. The use of ultrasounds with 50% ethanol resulted in an enrichment of PUFA (from 40% to 50%), lower cholesterol concentrations (from 0.47% to 0.26%) as well as improvements of the PUFA/SFA ratio, AI and TI of the extracts. Concerning other innovative modes of processing, a novel food application was performed to the European Food Safety Authority (EFSA) in 2023 for the evaluation of UV-treated powder of *T. molitor* larvae to increase the levels of vitamin D3 in the final food powder (Turck et al., 2023). The applicant reported that the values of vitamin D3 increased from around 2 μg/100 g of powder to more than 50 μg/100 g of powder after UV treatment. The EFSA panel concluded the safety of this novel food. Therefore, all these examples evidence how human intervention by the different technological processing steps performed after insect rearing can also impact the lipid composition of edible insects as *T. molitor*, modulating the nutritional quality of the insects.

4.3 LIPIDS AND FATTY ACIDS OF *ACHETA DOMESTICUS*

A. domesticus belongs to family of Gryllidae of the order Orthoptera. Its common name is house cricket. In general, crickets are within the most popular edible insects worldwide, but especially house cricket is one of the most extended farmed species for food and feed purposes. The first authorisation as novel food for *A. domesticus* was obtained in Europe in 2022 (European Commission, 2022b) as dried, powder and frozen form, and in 2023, it was also authorized as partially defatted powder (European Commission, 2023a).

TABLE 4.3 Lipid Content of *Acheta domesticus* According to References from the Scientific Literature.

LIPID CONTENT (G/100 G DM)	*REFERENCE*	*LIPID CONTENT (G/100 G DM)*	*REFERENCE*
11.2	Pranil et al. (2024)	21.3	Perez-Santaescolastica et al. (2023)
22.8	Orkusz et al. (2023)	15.4	Brogan et al. (2023)
23.1	Grdeń and Sołowiej (2022)	15.1	Khatun et al. (2021)
22.4	Tzompa-Sosa et al. (2021)	29.0	Oonincx et al. (2020)

House crickets are consumed as adult animals, similarly to *Locusta migratoria*, and unlike the larvae form of other insects as *T. molitor, H. illucens* or *A. diaperinus*. Despite this, the lipid content of house cricket can be also considered high, being in the range of 10–30% (Table 4.3). As already explained, the diet and also the sex are important exogenous and endogenous factors, respectively, that can modulate the lipid content. On one hand, diverse studies have described a higher lipid content for female crickets compared to male animals (Gutiérrez et al., 2022; Kulma et al., 2019). On the other hand, Pastell et al. (2021) evidenced a variable range of lipid content of 25–33% depending on the diet, showing that the house crickets fed with the highest carbohydrate-rich diet had the highest lipid content. Oonincx et al. (2020) found that the inclusion of flaxseed oil in the diet of house crickets increased the lipid content of the animals from 29% up to 32%.

Concerning the effect of the exogenous variability factor of technological processing of farmed crickets, the available information on lipid content is still scarce. Interestingly, Singh et al. (2020) evidenced a significant effect of the slaughtering method of the animals on the lipid content. Thus, the highest value of 16.2% of lipids was found for slaughtering by introduction in sealed plastic bags, compared to the minor lipid content of 12.9% found for the combination of CO_2 plus blanching for slaughtering. Concerning the effect of the subsequent processing of drying of the crickets, Lucas-González et al. (2019) also demonstrated an impact on the lipid content, with thermal drying crickets containing higher lipid content (24.9%) than non-thermal drying crickets by freeze-drying (19.3%).

The fatty acid profile of *A. domesticus* is mainly characterized by SFA and PUFA, both >35% of total fatty acids, together with a proportion of MUFA higher than 20% (Table 4.4). This composition is mainly due to the major fatty acids as linoleic acid (>30%), palmitic acid (>25%) and oleic acid (>20%). Therefore, its composition is quite different to *T. molitor*, being closer to other cricket species such as *Gryllodes sigillatus* (Section 8), *Gryllus assimilis* (Section 9) or locust as *L. migratoria* (Section 4).

Concerning its nutritional quality, the ratio PUFA/SFA close to 1 makes a good general quality of the lipid profile of *A. dometicus*. However, only minor n-3 values as linolenic acid can be found in house crickets oils (<2%). Furthermore, the ratio n6/n3 can be considered high (>25) although lower than the remarked value for *T. molitor* (Section 2).

Nevertheless, concerning the effect of the diet on the fatty acid profile, Oonincx et al. (2020) successfully showed that it was possible to increase the content of n-3 fatty acids of these animals by the substrate. Thus, the inclusion of different levels of flaxseed oil to the cricket diets increased progressively the content of n-3 linolenic acid, from 0.8% up to 12.7% at inclusion levels of up to 4% in the diets. This effect was reached without

TABLE 4.4 Fatty Acid Profile (g/100 of Fatty Acids) of *Acheta domesticus* According to Different References Found in the Scientific Literature.

c10	–	–	–	–	–
c12	0.1	–	–	–	0.1
c14	0.6	0.4	–	0.7	0.7
c15	–	–	–	–	–
c16	29.6	25.0	24.3	26.1	27.8
c17	0.3	–	–	0.3	–
c18	10.7	10.7	9.2	9.1	8.2
c20	0.4	0.6	–	0.2	–
c21	–	–	–	–	–
c22	–	0.2	–	–	–
c14:1	–	–	–	–	–
C16:1	0.8	0.6	–	0.4	1.0
c18:1	13.7	24.3	23.3	23.9	29.8
C20:1	–	–	–	–	–
C18:2	41.8	32.8	36.7	35.0	28.7
C18:3n6	–	–	–	–	–
C18:3n3	1.9	0.9	2.6	1.5	0.8
C20:2	–	–	–	–	–
C20:4	–	–	–	–	–
SFA	41.7	37.0	33.5	36.4	36.8
MUFA	14.6	24.9	23.3	24.4	30.8
PUFA	43.7	33.7	39.2	36.6	29.5
PUFA/SFA	1.0	0.9	1.2	1.0	0.8
n3	1.9	0.9	2.5	1.5	0.8
n6	41.8	32.8	36.7	35.0	28.7
n6/n3	22.1	37.3	14.4	22.8	35.9
References	Loypimai et al. (2024)	Orkusz et al. (2023)	Perez-Santaescolastica et al. (2023)	Tzompa-Sosa et al. (2021)	Oonincx et al. (2020)

compromising the normal growth of the crickets. As consequence, the total PUFA content increased (from 29.8% to 38.4%) and the n-6/n-3 ratio decreased (from 36 to 2), reaching much most favourable fatty acid profile from the nutritional point of view.

By technological processing, Otero et al. (2020) also demonstrated that is possible to improve the fatty acid profile of *A. domesticus*. Thus, the extraction assisted by ultrasounds by 50% ethanol allowed to obtain a lipid product from house crickets with lower SFA and higher PUFA contents, subsequently decreasing the estimated AI and TI of the fatty acid profile. Concerning other processing effects, the method of drying of crickets also seems to affect the fatty acid profile. Thus, Lucas-González et al. (2019) evidenced a lower PUFA and SFA when thermal drying was compared to non-thermal drying.

Respect to the composition in lipid compounds, the major lipid class for *A. domesticus* has been reported for neutral lipids, close to 45%, followed by PLs and glycolipids.

Similarly, Brogan et al. (2023) also described an abundance of TAGs around 40%, followed by FFA and PLs, both close to 30%. Therefore, despite TAGs is the major lipid class, it is interesting to remark that, compared with other insect species, the level of TAGs seems to be considered lower in house crickets, since values higher than 60% and even close to 90% of TAGs can be found in other species, as those described in this chapter. On the contrary, it seems that the level of polar lipids is interestingly higher in the case of *A. domesticus*. In this sense, Tzompa-Sosa et al. (2021) performed a more exhaustive lipidomic analysis of house cricket. In case of non-polar lipids, these authors found 94 different compounds within the families of TAG, DAG, waxes and sterols, the TAG being the major one. Additionally, for polar lipids, these authors described up to 211 different PLs classified as phosphatidylcholine, which was the most abundant, followed by phosphatidylserine, phosphatidylethanolamine, phosphatidylinositol, sphingomyelin, phosphatidylglycerol, hexosylceramides and lactosylceramide. Gutiérrez et al. (2022) also performed an interesting lipidomic analysis of *A. domesticus*, finding up to 811 different lipid species. Additionally, these authors found relevant differences in the lipidome between male and female, which were also, respectively, affected by a different mode depending on the diet, especially in case of female animals. Thus, female crickets fed with a balanced diet mainly contained TAGs and glycerophosphocholines, mostly represented by diacylglycerophosphocholines. However, the use of a protein-enriched diet had a higher abundance of TAGs on female animals. On the other hand, male crickets mainly contained glycerophosphoserines and sphingolipids.

Concerning lipid classes, it is important to remark the relevance of the method of extraction of lipids and the impact on lipid classes when insect lipids are analysed. This still unsolved issue has been already described for diverse insect species, including *A. domesticus* (Ramos-Bueno et al., 2016). Therefore, this factor should be cautiously considered when this component is characterized for most edible insects and when comparison of results between assays is performed. For example, Tzompa-Sosa et al. (2014) demonstrated that polar lipids were present in the lipid extract when organic solvents were used for extraction, whereas when aqueous extraction was compared, only non-polar lipids were present in the lipid extract.

Minor lipid compounds can also be found in *A. domesticus*. Otero et al. (2020) reported values of cholesterol close to 0.7% of total lipids, but higher values up to 1.5% of cholesterol have been described (Tzompa-Sosa et al., 2021). These authors also found values of 0.3% of phytosterol, with stigmasterol being the most abundant phytosterol. Additionally, it seems that house crickets are relevant sources of vitamin A, with values around 1.5 μg/100 g of cricket, together with β-carotene (>100 μg/100 g) (Pranil et al., 2024).

Other minor lipid compounds of bioactive interest have been described for *A. domesticus*. Pranil et al. (2024) demonstrated the coextraction of phenolic compounds in the oil of *A. domesticus*, but with variable values depending on the used method of extraction. Thus, conventional Soxhlet with hexane resulted in around 0.16 mg gallic acid equivalent/g compared to ultrasound assisted extraction, which yielded oils with total phenolic contents around 0.26 mg gallic acid equivalent/g. Much higher total phenolic compounds were found in the oil of house crickets when high hydrostatic pressure was used to assist the oil extraction, with values in the range of 0.45–0.75 mg gallic acid equivalent/g oil, depending on the pressure and temperature (Ugur et al., 2021). Therefore, despite these interesting evidence, more specific characterisation and quantification studies are still needed, since the available information of the minor lipid fraction from *A. domesticus* is really scarce.

4.4 LIPIDS AND FATTY ACIDS OF *LOCUSTA MIGRATORIA*

Locusta migratoria belongs to family of Acrididae of the order Orthoptera. Its common name is migratory locust. Within the general common group of locusts, this specie is the most popular for human and animal feed (Egonyu et al., 2021). The authorisation as novel food for *L. migratoria* was obtained for the first time in Europe in 2021 (European Commission, 2021b).

Similarly to most edible insects described in this chapter, the lipid content of migratory locust can also be considered high, since values in the range of 10–30% have been described (Table 4.5). In general, it has been evidenced that this content is higher in adult animals compared to the nymph state of these insects (Mazı, 2024). Additionally, concerning the effect of the diet of the lipid content, the available information is really scarce for *L. migratoria* compared to the intense research in this topic for other species, as the case of *T. molitor* or *H. illucens*. Oonincx et al. (2011) described that a substrate based on wheat bran and carrots increased the lipid content and decreased that of protein, compared to that just based on grass.

Concerning the fatty acid profile, the lipid fraction of the adults from migratory locust mainly consists of UFA (close to 60%), with a similar mean value of MUFA and PUFA around 30% for both groups (Table 4.6). The SFA content is close to 35%. This composition is due to the main fatty acids as oleic acid (close to 30%), palmitic acid (around 25%), linoleic acid (close to 20%), n-3 linolenic acid (>10%) and stearic acid (close to 10%). Similar to the rest of insect species, the ratio PUFA/SFA is close to 1. However, the most remarkable nutritional quality is that this is one the insect species with a notable content of n-3, as linolenic acid, together with a lower content of n-6 fatty acids, compared to other insect species. Due to this, this insect shows a n-6/n-3 ratio more favourable than other insect species from the nutritional point of view (estimated mean ratio 1.6:1, according to

TABLE 4.5 Lipid Content of *Locusta migratoria* According to References from the Scientific Literature.

LIPID CONTENT (G/100 G DM)	*REFERENCE*	*LIPID CONTENT (G/100 G DM)*	*REFERENCE*
23.0	Mazı (2024)	11.4	Brogan et al. (2023)
23.8	Purschke et al. (2018)	23.2	Althwab et al. (2021)
30.5	Fombong et al. (2021)	22.5	Verheyen et al. (2018)
32.3	Perez-Santaescolastica et al. (2023)	20.3	Mohamed (2015)

TABLE 4.6 Fatty Acid Profile (g/100 of Fatty Acids) of *Locusta migratoria* According to Different References Found in the Scientific Literature.

C10	0.01	–	–	–	0.1
c12	0.2	–	–	–	0.3
c14	1.4	1.7	–	2	2
c15	0.08	–	–	–	–
c16	23.5	28.4	23	24	26.8
c17	0.2	–	–	–	0.5
c18	9.4	8.3	11	12	8.6
c20	0.3	–	–	–	0.2
c21	0.4	–	–	–	–
c22	0.05	–	–	–	0.2
c14:1	0.01	–	–	–	–
C16:1	0.8	–	–	–	1.0
c18:1	35.8	31.6	22	30	29.1
C20:1	0.2	–	–	–	0.1
C18:2	17.3	17.0	23	17	17.9
C18:3n6	0.05	–	–	–	–
C18:3n3	10.2	10.5	13	14	11.5
C20:2	0.05	–	–	–	–
C20:4	0.2	–	–	–	0.2
SFA	35.4	38.4	34	38	38.7
MUFA	36.8	31.6	22	30	30.2
PUFA	27.8	27.4	36	31	29.6
PUFA/SFA	0.8	0.7	1.1	0.8	0.8
n3	10.2	10.5	13	14	11.5
n6	17.6	17.4	23	17	18.1
n6/n3	1.7	1.6	1.8	1.2	1.6
References	Mazı et al. (2024)	Perez-Santaescolastica et al. (2023)	Brogan et al. (2023)	Verheyen et al. (2018)	Ramos-Bueno et al. (2016)

values in Table 4.6), being closer to the desirable range of 2:1–4:1 to reduce inflammation and associated health risks.

The lipid classes of migratory locust have not been extensively described. Perez-Santaescolastica et al. (2023) found a major lipid composition on neutral lipids (24.4%), and minor content in PLs (4.8%) and glycolipids (3.0%). Similarly, by thin layer chromatography, Brogan et al. (2023) described a lipid fraction that consisted of 24.6% of TAGs but found a great difference due to a content of 41.3% of FFAs and 24.6% of PLs. On the contrary, Verheyen et al. (2018) described an FFA content for *L. migratoria* of 2.2%. Therefore, even more evident than *A. domesticus*, the content of TAGs for *L. migratoria* seems to be low respect to a relevant level of polar lipids. In this sense, Rose et al. (2021) evidenced again the issue on the variability on lipid characterisation related to the specific used method for extraction. These authors found that the TAG content of migratory locust was in the range of 18.9–47.0%, depending on the organic solvent used for extraction. Furthermore, a great variability was also observed by these authors for PLs (14.5–30.6%) and FFAs (19.3–33.9%). These authors also described important differences on lipid extraction efficiency depending on the solvent. Therefore, together with the issue of the mode of lipid extraction, the characterisation of migratory locus oils seems to be still quite uncomplete, but the results suggest an unconventional composition respect to other animal fats, which motivate the needing of a more exhaustive characterisation of the oil for these insects.

Concerning minor lipid compounds, cholesterol has been described at levels of 8.8% (Brogan et al., 2023) of total lipids. However, Rosc et al. (2021) found values in the range of 8.3 up to 18.9% of the total lipid fraction, depending on the used solvent for extraction of the lipids. Both values were determined by thin layer chromatography densitometry. However, since these given high values of cholesterol would be an important issue for *L. migratoria* from the nutritive quality point of view, this remark again the more exhaustive characterisation needed for locust lipids. Ramos-Bueno et al. (2016) reported a more reasonable value of cholesterol for any animal fat, being close to 80 mg/100 g of dry insect (0.3% of total lipids, estimated according to the mean lipid content of Table 4.5). It has been described that *L. migratoria* also contains carotenoids, such as β-carotene, α-carotene as well as β-cryptoxanthin, lutein and zeaxanthin (Oonincx et al., 2011). Furthermore, it seems that substrates may have an impact on these minor compounds. Thus, wheat bran decreased the α-carotene content. On the contrary, carrots effectively caused higher β-carotene concentrations, whereas α-carotene content did not change.

4.5 LIPIDS AND FATTY ACIDS OF *ALPHITOBIUS DIAPERINUS*

A. diaperinus (lesser mealworm) is another member of the family Tenebrionidae (order Coleoptera). The frozen, paste, dried and powder forms of lesser mealworm larvae were approved for the first time in 2023 as a novel food for their commercialisation in Europe (European Commission, 2023b).

The lipid content of *A. diaperinus* larvae is high, presenting amounts generally above 20% (Table 4.7). In general, this content is higher in larvae compared to the pupal state (Kurecka et al., 2021).

Regarding its fatty acid composition (Table 4.8), SFA and MUFA are predominant, and their concentration is similar (30–41%). Palmitic (25–30%) and oleic acid (29–36%) are the most abundant fatty acids. The amount of PUFA is usually in the range of 20–36%, and this fraction is mainly composed of linoleic acid, which constitutes more than 90% of PUFAs.

This fatty acid composition entails AI and TI indices below one, as well as HH ratio higher than the unit (Roncolini et al., 2020). However, the concentration of n-3 fatty acids in the fat of *A. diaperinus* larvae is low, and it is almost exclusively represented by linolenic acid (less than 2.3% of total fatty acids). Therefore, in general, the nutritional quality of the fatty acid profile of *A. diaperinus* seems quite similar to that of *T. molitor.*

TABLE 4.7 Lipid Content of *A. diaperinus* Larvae According to References from the Scientific Literature.

LIPID CONTENT (G/100 G DM)	*REFERENCE*	*LIPID CONTENT (G/100 G DM)*	*REFERENCE*
10.8	Mattioli et al. (2024)	26.2	Roncolini et al. (2020)
28.5	Accardo et al. (2022)	24.1	Kurecka et al. (2021)
24.5	Renna et al. (2022)	29.0	Adamkova et al. (2016)
14.4–28.0	Soetemans et al. (2020)	24.3	Van Broekhoven et al. (2015)

TABLE 4.8 Fatty Acid Profile (g/100 of fatty acids) of *A. diaperinus* Larvae According to Different References Found in the Scientific Literature.

C10:0	–	–	0.42	–	–
C12:0	0.34	–	0.03	0.03	0.1
C14:0	1.6	1.1	0.78	0.77	1.4
C15:0	–	0.2	–	0.21	0.3
C16:0	26.7	27.2	29.3	25.0	26.4
C17:0	–	–	–	0.42	0.7
C18:0	10.6	6.8	9.11	7.2	10.9
C20:0	0.32		0.56	0.31	0.6
C22:0	–	–	0.25	0.08	–
C14:1	0.03	–	–	0.02	–
C16:1	1.7	–	0.32	0.29	1.1
C18:1	31.4	34.3	35.6	29.0	35.9
C20:1	0.48	–	–	0.36	0.4
C18:2 n-6	20.3	28.0	20.5	33.7	20.2
C20:2 n-6	–	–	–	0.19	0.2
C18:3α, n-3	0.38	2.3	0.76	1.9	0.4
C20:4 n-6	0.55	–	–	–	0.4
C22:5 n-3	0.02	–	–	–	–
SFA	39.2	35.4	40.7	34.0	40.6
MUFA	33.0	34.3	37.3	30.1	37.8
PUFA	21.2	30.3	21.3	35.8	21.6
PUFA/SFA	0.54	0.9	0.5	1.1	0.5
n-3	0.40	2.3	0.8	1.9	0.4
n-6	20.8	28.0	20.5	33.7	20.6
n-6/n-3	53.8	12.2	27.0	17.9	51.5
References	Mattioli et al. (2024)	Accardo et al. (2022)	Renna et al. (2022)	Roncolini et al. (2020)	Adamkova et al. (2016)

Soetemans et al. (2020) demonstrated the impact of the diet on the *A. diaperinus* larvae lipids. The insects were reared on mixtures of wheat middlings, rice bran, rapeseed meal, distillers dried grains, corn gluten free, carrot and brewery grains. The lipid content of larvae varied between 14% and 28%, and the concentration of essential fatty acids ranged from 2.3% to 8.9% depending on the diet. Oonincx et al. (2020) investigated the inclusion of flaxseed oil, which is rich in α-linolenic acid. The addition of this substrate led to an increase in the lipid content and concentration of α-linolenic acid (10-fold increase), as well as a reduction of the n-6/n-3 ratio (from 21.7 to 2.4). These changes in the lipid composition were proportional to the amount of flaxseed oil added. Van Broekhoven et al. (2015) studied larvae reared on diets with different concentrations of starch and protein. Diets were composed of by-products from beer brewing, bread and cooking baking, potato processing and bioethanol production. The lipid content of the larvae decreased from 24.3% to 13.4% when the experimental diet low in protein and high in starch was tested. The fatty acid composition changed with the experimental diets, providing long-chain PUFA that were not present in the fat of larvae from the control diet. However, the n-6/n-3 ratio remained constant, or it was even higher compared to that obtained from the control diet.

With respect to the major and minor lipid classes of *A. diaperinus*, they have been scarcely determined. Tzompa-Sosa et al. (2014) carried out a qualitative analysis of the main lipid species by thin layer chromatography (TLC). As expected, TAGs were the most abundant lipid class, followed by FFA, based on the size of the spot on the TLC plate. MAGs, DAGs, PLs and cholesterol were other lipid species found in lower concentrations. As already mentioned for other insect species in this chapter, the lipid composition of *A. diaperinus* is also strongly influenced by the defatting method. In the same study, the authors observed that the lipids obtained by aqueous extraction lacked polar lipids, such as PLs, MAGs and DAGs, and contained less FFA than lipids obtained by Folch and Soxhlet extraction using petroleum ether. Gharibzahedi and Altintas (2023) explored the performance of hexane, ethanol, isopropanol and their mixtures. The use of ethanol/isopropanol and ethanol/hexane (1:1 v/v) resulted in larger extraction yields and led to lipids with higher concentrations of tocopherols (171.1–143.2 mg/kg of γ-tocopherol and 7.7–7.2 mg/kg of α-tocopherol), carotenoids (0.8–0.7 mg/g), vitamin D_2 (0.07–0.08 μg/g), campesterol (167.4–162.5 mg/kg), β-sitosterol (311.2–277.9 mg/kg), phosphatidylinositol and phosphatic acid (8.1–8.0 mg/g). In another study, Gharibzahedi and Altintas (2022) found that the ultrasound-assisted extraction with ethanol/isopropanol (1:1 v/v) was more efficient than conventional extraction with hexane for obtaining a lipid fraction with higher carotenoid concentration (0.8 mg/g) and antioxidant activity. Furthermore, Mattioli et al. (2024) found low concentrations of tocols (0.2 mg/kg) in a *A. diaperinus* lipid obtained by Soxhlet with petroleum ether.

4.6 LIPIDS AND FATTY ACIDS OF *HERMETIA ILLUCENS*

In recent years, among bioconverter insects, particular attention has been given to *Hermetia illucens*, popularly known as black soldier fly, which belongs to family of Stratiomyidae

TABLE 4.9 Lipid Content of *Hermetia illucens* Larvae According to References from the Scientific Literature.

LIPID CONTENT (G/100 G DM)	*REFERENCE*	*LIPID CONTENT (G/100 G DM)*	*REFERENCE*
26.0	Makkar et al. (2014)	25.7 – 38.4	Zulkifli et al. (2022)
16.8 – 29.1	Fornari et al. (2023)	30.1 – 34.3	Shumo et al. (2019)
29.4	Onsongo et al. (2018)	33.7	Matthäus et al. (2019)
29.8	Rawski et al. (2020)	35.9	Loho and Lo (2023)
29.1 – 30.1	Li et al. (2011)	35.1 – 51.5	Alifian et al. (2019)

(subfamily Hermetiinae) in the order Diptera. Since 2017, *H. illucens* larvae was authorized for the first time in animal feeding, primarily for aquaculture, according to Commission Regulation (EU) No. 2017/893. This authorisation was expanded to include poultry and pig feed in 2021, under Commission Regulation (EU) No. 2021/1372. Although *H. illucens* is not yet authorized for human consumption in Europe, it is expected that in the coming months, EFSA will issue more favorable opinions on related requests, as those ongoing assessments about dried defatted powder (Novel Foods EFSA-Q-2023–00703) or even for refined fat (Novel Foods EFSA-Q-2023–00847). In fact, this last case is the first safety assessment that is being performed for any insect oil by EFSA.

As shown in Table 4.9, *H. illucens* larvae are particularly rich in lipids, ranging from 15% to 50% of their weight, making them a natural source of biologically active hydrophobic substances (Franco et al., 2021). Lipid content is remarkably affected by the composition of the rearing substrate (Spranghers et al., 2017), but in general, the crude fat content in *H. illucens* is much higher than other insects, soybean, or fish meal (Franco et al., 2021). The number of studies showing the impact of the diet on the lipid content and composition of *H. illucens* can be considered the most extensive among all the edible insect species. As examples, Makkar et al. (2014) reported a lipid content ranging between 15% and 25% for larvae fed on chicken manure, 28% on pig manure, 35% on cattle manure, and 42–49% on oil-rich food waste. Additionally, it seems that low protein and high carbohydrate substrates allow a higher lipid accumulation in larval biomass (Pimentel et al., 2017) because of the conversion of carbohydrate into lipids by larvae (Li et al., 2015).

The fatty acid composition of *H. illucens* fat is also significantly influenced by the type of growth substrates(Mai et al., 2019). This would explain the great variability of data found in the scientific literature, as shown in Table 4.10.

Irrespective of diet, SFA are the main group of fatty acids in *H. illucens* larvae (Table 4.10). This is the first major difference of this insect compared to the rest of edible insects in this chapter. Thus, total SFA content in the range of 50–85% can be found in these larvae. Specifically, lauric acid is the most abundant fatty acid, reaching a content up to 50% of the isolated fats, followed by palmitic acid (10–20%) and oleic acid (6–30%). The presence of high levels of lauric acid, despite its limited amounts in the diets, strongly indicates that this fatty acid is synthesized by the larvae, presumably from the carbohydrates in the substrate (Spranghers et al., 2017). It should be remarked that the extremely high values of lauric acid make *H. illucens* unique among the different species of edible insects, since the presence of this fatty acid can be negligible in most species. Lauric acid and its derivative, monolaurin, are recognized as excellent antimicrobial lipids, exhibiting antibacterial, antifungal, and antiviral activity with broad-spectrum inhibition (Nitbani

TABLE 4.10 Fatty Acid Profile (g/100 of Fatty Acids) of *Hermetia illucens* Larvae According to Different References Found in the Scientific Literature.

C10:0	1.4	0.9	1.5	0.9	0.7	1.0	0.8
C12:0	45.3	46.0	54.8	41.4	26.2	50.3	31.9
C14:0	14.6	8.7	8.9	6.7	6.2	10.2	5.5
C15:0	0.6	0.2	–	0.1	0.2	–	–
C16:0	19.0	12.2	11.4	16.5	14.8	13.4	21.9
C17:0	0.6	0.2	–	0.1	0.2	0.4	–
C18:0	3.5	2.6	1.6	1.5	2.8	2.4	2.7
C20:0	0.1	0.1	–	0.1	0.1	–	–
C21:0	–	–	–	–	–	–	–
C22:0	–	–	–	–	–	–	–
C14:1	0.2	–	–	–	–	0.4	–
C16:1	3.7	1.9	2.1	2.4	2.3	2.4	2.3
C18:1	10.6	11.2	6.4	18.5	30.8	8.6	20.2
C20:1	–	0.1	–	0.1	0.1	–	–
C18:2 n-6	0.2	14.1	12.1	10.2	13.7	7.1	13.0
C18:3γ, n-6	–	–	–	–	–	0.1	1.7
C18:3α, n-3	–	1.7	1.1	1.6	1.2	3.2	–
C20:2 n-6	–	–	–	–	0.3	–	–
C20:4 n-6	–	0.1	–	–	0.1	–	–
C20:5 n-3	–	–	–	–	0.1	–	–
SFA	85.2	70.9	78.1	67.2	51.3	77.7	62.8
MUFA	14.5	13.2	8.5	21.0	33.3	18.5	35.5
PUFA	0.2	15.9	13.1	11.8	15.4	3.3	1.7
PUFA/SFA	0.0	0.22	0.17	0.18	0.30	0.04	0.03
n-3	0.0	1.7	1.1	1.6	1.4	3.2	0.0
n-6	0.2	14.2	12.1	10.2	14.1	3.3	1.7
n-6/n-3	–	8.4	11.5	6.4	10.3	1.0	–
References	Ushakova et al. (2016)	Daszkiewicz et al. (2022)	Rawski et al. (2020)	Rabani et al. (2019)	Tognocchi et al. (2023)	Fornari et al. (2023)	Mai et al. (2019)

et al., 2022). Additionally, it is often used in cosmetics, body care products, soaps and detergents, plastics, and textiles (Suryati et al., 2023), as well as in the food industry for its potential health benefits as MCFA, including the promotion of healthy cholesterol levels (Dayrit, 2015). Furthermore, Li et al. (2022) reported that *H. illucens* larvae could produce and retain 12:0, 14:0, and 16:1c9 in vivo, but 16:0, 18:1c11, 18:2 n-6 and 18:3 n-3 could only be partly incorporated from diets, and the process may be enhanced by a higher amount of dietary fat. It should also be noted that small amounts of conjugated linoleic acid isomers (CLA, 18:2c9t11) were found in the fat of *H. illucens* larvae and pupae, even though these lipids were not present in the substrate (Tognocchi et al., 2023). Previous findings about the presence of CLA in *H. illucens* are scarce but increasing in the last years (Renna et al., 2024). CLA is a popular bioactive fatty acid that has been shown to exert various potent physiological functions such as antitumor, antiobese, antidiabetic and antihypertensive properties (Koba and Yanagita, 2014). Likely, the advance in the characterisation of insect lipids would probably contribute to giving novel unique attributes to *H. illucens*, as lauric acid and CLA are, since the presence of CLA has not been described for most of the edible insects. Low amounts of azelaic acid (and its esters) and sebacic acid have also been reported (Ushakova et al., 2016).

The PUFA/SFA ratio of *H. illucens* larvae is extremely low (0–0.3) due to the predominant fatty acid, lauric acid, which is saturated. Although fats with these values are not recommended from a cardiovascular perspective, a different effect may occur in this case, as lauric acid can be metabolized as an MCFA. These fatty acids are preferentially transported via the portal vein to the liver and are metabolized as rapidly as glucose in the body, with minimal tendency to be stored as fat (Osborn and Akoh, 2002). Furthermore, the fat of *H. illucens* has n-6/n-3 values ranging from 1 to 12, which are intermediate compared to other insect species and do not suggest potential anti-inflammatory effects. Thus, these values are higher than in *L. migratoria* but significantly lower than in *T. molitor* and *A. domesticus*.

Concerning interesting findings about the impact of the diet, larvae fed with fish offal and algae was found to incorporate significant amounts of eicosapentaenoic acid (EPA) (20:5 n-3) and docosahexaenoic acid (DHA) (22:6 n-3) (Liland et al., 2017), but when the larvae gain weight, the percentage of these fatty acids decreases. Hence, in general, larvae with a higher weight contained a higher percentage of saturated fatty acids (SFAs) and a lower percentage of UFAs (Ewald et al., 2020).

As example of studies about the lipid classes of *H. illucens*, Li et al. reported that it mainly consisted of TAGs (72.6%). Additionally, phosphatidylcholine (8.6%), sphingomyelin (4.6%), phosphatidylethanolamine (2.2%), DAGs (2.4%), ceramide (2.1%), methyl phosphatidylcholine (1.2%), bis-methyl phosphatidic acid (1.2%), phosphatidic acid (0.01%), phosphatidylglycerol (0.32%), phosphatidylinositol (0.14%), phosphatidylserine (0.13%), acyl carnitine (1.4%) and coenzyme (0.40%) were also described as minor lipids (Li et al., 2021). Tognocchi et al. (2023) reported similar values for glycerophospholipids and sphingolipids from *H. illucens* larval fat. However, this study found a lower phosphatidylcholine content (1.9%) and a higher level of phosphatidylethanolamine (9.5%). These authors also described the absence of sphingomyelin in the pupa and an approximately two-fold increase in the main species of glycerophospholipids in the pupa compared to the larval content (Tognocchi et al., 2023). Therefore, it seems that PLs can be a relevant and variable fraction of lipids in case of *H. illucens*, as already described for other edible insects.

Typically, the content of FFAs in larva fat ranges from 3% to 5%, according to several authors (Li et al., 2011; Mai et al., 2019; Bogevik et al., 2022). However, it should be noticed that the technological processing of insects can significantly influence this content. This has been pointed out for most insect species, but this phenomenon seems to be extremely marked and has been more deeply studied, in case of *H. illucens*. Specifically, it has been evidenced that the initial step of slaughtering of the larvae can favor the hydrolysis of the fat due to active endogenous lipases (Hurtado-Ribeira et al., 2024). This hydrolysis is mainly related to the use of non-thermal methods of slaughtering; hence, the lipases remain active. Thus, during the slaughtering by freezing, typically used in edible insects, an intense lipolysis seems to take place. Hence, Hurtado-Ribeira et al. (2023b) showed that the combination of freezing for slaughter (instead of blanching) and freeze-drying (instead of oven drying), both non-thermal methods of processing, increased endogenous larval lipolysis, leading to a higher content of FFAs (21%). Additionally, this content of FFA can be even higher when the frozen larvae are kept storage few days after slaughter by freezing (Hurtado-Ribeira et al., 2024). Although the quality of this fat is considered poor due to the high acidity, it exhibits high antibacterial activity due to the high concentration of lauric acid in its free and MAG forms (Hurtado-Ribeira et al., 2023a).

With respect to minor lipid compounds, *H. illucens* has another relevant difference with respect to most edible insects and with animals, in general. Insects generally contain mostly cholesterol, even when their diet is rich in phytosterols due to the dealkylation process that converts phytosterols into cholesterol (Behmer and Elias, 2000). However, *H. illucens* has negligible levels of cholesterol. On the contrary, the phytosterol content in *H. illucens* larvae is relatively high, ranging from 0.4% to 4% total lipids, and exceeds the levels found in other insect species, such as *Tenebrio molitor*, with less than 1% (Matthäus et al., 2019; Hurtado-Ribeira et al., 2023b; Tognocchi et al., 2023). In fact, it should be remarked that the sterol fraction can reach quantities comparable or even higher to concentrations found in plant oils typically known as sources of phytosterols (Liland et al., 2017). Regarding the sterol profile, studies consistently showed that β-sitosterol is the predominant sterol (57.7–65.1%), with campesterol (12.3–32.0%) and δ-avenasterol (4.1–16.1%) following in abundance. Significant amounts of brassicasterol (0.5–4.3%) and lower quantities of other sterols have also been observed (Matthäus et al., 2019; Boukid et al., 2021; Tognocchi et al., 2023). This bioaccumulation of phytosterols would give an attractive value to *H. illucens* lipids due to the well-known health properties of phytosterols, mainly as hypocholesterolemic.

H. illucens larvae fat also contains small amounts of tocopherols and tocotrienols. Matthäus et al. (2019) reported a total amount of 64.7 μg/g fat, with α-tocopherol and ß-tocotrienol as predominant members of this group (30.4 and 19.8 μg/g, respectively). ß-tocopherol (9.7 μg/g), γ-tocopherol (2.2 μg/g), and α-tocotrienol (2.7 μg/g) were found in smaller amounts. However, other authors have reported greater and highly variable amounts of vitamin E isomers, ranging from 53.3 μg/g dry matter (Liland et al., 2017) to 614.6 μg/g dry matter (Zhang et al., 2023), identifying α-tocopherol and α-tocotrienol as the predominant compounds in each study. Additionally, low amounts of β-carotene, lutein and zeaxanthin have been found in *H. illucens* larvae, measuring <0.2, 0.6 and 1.3 μg/g of insect, respectively (Finke, 2013).

4.7 LIPIDS AND FATTY ACIDS OF *MUSCA DOMESTICA*

Musca domestica L., also known as common fly or house fly, is a member of dipteran brachyceran insect in the family Muscidae (Moon, 2019). It is the most common and widespread fly in urban and rural environments (Geden et al., 2021). The ubiquitous presence, well known and transferable culture practices, and the fact that they feed on waste makes house fly larvae a highly productive and ecologically efficient alternative to the black soldier fly to produce mainly protein for animal feed (Miranda et al., 2020). Accordingly,

TABLE 4.11 Lipid Content of *Musca domestica* Larvae According to References from the Scientific Literature.

LIPID CONTENT (G/100 G DM)	*REFERENCE*	*LIPID CONTENT (G/100 G DM)*	*REFERENCE*
14.1	Pieterse and Pretorius (2014)	16 – 21	Ganda et al. (2022)
15.1	Niu et al. (2017)	21.7	Ali et al. (2024)
18.9	Makkar et al., 2014)	21.7	Sánchez et al. (2021)
19.7	Hussein et al. (2017)	16.4 – 27.4	Ukanwoko and Olalekan (2015)
15–20	Boatta et al. (2023)	20.8 – 29.7	Aniebo and Owen (2010)

like other insect species previously mentioned, *M. domestica* has been authorized for use in aquaculture since 2017 and for poultry and pig feeding since 2021, under current European legislation (Commission Regulation 2021/1372); however, it is still not authorized for human food in Europe (Delgado et al., 2023). Maggots have a rapid reproductive rate, high feeding value and easy to process and stored for sustainable use and are available throughout the year (Moreki et al., 2012).

The lipid content of *M. domestica* has generally been determined using ether extraction procedures, and it ranges from 14% to 30% of dry matter, as shown in Table 4.11.

Additionally, it has been reported that the age and drying method of the insect can affect its proximal composition. Hence, lipid content increased from 20.8% at the age of 2 days to 25.3% at the age of 4 days. Regarding the drying, oven-dried maggots had mean higher protein content (50.9%) and less lipids (22.8%) than sun dried maggots (47% and 26.4%, respectively) (Aniebo and Owen, 2010).

Regarding fatty acid composition, the *Musca domestica* larvae fat exhibits a high content of SFA (36–43%) and MUFA (38–58%), according to the literature detailed in Table 4.12.

As for most insects, the SFA are mainly represented by palmitic acid, which is the predominant fatty acid, with values ranging from 25% to 31%, while the MUFA consists mainly of oleic acid (16–25%) and palmitoleic acid (13–33%), with the latter being remarkable for its content, as it is a relatively uncommon fatty acid in nature. The level of this n-7 fatty acid in *M. domestica* larvae is much higher (4–15 times) than that in *T. molitor* and *A. domesticus*. Thus, in general, edible insects are not sources of palmitoleic acid. Some of the scarce sources known for richness in palmitoleic acid are macadamia nut oil (Hu et al., 2019) or sea buckthorn oil (Solà Marsiñach and Cuenca, 2019), which contain quite similar values to *M. domestica*. Palmitoleic acid is widely used in skincare for its moisturizing effects and as supplement to aid lipid metabolism, reduce inflammation, and prevent diabetes and cardiovascular diseases. (Hu et al., 2019). Therefore, due to the increasing interest on the biological and cosmetic properties of this unusual fatty acid, *M. domestica* can be pointed out as a novel alternative source for palmitoleic acid-rich products. Furthermore, the fat of *M. domestica* has a variable PUFA content (4–22%), predominantly composed of linoleic acid, with small amounts, less than 2%, of α-linolenic acid. It should be remarked that the linoleic acid content is much lower in *M. domestica* larvae compared with those in the other edible insects (Makkar et al., 2014), being closer

TABLE 4.12 Fatty Acid Profile (g/100 of Fatty Acids) of *Musca domestica* Larvae According to Different References Found in the Scientific Literature.

C10:0	–	–	–	–	–
C12:0	1.2	–	–	0.5	–
C14:0	6	5.5	4.8	8.3	3.7
C15:0	2.5	–	3	–	
C16:0	27.1	31.1	25.3	33.4	26.9
C17:0	0.7	–	–	–	–
C18:0	5	3.4	2.4	2.4	9
C20:0	0.1	–	–	–	–
C21:0	–	–	–	–	–
C22:0	–	–	–	–	–
C14:1	0.6	–	2	2.5	–
C16:1	17.4	13.4	33.3	23.4	23.2
C18:1	24.5	24.8	22.2	18.3	23.1
C20:1	–	–	–	–	–
C18:2 n-6	13.5	19.8	2.8	10.2	9.1
C18:3γ, n-6	–	–	–	0.2	–
C18:3α, n-3	0.4	2	0.6	0.8	1.9
C20:2 n-6	–	–	–	–	–
C20:4 n-6	–	–	0.05	–	–
SFA	42.6	40	35.5	44.5	39.6
MUFA	42.5	38.2	57.5	44.3	46.3
PUFA	13.9	21.8	3.45	11.2	11
PUFA/SFA	0.33	0.55	0.10	0.25	0.28
n-3	0.4	2.0	0.6	0.8	1.9
n-6	13.5	19.8	2.85	10.4	9.1
n-6/n-3	33.8	9.9	4.8	13.0	4.8
References	Niu et al. (2017)	Makkar et al. (2014)	Hussein et al. (2017)	Boatta et al. (2023)	Stefanov et al. (2002)

to the content of *H. illucens* (Table 4.10). Therefore, concerning the nutritional quality, the fat of *Musca domestica* does not exhibit favorable PUFA/SFA values, ranging from 0.1 to 0.5, due to its high SFA content, mainly palmitic acid, and low PUFA levels. The n-6/n-3 values are in a similar range to *H. illucens*, although with higher variability.

The studies by Makkar et al. (2014) and Stefanov et al. (2002) reported very small differences when comparing the fatty acid composition of the larval and adult stages. However, the fatty acid profile can be largely influenced by substrate composition (Makkar et al., 2014). Hence, Hussein et al. (2017) showed that providing fish oil changed larvae fatty acid composition, increasing the proportion of PUFA by 4.2-fold (from 3.9% to 16.1%). This increase in PUFA was mainly in the form of EPA (20:5 n-3) and DHA (22:6 n-3), being 8.7% and 0.8%, respectively. Relative to the increase in PUFA, the MUFA and SFA contents dropped by 0.8-fold (from 57.5% to 48.5%) and 8.4% (from 38.6% to 35.4%), respectively.

The fat of *M. domestica* larvae contains TAGs as main lipids, together with PLs accounting around 16% of the mass of the total lipids (Stefanov et al., 2002). According to Crone and Bridges, the major components of the PL fraction of *M. domestica* are phosphatidylethanolamine (65% of the total PLs), phosphatidylcholine (17%), phosphatidylserine (3.5%), phosphoinositide (3%) and sphingolipid (3.5%) (Crone and Bridges, 1963). Therefore, despite the available information being scarce, *M. domestica* seems to be an interesting source of PLs, as other insects reviewed in this chapter.

Regarding minor lipids, Stefanov et al. (2002) found that cholesterol, sitosterol and campesterol were the predominant sterols in *M. domestica* larvae, demonstrating that composition of sterols could depend on the diet of the larvae before pupation. Likewise, Gołębiowski et al. (2013) found differences in the sterol composition between the cuticular and internal lipids of *M. domestica* larvae, pupae, males and females. Regarding internal lipids, the major internal sterol in all stages was cholesterol, with relative contents ranging from 54.8% in larvae to 75.1% in females. Another important internal sterol was campesterol, which constituted from 15.0% of the total sterols in females to 19.9% in pupae. However, in larvae, the two major internal sterols were cholesterol (54.8%) and sitosterol (28.7%). According to these authors, sitosterol was the major cuticular sterol constituting 74.1% and 57.9% in larvae and pupae, respectively. In contrast, cholesterol was the major sterol in the cuticular lipids of males and females comprising 79.9% and 76.3% of the sterol fraction, respectively (Gołębiowski et al., 2013). However, similar to *H. illucens*, in *M. domestica*, cholesterol levels can be found unusually low, while phytosterols are present in significant amounts, indicating that phytosterol dealkylation occurs more slowly in these larvae (Stefanov et al., 2002). Furthermore, sterol esters, common for plants, marine invertebrates and vertebrates, were practically absent in *M. domestica* (Stefanov et al., 2002).

In addition, *M. domestica* showed selective accumulation of different vitamin E isomers. For instance, α-tocopherol was detected in larvae, pupae and adults, with larvae containing 616.5 μg/g dry matter. The content of α-tocotrienol can vary during growth, being abundant in larvae, with a maximum value of nearly 200.0 μg/g dry matter (Zhang et al., 2023). Regarding other minor lipids, the fat from *M. domestica*, both cuticular and internal, contains fatty alcohols in the range of 2–5 μg/g insect body, which are associated with antimicrobial activity. The main fatty alcohols in larvae and pupae are 12:0 and 24:0, respectively (Gołębiowski et al., 2012). Besides, β-carotene, lutein and zeaxanthin have been detected in adults of *M. Domestica* in small amounts, less than 1 μg/g insect (Finke, 2013).

4.8 LIPIDS AND FATTY ACIDS OF *GRYLLODES SIGILLATUS*

G. sigillatus (banded cricket) belongs to the family Gryllidae (order Orthoptera). At present, the dry form of the adult insect is pending for authorisation as a novel food by the European Commission (European Commission, n.d.). Nevertheless, this specie is already approved for animal feeding since 2017.

TABLE 4.13 Lipid Content of *G. sigillatus* According to References from the Scientific Literature.

LIPID CONTENT (G/100 G DM)	*REFERENCE*	*LIPID CONTENT (G/100 G DM)*	*REFERENCE*
21.5	Kepinska-Pacelik et al. (2023)	23.5	Ribeiro et al. (2019)
15.4	Tran et al. (2023)	21.6	Kilburn et al. (2020)
26.9	Renna et al. (2022)	19.5	Jozefiak et al. (2018)
32.4	Tzompa-Sosa et al. (2021)	18.2	Zielinska et al. (2015)

In general, the lipid content of *G. sigillatus* is higher than 20% (Table 4.13), being one of the cricket species with the highest lipid content (Zafar et al., 2024). In addition to factors, such as dietary components, environmental factors and regional variations between species, the solvent selection is an important parameter to consider in fat quantification, which leads to high variability in results, as already explained along this chapter. In this regard, Ribeiro et al. (2019) obtained lipid extraction yields 40% higher with ethanol than using ether petroleum, hexane and diethyl ether from *G. sigillatus*.

Regarding the fatty acid composition of *G. sigillatus*, in broad terms, somewhat higher concentrations of SFA are present (>30%). The main fatty acids are palmitic (20–30%), oleic (20–30%) and linoleic acid (30–35%). These three fatty acids, together with stearic acid (7–15%), comprise more than 90% of total fatty acids. *G. sigillatus* also contains linolenic acid, but its concentration is low, generally below 2.5%. With respect to its nutritional quality, PUFA/SFA ratios close to 1 and n-6/n-3 ratios above ten are present in the fat of this insect. Thus, in general, the fatty acid profile can be considered similar to that of *A. domesticus* (Section 3).

Cricket species, such as house cricket (*Acheta domesticus*) and Jamaican field cricket (*Gryllus assimilis*), have demonstrated to be able to bioaccumulate the long-chain n-3 fatty acid EPA in small amounts from the diet (Barroso et al., 2014; Oonincx et al., 2020). In addition, previous studies have suggested that some cricket species may be able to elongate and desaturate linoleic and linolenic acid to EPA (Jurenka et al., 1988; Blomquist et al., 1991). In the case of *G. sigillatus*, Kepinska-Pacelik et al. (2023) and Drosdowech et al. (2024) detected small amounts of EPA and DHA (Table 4.14), but the composition of the diet supplied to the insects was not specified in both studies.

Concerning other lipid species of *G. sigillatus*, Tzompa-Sosa et al. (2021) carried out a determination of its lipid profile. TAGs were the most predominant lipid species in the fat. Phosphatidylcholine was the most abundant PL (almost 45% of total PLs analysed), followed by phosphatidylserine, phosphatidylethanolamine and phosphatidylinositol (13–15% of total PLs). Other PLs identified were sphingomyelin, phosphatidylglycerol, hexosylceramides and lactosylceramide.

With respect to minor lipids, cholesterol, stigmasterol, ß-sitosterol, desmosterol, campesterol, cholestane-3-ol and cholest-7-en-3-ol were the main sterols (1.5% in the fat). The most abundant was cholesterol (84.2% of total sterols), followed by ß-sitosterol (6.4%), stigmasterol (5.0%) and campesterol (4.3%). The presence of carotenoids in *G. sigillatus* has also been described. In this sense, Ogilvy et al. (2012) studied the influence of the diet on the concentration of carotenoids. Insects fed on a fish-based diet had a

TABLE 4.14 Fatty Acid Profile (g/100 of Fatty Acids) of *G. sigillatus* According to Different References Found in the Scientific Literature.

C10:0	–	–	0.34	–	–
C12:0	0.06	0.06	0.13	–	0.10
C14:0	0.67	0.54	0.71	0.80	1.7
C15:0	–	–	–	–	0.24
C16:0	27.1	23.4	25.7	24.6	23.5
C17:0	–	0.2	–	0.22	0.32
C18:0	14.2	15.5	7.9	10.6	7.4
C20:0	–	0.44	0.81	0.29	0.4
C21:0	–	–	–	–	0.13
C22:0	–	–	0.14	–	0.07
C16:1	0.64	0.75	1.5	1.8	3.8
C18:1	22.4	24.7	27.6	25.9	29.1
C20:1	0.23	0.15	–	0.05	1.0
C18:2 n-6	30.3	30.6	32.0	33.7	29.8
C18:3α, n-3	1.7	2.4	0.77	1.6	2.1
C18:3γ, n-6	–	–	0.04	–	–
C20:5 n-3	0.59	0.30	0.08	–	–
SFA	43.1	40.1	35.9	36.5	33.7
MUFA	23.3	26.1	30.2	27.8	34.3
PUFA	32.8	33.8	33.0	35.3	31.9
PUFA/SFA	0.76	0.84	0.92	0.97	0.95
n-3	2.5	3.2	0.9	1.6	2.1
n-6	30.3	30.6	32.0	33.7	29.8
n-6/n-3	12.1	9.6	37.7	21.1	14.0
References	Drosdowech et al. (2024)	Kepinska-Pacelik et al. (2023)	Renna et al. (2022)	Tzompa-Sosa et al. (2021)	Zielinska et al. (2015)

higher concentration of total carotenoids (around 22 μg/g of insect) than those reared on a vegetable-based diet. On the contrary, the dietary pattern based on wheat bran led to very low carotenoid concentrations.

4.9 LIPIDS AND FATTY ACIDS OF *GRYLLUS ASSIMILIS*

Gryllus assimilis is a member of the Gryllidae family (order Orthoptera). Commonly known as the Jamaican field cricket and occasionally referred to as the silent cricket, among other names. This cricket is authorized for animal feeding since 2017.

The average lipid content determined for this species is about 20% (Table 4.15). This total value is within the ranges reported for other cricket species, such as *A. domesticus*,

TABLE 4.15 Lipid Content of *G. assimilis* Larvae According to References from the Scientific Literature.

LIPID CONTENT (G/100 G DM)	*REFERENCE*	*LIPID CONTENT (G/100 G DM)*	*REFERENCE*
11.9	Mlcek et al. (2018)	21.6	Khatun et al. (2021)
21.8	Soares Araújo et al. (2019)	33.0	Fernandez et al. (2022)
20.0	Da Rosa et al. (2019)	22.2	Oliveira et al. (2024)

G. sigillatus, and the two-spotted cricket (*Gryllus bimaculatus*). However, as for most insects, factors such as diet, developmental stage, sex, and the origin of these insects can influence the lipid content, with a minimum value of 11% described for crickets from the Czech Republic (Mlcek et al., 2018) and a maximum value of 33% found for crickets from Honduras (Fernandez et al., 2022). Furthermore, it has been reported that different processing methods, such as oven drying and blanching, do not significantly alter the overall lipid composition, although they may affect lipid oxidation (Khatun et al., 2021), as observed for most insects species.

The fatty acid composition (Table 4.16) of the field cricket is approximately balanced between PUFA (15.5–34.8%), SFA (35.3–44.6%), and MUFA (26.0–35.2%). The main fatty acids are oleic acid (>25%), palmitic acid (>20%), linoleic acid (>20%) and stearic acid (close to 10%). The n-6/n-3 fatty acid ratio of this cricket species is considerably high (>18:1), as already described for other crickets and insects species. This elevated ratio is attributable to the high concentration of linoleic acid in the feed, which can be accumulated in the crickets. Therefore, as recommended for different species showing these high ratios, it would be advisable to decrease the n-6/n-3 fatty acid ratio, which could be achieved in *G. assimilis* by incorporating n-3 PUFA into the cricket diet (Oonincx et al., 2020).

TAGs are the predominant class of lipids. Regarding polar lipids, Ochiai and Komiya (2021) found, in a comparative study of different cricket species, a PL content of 0.23% of dry matter in *G. assimilis*, which is less than the 0.28% and 0.35% found in *A. domesticus* and *G. bimaculatus*, respectively. Tzompa-Sosa et al. (2021) reported that the most abundant PL was phosphatidylcholine, followed by phosphatidylserine, phosphatidylethanolamine and phosphatidylinositol, which together represent approximately 83% of the total relative abundance of PLs. Other polar lipids identified were sphingomyelin, phosphatidylglycerol, hexosylceramide and lactosylceramide.

The identification of minor lipids conducted by matrix-assisted laser desorption/ionization time-of-flight mass spectrometry (MALDI-TOF MS) by Tzompa-Sosa et al. (2021) showed a total sterol content of 0.15%, including 0.13% cholesterol, 0.04% campesterol, 0.01% stigmasterol and 0.07% β-sitosterol, along with other phytosterols at 0.13%, expressed as a percentage of these compounds in the oil. Similarly, Ochiai and Komiya (2021) found that the cholesterol content in *G. assimilis* was relatively high (5.4 μg/g of total lipids) but did not differ significantly from that of other species, such as *A. domesticus* or *G. bimaculatus*. Kulma et al. (2022) found α-carotene ranging from 5.4 to 15.2 μg/g dry matter, and β-carotene ranging from 12.8 to 37.2 μg/g dry matter in *G. assimilis* fed with varying levels of carrot supplementation in their diet. However, this study suggested that the capacity of these insects to store carotenes is limited.

TABLE 4.16 Fatty Acid Profile of *G. assimilis* (g/100 of Fatty Acids) According to Different References Found in the Scientific Literature.

C10:0	0.03	–	0.07	–	–
C12:0	0.12	0.10	0.17	–	0.19
C14:0	1.3	1.0	0.23	–	1.8
C15:0	0.37	–	0.28	–	0.23
C16:0	25.9	33.1	25.5	29.0	22.2
C17:0	0.57	–	0.46	–	0.27
C18:0	14.1	9.0	11.9	8.0	12.0
C20:0	0.56	0.45	0.6	–	–
C21:0	0.03	–	–	–	–
C22:0	0.57	–	–	–	–
C14:1	0.06	–	0.77	–	–
C16:1	1.9	3.3	1.6	–	4.7
C18:1	25.0	30.3	27.1	26.0	30.6
C20:1	0.24	–	–	–	–
C18:2 n-6	26.1	14.9	32.6	31.0	23.4
C20:2 n-6	0.4	–	–	–	–
C18:3α, n-3	1.6	–	1.3	2.0	3.3
C20:4 n-6	0.21	–	0.61	–	–
C20:5 n-3	0.38	–	0.34	–	–
SFA	43.7	44.6	35.4	37.0	36.6
MUFA	27.5	35.1	29.9	26.0	35.3
PUFA	28.8	15.6	34.9	33.0	27.1
PUFA/SFA	0.66	0.35	0.99	0.89	0.74
n-3	2.0	–	1.6	2.0	3.3
n-6	26.8	14.9	33.2	31.0	23.4
n-6/n-3	13.5	–	20.7	15.5	7.2
References	Mlcek et al. (2018)	Soares Araújo et al. (2019)	Khatun et al. (2021)	Ochiai et al. (2021)	Hassen et al. (2024)

4.10 LIPIDS AND FATTY ACIDS OF *BOMBYX MORI*

Bombyx mori belongs to family of Bombycidae in the order Lepidoptera. Its common name is mulberry silkworm. Within the silkworm species, this is the most popular one. The major industry of silkworms is not for food or feed, but their traditional production is for the commercial use of the silk, known as sericulture. However, it is noteworthy to remark that during the process of recuperation of the silk, the pupae inside the silk cocoon is released as a by-product (Yeasmin et al., 2024). In last decades, the silkworm pupae is gaining more and more relevance due to the nutritional potential that is being demonstrated

TABLE 4.17 Lipid Content of *Bombyx mori* According to References from the Scientific Literature.

LIPID CONTENT (G/100 G DM)	*REFERENCE*	*LIPID CONTENT (G/100 G DM)*	*REFERENCE*
30.4	Yeasmin et al., 2024	33.3	Brogan et al. (2023)
27.8	Ferdousi et al. (2023)	33.0	Zhao et al. (2022)
26.9	Tangsanthatkun et al. (2022)	28.3	Wang et al. (2020)

for these by-products. This is due not only to the protein content (40–70%) but also to the lipid one, together with other micronutrients and minor bioactive compounds of interest (del Hierro et al., 2024). Therefore, despite that *B. mori* pupae is not currently authorized for human food in Europe, it was already authorized for animal feed in aquaculture, poultry and pig feeding since 2021 (Commission Regulation 2021/1925, 2021).

Concerning the lipid composition of *B. mori*, it is interesting to remark that this specie has diverse varieties, and most of them are industrially produced for silk. Therefore, the different varieties also determine different lipid compositions (Zhai et al., 2024). For a deeper information in this sense, the review article of Zhai et al. (2024) about 15 different varieties of silkworm is recommended. In the following paragraphs, general information about the lipid composition of *B. mori* will be given, without distinguish on the specific variety, since the specific data of the variety is not always given in the studies.

The lipid content of *B. mori* is high, since values in the range of 20–30% have been described (Table 4.17). Conventionally, this lipid fraction has been used for non-food products such as cosmetics, pharmaceuticals, paints, biofuels or plastics (Yeasmin et al., 2024). However, it can be considered that the commercial production of silkworm oil as edible oil just started to emerge in the last few years.

Concerning the fatty acid profile, silkworm pupae have a rare but interesting composition. It mainly consists of UFAs (close to 70%), with a mean value of MUFA around 30% and PUFA higher than 35% (Table 4.18). The SFA content is close to 30%.

This composition is due to the main fatty acids as n-3 linolenic acid (in the range of 25–35%), oleic acid (close to 30%) and palmitic acid (around 25%), together with minor levels of stearic acid and linoleic acid. The high content of α-linolenic acid in silkworm is really attractive since this fatty acid is almost exclusively found in vegetable sources, as linseed oil being the most representative, whereas animals containing this fatty acid are less known. The presence of α-linolenic acid is mainly related to the diet of silkworms with mulberry leaves, also characterized by a high content of α-linolenic acid. However, the content of lipids and this specific fatty acid in silkworms is much higher than that in mulberry leaves, which suggests that specific lipids are synthesized de novo by silkworms (Zhai et al., 2024). Due to this particular fatty acid profile, the n6/n3 ratio of silkworms is exceptionally low (close to 0.2), showing a really interesting oil from the health point of view. This would make silkworm oil unique, since similar ratios are only found for marine animals or algae but due to long-chain PUFA, and the analogous vegetable source of α-linolenic acid, linseed oil, has n-6/n-3 ratio higher than 1.

The fat body of silkworms mainly consists of TAGs (70%), being this lipid class the major in *B. mori* (90%) (Zhai et al., 2024). PLs are also relevant in this oil, with a

TABLE 4.18 Fatty Acid Profile (g/100 of Fatty Acids) of *Bombyx mori* According to Different References Found in the Scientific Literature.

c10	–	–	–	–	–
c12	0.03	–	0.13	–	0.05
c14	0.16	–	0.49	–	0.16
c15	0.03	–	–	–	–
c16	26.77	24.9	21.27	21	26.0
c17	0.10	–	–	–	–
c18	6.38	5.4	6.60		6.78
c20	0.19	–	–	–	–
c21	–	–	–	–	–
c22	0.02	–	–	–	–
c14:1	–	–	–	–	–
C16:1	1.24	0.8	–	–	0.98
c18:1	35.28	24.3	28.97	31	36.84
C20:1	0.02	–	–	–	–
C18:2	4.13	6.3	4.73	6	4.25
C18:3n6	–	–	–	–	–
C18:3n3	24.6	36.0	37.81	33	24.85
C20:2	0.01	–	–	–	–
C20:4	0.03	–	–	–	–
SFA	33.7	30.3	28.5	21	33.0
MUFA	36.5	25.1	29.0	31	37.8
PUFA	28.8	42.5	42.5	39	29.1
PUFA/SFA	0.9	1.4	1.5	1.9	0.9
n3	24.6	36.2	37.8	33	24.9
n6	4.1	6.3	4.7	6	4.3
n6/n3	0.2	0.2	0.1	0.2	0.2
References	Chantankun et al. (2024)	Kotake-Nara et al. (2002)	Susirirut et al. (2023)	Brogan et al. (2023)	Tangsanthatkun et al. (2022)

proportion of lipids close to 10% and with phosphatidylethanolamine as the major PL (Kotake-Nara et al., 2002). Similarly, Brogan et al. (2023) described values close to 6% of oil. It is noteworthy to remark the recent study of Hu et al. (2024) that performed a detailed lipidomic analysis of silkworm pupae. They found up to 241 lipid molecular species. In agreement with classical characterisation, TAGs were confirmed as the major lipid class, together with phosphoethanolamine and phosphocoline. Additionally, a considerable abundance of sphingomyelin, lysophosphatidylcholine, lysophosphatidylethanolamine, ceramide, diacylglycerol, lysophosphatidiylinositol, lysophosphatidylserine, cardiolipin, fatty acid esters, phosphatidylglycerol, phosphatidylinositol, lysophosphatidylglycerol, phosphatidylserine and sphingobase was also detected. Additionally, Yeasmin et al. (2024) described values of FFAs of 3.1%, whereas this value was higher in the study of Wang et al. (2020), close to 8%.

Within minor lipid compounds, cholesterol is the major sterol found in silkworm oil, together with phytosterols (Zhai et al., 2024). Tangsanthatkun et al. (2022) detailed cholesterol values of 0.1% of oil, but such value was higher according to Wang et al. (2020), being closer to 1.2%. Concerning total phytosterols, Wang et al. (2020) found values close to 0.4% of oil. The presence of sterol esters (0.02–0.08%) and tocopherol (0.08–0.14%) has also been described as minor compounds in *B. mori* oils. Specifically, Wang et al. (2020) found contents of total tocopherols at levels close to 37 mg/100 g of oil, being α-tocopherol the major one, with values around 20 mg/100 g of oil. Tangsanthatkun et al. (2022) described lower contents of α-tocopherol, with around 9 mg/100 g of oil. These contents have been found to be variable depending on the sex of the pupae, being higher for female than male, as it is typically found in diverse insect species (Kotake-Nara et al., 2002). Other fat-soluble vitamins and glycolipids are also found in these oils (Zhai et al., 2024). Thus, within the group of carotenoids, neoxanthin, violaxanthin and lutein have been identified (Kotake-Nara et al., 2002). Specifically, Tangsanthatkun et al. (2022) described contents of β-carotene close to 0.8 mg/100 g of oil.

4.11 CONCLUDING REMARKS

Most popular edible insect species contain significant lipid contents, with a diverse but attractive lipid composition influenced by a wide range of factors. Specifically, the species, sex and developmental stage are the main endogenous factors that determine differences on lipid composition. Concerning exogenous factors that can be intervened by the human practices to modulate the lipid content and composition of insects, the used diets for insect feeding is demonstrated as the most relevant one. Additionally, the subsequent technological processing can also offer interesting possibilities, including the slaughtering, drying, but specially the defatting steps. Specially, this last step of lipid extraction is relevant from the industrial but also from the analytical point of view. On one hand, advances and innovative tools of lipid extraction such as ultrasound or pressurized extraction can lead to products with modified fatty acid profiles. On the other hand, the analytical study of insect lipids implies the extraction with organic solvents, but the evidence clearly is demonstrating that depending on the solvent, the lipid content and lipid composition of insects can be extremely different. Therefore, this is an issue that needs to be solved in further studies to clearly have the real lipid composition of the insects.

Despite all these circumstances and despite that still the available information can be considered scarce, general conclusions can be extracted. Most edible insects are characterized by a medium to high UFA profile, with the oleic acid or linoleic acid as the major UFAs. This is the case of species as *T. molitor*, *L. migratoria* and *A. diaperinus*, and all cricket species as *A. domesticus*, *G. sigillatus* and *G. assimilis*. Among them, the oleic acid is especially relevant in case of *T. molitor*. Concerning SFAs, for all these species, pamitic acid is the major one. Due to these profiles, most insects have a n6/n3 ratio far from the recommended for health, but it can be considered in between vegetable sources and animal sources. In this sense, the potential of diet intervention strategies to increase n3 or decrease n6 fatty acids is being intensively explored for most edible insects, with promising results due to the plasticity of most insects. However, there are other quite different species interesting due to their unique fatty acid profile. This is the case of *L. migratoria* but mainly *B. mori*, with noticeable values of n3 linolenic acid, being within the major fatty acids in case of *B. mori*, at levels close to linseed oil. Also, the larvae of *H. illucens* are exceptional due to extremely high lauric acid content, comparable to vegetable sources as coconut or palm oils. The case of *M. domestica* is also unique due to the major content of palmitoleic acid, a rare fatty acid in nature, but at comparable levels to the conventional source as macadamia nuts. Respect to lipid classes, despite TAGs is the major one for edible insects, the content of partial glycerides and especially PLs is also noticeable. Furthermore, minor lipid compounds, mainly as sterols, tocopherols and carotenoids, are described in most edible species.

In conclusion, edible insects represent a valuable source of lipids with unique compositions and promising health benefits, influenced by both intrinsic species characteristics and modifiable external factors such as diet and processing methods. The ability to adjust the fatty acid profile through targeted feeding strategies and advanced extraction technologies highlights the versatility and potential of insect lipids for both food and feed applications. Future research focused on optimizing lipid extraction methods, and refining dietary interventions could unlock even greater nutritional and functional benefits, positioning edible insects as an increasingly important component of sustainable and health-focused food systems.

REFERENCES

Accardo, F., Miguéns-Gómez, A., Lolli, V., Faccini, A., Ardévol, A., Terra, X., Caligiani, A., Pinent, M., & Sforza, S. (2022). Molecular composition of lipid and protein fraction of almond, beef and lesser mealworm after in vitro simulated gastrointestinal digestion and correlation with the hormone-stimulating properties of the digesta. Food Research International, 158, 111499.

Adamkova, A., Kourimska, L., Borkovcova, M., Kulma, M., & Mlcek, J. (2016). Nutritional values of edible Coleoptera (Tenebrio molitor, Zophobas morio and Alphitobius diaperinus) reared in the Czech Republic. Potravinarstvo, 10, 663–671.

Alfiko, Y., Xie, D., Astuti, R. T., Wong, J., & Wang, L. (2022). Insects as a feed ingredient for fish culture: Status and trends. Aquaculture and Fisheries, 7(2), 166–178.

Ali, M. H., Ferdousi, L., Shaikh, M. E. A., Begum, M., Salma, M., Hossain, M. S., Juliana, F. M., Faruki, M. S. I., Ahmed, S., & Reza, M. S. (2024). Exploring the impact of culture media on life history characteristics of the house fly, Musca domestica L. (Diptera: Muscidae). Scientific African, 24, e02133.

Alifian, M. D., Sholikin, M. M., Evvyerni, & Nahrowi, D. (2019). Potential fatty acid composition of Hermetia illucens oil reared on different substrates. IOP Conference Series: Materials Science and Engineering, 546(6), 062002.

Althwab, S. A., Alhomaid, R. M., Ali, R. F., Mohammed El-Anany, A., & Mousa, H. M. (2021). Effect of migratory locust (Locusta migratoria) powder incorporation on nutritional and sensorial properties of wheat flour bread. British Food Journal, 123(11), 3576–3591.

Aniebo, A. O., & Owen, O. J. (2010). Effects of age and method of drying on the proximate composition of housefly larvae (Musca domestica Linnaeus) meal (HFLM). Pakistan Journal of Nutrition, 9(5), 485–487.

Antunes, A. L. M., Mesquita, B. M. A. D. C., Fonseca, F. S. A. D., Carvalho, L. M. D., Brandi, I. V., Carvalho, G. G. P. D., & Coimbra, J. S. D. R. (2024). Extraction and application of lipids from edible insects. Critical Reviews in Food Science and Nutrition, 1–9.

Barroso, F. G., de Haro, C., Sánchez-Muros, M. J., Venegas, E., Martínez-Sánchez, A., & Pérez-Bañón, C. (2014). The potential of various insect species for use as food for fish. Aquaculture, 422–423, 193–201.

Behmer, S. T., & Elias, D. O. (2000). Sterol metabolic constraints as a factor contributing to the maintenance of diet mixing in grasshoppers (Orthoptera: Acrididae). Physiological and Biochemical Zoology, 73(2), 219–230.

Blomquist, G. J., Borgeson, C. E., & Vundla, M. (1991). Polyunsaturated fatty acids and eicosanoids in insects. Insect Biochemistry, 21, 99–106.

Boatta, F., Jansen, W. L., Beukeboom, L. W., & Ellers, J. (2023). Lipid quantity and composition vary between three European house fly strains. Entomologia Experimentalis et Applicata, 171(4), 248–257.

Bogevik, A. S., Seppänen-Laakso, T., Samuelsen, T. A., & Thoresen, L. (2022). Fractionation of oil from black soldier fly larvae (Hermetia illucens). European Journal of Lipid Science and Technology, 124(5), 2100252.

Boukid, F., Riudavets, J., del Arco, L., & Castellari, M. (2021). Impact of diets including agro-industrial by-products on the fatty acids and sterol profiles of larvae biomass from Ephestia kuehniella, Tenebrio molitor and Hermetia illucens. Insects, 12(8), 672.

Brogan, E. N., Park, Y. L., Shen, C., Matak, K. E., & Jaczynski, J. (2023). Characterization of lipids in insect powders. LWT, 184, 115040.

Chantakun, K., Petcharat, T., Wattanachant, S., Ab Karim, M. S. B., & Kaewthong, P. (2024). Fatty acid profile and thermal behavior of fat-rich edible insect oils compared to commonly consumed animal and plant oils. Food Science of Animal Resources, 44(4), 790–804.

Commission Regulation (EU) 2017/893 of 24 May 2017 amending Annexes I and IV to Regulation (EC) No. 999/2001 of the European Parliament and of the Council and Annexes X, XIV and XV to Commission Regulation (EU) No 142/2011 as regards the provisions on processed animal protein.

Commission Regulation (EU) 2021/1372 of 17 August 2021 amending Annex IV to Regulation (EC) No. 999/2001 of the European Parliament and of the Council as regards the prohibition to feed non-ruminant farmed animals, other than fur animals, with protein derived from animals (1), as corrected by OJ L 398, 11.11.

Commission Regulation (EU) 2021/1925 of 5 November 2021 amending certain Annexes to Regulation (EU) No 142/2011 as regards the requirements for placing on the market of certain insect products and the adaptation of a containment method.

Costa, S., Pedro, S., Lourenço, H., Batista, I., Teixeira, B., Bandarra, N. M., Murta, D., Nunes, R., & Pires, C. (2020). Evaluation of Tenebrio molitor larvae as an alternative food source. NFS Journal, 21, 57–64.

Crone, H. D., & Bridges, R. G. (1963). The phospholipids of the housefly, Musca domestica. Biochemical Journal, 89(1), 11–21.

Da Rosa, M. C., & Cruz Silveira Thys, R. (2019). Cricket powder (Gryllus assimilis) as a new alternative protein source for gluten-free breads. Innovative Food Science and Emerging Technologies, 56, 102180.

Daszkiewicz, T., Murawska, D., Kubiak, D., & Han, J. (2022). Chemical composition and fatty acid profile of the Pectoralis major muscle in broiler chickens fed diets with full-fat black soldier fly larvae meal. Animals, 12(4), 464.

Dayrit, F. M. (2015). The properties of lauric acid and their significance in coconut oil. Journal of the American Oil Chemists' Society, 92(1), 1–15.

del Hierro, J. N., Hernández-Ledesma, B., & Martin, D. (2024). Bioactive peptides released from edible insects during gastrointestinal digestion. In Protein Digestion-Derived Peptides (pp. 387–407). Academic Press.

Delgado, L., Garino, C., Moreno, F. J., Zagon, J., & Broll, H. (2023). Sustainable food systems: EU regulatory framework and contribution of insects to the Farm-to-Fork strategy. Food Reviews International, 39(9), 6955–6976.

Downer, R., & Matthews, J. (1976). Patterns of lipid distribution and utilisation in insects. Integrative and Comparative Biology, 16, 733–745.

Dragojlovic, D., Duragic, O., Pezo, L., Popovic, L., Rakita, S., & Tomicic, Z. (2022). Comparison of nutritional profiles of superworm (Zophobas morio) and yellow mealworm (Tenebrio molitor) as alternative feeds used in animal husbandry. Animals, 12, 1277.

Drosdowech, S., Chiasson, M., Ma, D., Huyben, D., & Rooney, N. (2024). Comparison of black soldier fly, cricket, and superworm on growth performance, nutrient utilization and fatty acid profiles of rainbow trout. *BioRxiv*, *2024*(07).

Egonyu, J. P., Subramanian, S., Tanga, C. M., Dubois, T., Ekesi, S., & Kelemu, S. (2021). Global overview of locusts as food, feed and other uses. Global Food Security, 31, 100574.

European Commission. (n.d.). Food safety: Summary of applications and notifications. Retrieved August 13, 2024, from https://food.ec.europa.eu/safety/novel-food/authorisations/summary-applications-and-notifications_en

European Commission. (2021a). Commission Implementing Regulation (EU) 2021/882 of 1 June 2021 authorizing the placing on the market of dried Tenebrio molitor larva as a novel food under Regulation (EU) 2015/2283 of the European Parliament and of the Council, and amending Commission Implementing Regulation (EU) 2017/2470. Official Journal of the European Union, 194, 16–21.

European Commission. (2021b). Commission Implementing Regulation (EU) 2021/1975 of 12 November 2021 authorising the placing on the market of frozen, dried and powder forms of Locusta migratoria as a novel food under Regulation (EU) 2015/2283 of the European Parliament and of the council and amending Commission Implementing Regulation (EU) 2017/2470. Official Journal of the European Union, 402, 10–16.

European Commission. (2022a). Commission Implementing Regulation (EU) 2022/169 of 8 February 2022 authorizing the placing on the market of frozen, dried and powder forms of yellow mealworm (Tenebrio molitor larva) as a novel food under Regulation (EU) 2015/2283 of the European Parliament and of the Council, and amending Commission Implementing Regulation (EU) 2017/2470. Official Journal of the European Union, 28, 10–16.

European Commission. (2022b). Commission Implementing Regulation (EU) 2022/188 of 10 February 2022 authorising the placing on the market of frozen, dried and powder forms of Acheta domesticus as a novel food under Regulation (EU) 2015/2283 of the European Parliament and of the Council, and amending Commission Implementing Regulation (EU) 017/2470.

European Commission. (2023a). Commission Implementing Regulation (EU) 2023/5 of 3 January 2023 authorising the placing on the market of Acheta domesticus partially defatted powder as a novel food and amending Implementing Regulation (EU) 2017/2470.

European Commission. (2023b). Commission Implementing Regulation (EU) 2023/58 of 5 June 2023 authorizing the placing on the market of the frozen, paste, dried and powder forms of Alphitobius diaperinus larvae (lesser mealworm) as a novel food and amending Implementing Regulation (EU) 2017/2470. Official Journal of the European Union, 5, 10–15.

Ewald, N., Vidakovic, A., Langeland, M., Kiessling, A., Sampels, S., & Lalander, C. (2020). Fatty acid composition of black soldier fly larvae (Hermetia illucens): Possibilities and limitations for modification through diet. Waste Management, 102, 40–47.

Ferdousi, L., Begum, M., Yeasmin, M. S., Uddin, J., Miah, M. A. A., Rana, G. M., & Siddique, M. A. B. (2023). Facile acid fermentation extraction of silkworm pupae oil and evaluation of its physical and chemical properties for utilization as edible oil. Heliyon, 9(1).

Fernandez, S. E., Marcia, J., Menjivar, R. D., Santos, R. J., Pinto, A. G., Montero-Fernandez, I., & Reyes, J. T. (2022). Physico-chemical and sensory characteristics of barbecue sauce as influenced by cricket flour (Gryllus assimilis). Chemical Engineering Transactions, 93, 205–210.

Fernández, V., Guzmán-Delgado, P., Graça, J., Santos, S., & Gil, L. (2016). Cuticle structure in relation to chemical composition: Re-assessing the prevailing model. Frontiers in Plant Science, 7, 427.

Finke, M. D. (2013). Complete nutrient content of four species of feeder insects. Zoo Biology, 32(1), 27–36.

Fombong, F. T., Kinyuru, J., Nganga, J., Ayieko, M., Tanga, C. M., Vanden Broeck, J., & Van Der Borght, M. (2021). Affordable processing of edible orthopterans provides a highly nutritive source of food ingredients. Foods, 10(1), 144.

Fornari, T., Vázquez, L., Villanueva-Bermejo, D., Hurtado-Ribeira, R., Martin-Hernández, D., & Martin, D. (2023). Effect of moisture and oil content in the supercritical CO_2 defatting of Hermetia illucens larvae. Foods, 12(3), 490.

Franco, A., Scieuzo, C., Salvia, R., Petrone, A. M., Tafi, E., Moretta, A., Schmitt, E., & Falabella, P. (2021). Lipids from Hermetia illucens, an innovative and sustainable source. Sustainability, 13(18), 10198.

Ganda, H., Zannou, E. T., Kenis, M., Abihona, H. A., Houndonougbo, F. M., Chrysostome, C. A. A. M., Chougourou, D. C., & Mensah, G. A. (2022). Effect of four rearing substrates on the yield and the chemical composition of housefly larvae (Musca domestica L., 1758). International Journal of Tropical Insect Science, 42(2), 1331–1339.

Ganina, M. D., Tyurin, M. V., Zhumatayeva, U. T., Lednev, G. R., Morozov, S. V., & Kryukov, V. Y. (2022). Comparative analysis of epicuticular lipids in Locusta migratoria and Calliptamus italicus: A possible role in susceptibility to entomopathogenic fungi. Insects, 13(8), 736.

Geden, C. J., Nayduch, D., Scott, J. G., Burgess, E. R., IV, Gerry, A. C., Kaufman, P. E., Thomson, J., Pickens, V., & Machtinger, E. T. (2021). House fly (Diptera: Muscidae): Biology, pest status, current management prospects, and research needs. Journal of Integrated Pest Management, 12(1).

Gharibzahedi, S. M. T., & Altintas, Z. (2022). Ultrasound-assisted alcoholic extraction of lesser mealworm larvae oil: Process optimization, physicochemical characteristics, and energy consumption. Antioxidants, 11, 1943.

Gharibzahedi, S. M. T., & Altintas, Z. (2023). Lesser mealworm (Alphitobius diaperinus L.) larvae oils extracted by pure and binary mixed organic solvents: Physicochemical and antioxidant properties, fatty acid composition, and lipid quality indices. Food Chemistry, 408, 135209.

Ghosh, S., Lee, S., Jung, C., & Meyer-Rochow, V. B. (2017). Nutritional composition of five commercial edible insects in South Korea. Journal of Asia-Pacific Entomology, 20, 686–694.

Gołębiowski, M., Cerkowniak, M., Boguś, M. I., Włóka, E., Przybysz, E., & Stepnowski, P. (2013). Developmental changes in the sterol composition and the glycerol content of cuticular and internal lipids of three species of flies. Chemistry & Biodiversity, 10(8), 1521–1530.

Gołębiowski, M., Dawgul, M., Kamysz, W., Boguś, M. I., Wieloch, W., Włóka, E., Paszkiewicz, M., Przybysz, E., & Stepnowski, P. (2012). Antimicrobial activity of alcohols from Musca domestica. Journal of Experimental Biology, 215(19), 3419–3428.

Grdeń, A. S., & Sołowiej, B. G. (2022). Macronutrients, amino and fatty acid composition, elements, and toxins in high-protein powders of crickets, Arthrospira, single cell protein, potato, and rice as potential ingredients in fermented food products. Applied Sciences, 12(24), 12831.

Gutiérrez, Y., Fresch, M., Scherber, C., & Brockmeyer, J. (2022). The lipidome of an omnivorous insect responds to diet composition and social environment. Ecology and Evolution, 12(11), e9497.

Hassen, H., Škvorová, P., Pokhrel, K., Kulma, M., Piątkowska, E., Kostogrys, R. B., Kouřimská, L., Tarko, T., & Franczyk-Żarów, M. (2024). Effect of diets with the addition of edible insects on the development of atherosclerotic lesions in ApoE/LDLR–/– mice. International Journal of Molecular Sciences, 25(13).

Hu, L., Hu, Y., Hong, A., Guo, J., Zhong, C., Cai, J., & Cai, Q. (2024). Comparison of lipid profiles of male and female silkworm (Bombyx mori) pupae through high-resolution mass spectrometry-based lipidomics and chemometrics. Food Chemistry, 459, 140396.

Hu, W., Fitzgerald, M., Topp, B., Alam, M., & O'Hare, T. J. (2019). A review of biological functions, health benefits, and possible de novo biosynthetic pathway of palmitoleic acid in macadamia nuts. Journal of Functional Foods, 62, 103520.

Hurtado-Ribeira, R., Franco, A., & Martin, D. (2024). Freezing storage combined with freeze-drying of black soldier fly (Hermetia illucens) larvae to produce oil rich in free lauric acid. Journal of Insects as Food and Feed, 1–14.

Hurtado-Ribeira, R., Silvan, J. M., Fornari, T., Vázquez, L., Martinez-Rodriguez, A. J., & Martin, D. (2023a). Modulation of lipolysis and subsequent antibacterial activity of fat from black soldier fly (Hermetia illucens) by the combined selection of slaughtering, drying and defatting methods of the larvae. Innovative Food Science & Emerging Technologies, 90, 103510.

Hurtado-Ribeira, R., Villanueva-Bermejo, D., García-Risco, M. R., Hernández, M. D., Sánchez-Muros, M. J., Fornari, T., Vázquez, L., & Martin, D. (2023b). Evaluation of the interrelated effects of slaughtering, drying, and defatting methods on the composition and properties of black soldier fly (Hermetia illucens) larvae fat. Current Research in Food Science, 7, 100633.

Hussein, M., Pillai, V. V., Goddard, J. M., Park, H. G., Kothapalli, K. S., Ross, D. A., Ketterings, Q. M., Brenna, J. T., Milstein, M. B., Marquis, H., Johnson, P. A., Nyrop, J. P., & Selvaraj, V. (2017). Sustainable production of housefly (Musca domestica) larvae as a protein-rich feed ingredient by utilizing cattle manure. PLoS One, 12(2), e0171708.

Jankauskiene, A., Aleknavicius, D., Andruleviciute, V., Mockus, E., Bartkiene, E., Jukniene, I., Kiselioviene, S., Zavistanaviciute, P., Zaborskiene, G., & Kabašinskiene, A. (2024). Nutritional composition and safety parameters of mealworms (Tenebrio molitor) reared on substrates derived from by-products. Applied Sciences, 14, 2744.

Jozefiak, Z., Kieronczyk, B., Rawski, M., Mazurkiewicz, J., Benzertiha, A., Gobbi, P., Nogales-Mérida, S., Swiatkiewicz, S., & Jozefiak, D. (2018). Full-fat insect meals as feed additive: The effect on broiler chicken growth performance and gastrointestinal tract microbiota. Journal of Animal and Feed Sciences, 27, 131–139.

Jurenka, R. A., Stanley-Samuelson, D. W., Loher, W., & Blomquist, G. J. (1988). De novo biosynthesis of arachidonic acid and 5,11,14-eicosatrienoic acid in the cricket Teleogryllus commodus. Biochimica et Biophysica Acta – Lipids and Lipid Metabolism, 963, 21–27.

Kepinska-Pacelik, J., Biel, W., Podsiadlo, C., Tokarczyk, G., Biernacka, P., & Bienkiewicz, G. (2023). Nutritional value of banded cricket and mealworm larvae. Foods, 12, 4174.

Khanal, P., Pandey, D., Naess, G., Gabrita, A. R. J., Fonseca, A. J. M., Maia, M. R. G., Timilsina, B., Veldkamp, T., Sapkota, R., & Overrein, H. (2023). Yellow mealworms (Tenebrio molitor) as an alternative animal feed source: A comprehensive characterization of nutritional values and the larval gut microbiome. Journal of Cleaner Production, 389, 136104.

Khatun, H., Claes, J., Smets, R., Winne, A., Akhtaruzzaman, M., & Borght, M. (2021). Characterization of freeze-dried, oven-dried and blanched house crickets (Acheta domesticus) and Jamaican field crickets (Gryllus assimilis) by means of their physicochemical properties and volatile compounds. European Food Research and Technology, 247, 1291–1305.

Kilburn, L. R., Carlson, A. T., Lewis, E., & Rossoni Serao, M. C. (2020). Cricket (Gryllodes sigillatus) meal fed to healthy adult dogs does not affect general health and minimally impacts apparent total tract digestibility. Journal of Animal Science, 98, 1–8.

Koba, K., & Yanagita, T. (2014). Health benefits of conjugated linoleic acid (CLA). Obesity Research & Clinical Practice, 8(6), e525–e532.

Kolobe, S. D., Manyelo, T. G., Malematja, E., Sebola, N. A., & Mabelebele, M. (2023). Fats and major fatty acids present in edible insects utilised as food and livestock feed. Veterinary and Animal Science, 100312.

Kotake-Nara, E., Yamamoto, K., Nozawa, M., Miyashita, K., & Murakami, T. (2002). Lipid profiles and oxidative stability of silkworm pupal oil. Journal of Oleo Science, 51(11), 681–690.

Kotsou, K., Chatzimitakos, T., Athanasiadis, V., Bozinou, E., Rumbos, C. I., Athanassiou, C. G., & Lalas, S. I. (2023). Enhancing the nutritional profile of Tenebrio molitor using the leaves of Moringa oleifera. Foods, 12, 2612.

Kotsou, K., Chatzimitakos, T., Athanasiadis, V., Bozinou, E., Athanassiou, C. G., & Lalas, S. I. (2023). Utilization of spent coffee grounds as a feed additive for enhancing the nutritional value of Tenebrio molitor larvae. Sustainability, 15, 16224.

Kröncke, N., Grebenteuch, S., Keil, C., Demtröder, S., Kroh, L., Thünemann, A. F., Benning, R., & Haase, H. (2019). Effect of different drying methods on nutrient quality of the yellow mealworm (Tenebrio molitor L.). Insects, 10, 84.

Kulma, M., Petříčková, M., Kurečka, M., Kotíková, Z., Táborský, J., Michlová, T., & Kouřimská, L. (2022). Effect of carrot supplementation on nutritional value of insects: A case study with Jamaican field cricket (Gryllus assimilis). Journal of Insects as Food and Feed, 8(6), 621–630.

Kulma, M., Kouřimská, L., Plachý, V., Božik, M., Adámková, A., & Vrabec, V. (2019). Effect of sex on the nutritional value of house cricket, Acheta domestica L. Food Chemistry, 272, 267–272.

Kurecka, M., Kulma, M., Petrickova, D., Plachy, V., & Kourimska, L. (2021). Larvae and pupae of Alphitobius diaperinus as promising protein alternatives. European Food Research and Technology, 247, 2527–2532.

Lawal, K. G., Kavle, R. R., Akanbi, T. O., Mirosa, M., & Agyei, D. (2021). Enrichment in specific fatty acids profile of Tenebrio molitor and Hermetia illucens larvae through feeding. Future Foods, 3, 100016.

Lawal, K. G., Kavle, R. R., Akanbi, T. O., Mirosa, M., & Agyei, D. (2022). Lipid nutritional indices, regioisomeric distribution and thermal properties of Tenebrio molitor and Hermetia illucens larvae fat. Journal of Asia-Pacific Entomology, 25, 101951.

Lee, S., Kim, M., Cho, H., & Lee, G. (2024). Determination of triacylglycerol composition in mealworm oil (Tenebrio molitor) via electrospray ionization tandem mass spectrometry with multiple neutral loss scans. Insects, 15, 365.

Li, M., Wang, G., Shang, R., Xu, Q., Zhang, J., Sun, R., & Li, L. (2021). Comparative lipid profile analysis of Hermetia illucens larvae fed food waste at different days of age using an LC–MS-based lipidomics approach. Journal of Insect Science, 21(5).

Li, Q., Zheng, L., Cai, H., Garza, E., Yu, Z., & Zhou, S. (2011). From organic waste to biodiesel: Black soldier fly, Hermetia illucens, makes it feasible. Fuel, 90(4), 1545–1548.

Li, W., Li, M., Zheng, L., Liu, Y., Zhang, Y., Yu, Z., Ma, Z., & Li, Q. (2015). Simultaneous utilization of glucose and xylose for lipid accumulation in black soldier fly. Biotechnology for Biofuels, 8, 117.

Li, X., Dong, Y., Sun, Q., Tan, X., You, C., Huang, Y., & Zhou, M. (2022). Growth and fatty acid composition of black soldier fly (Hermetia illucens) larvae are influenced by dietary fat sources and levels. Animals, 12(4), 486.

Liland, N. S., Biancarosa, I., Araujo, P., Biemans, D., Bruckner, C. G., Waagbø, R., Torstensen, B. E., & Lock, E.-J. (2017). Modulation of nutrient composition of black soldier fly (Hermetia illucens) larvae by feeding seaweed-enriched media. PLoS One, 12(8), e0183188.

Loho, L., & Lo, D. (2023). Proximate and fatty acid analysis of black soldier fly larvae (Hermetia illucens). IOP Conference Series: Earth and Environmental Science, 1169(1), 012082.

Loypimai, P., Moontree, T., Pranil, T., & Moongngarm, A. (2024). A comparative study of nutritional components of Gryllus bimaculatus and Acheta domesticus cricket powders prepared using different drying methods. Journal of Food Measurement and Characterization, 18(5), 3974–3983.

Lucas-González, R., Fernández-López, J., Pérez-Álvarez, J. A., & Viuda-Martos, M. (2019). Effect of drying processes on the chemical, physicochemical, techno-functional and antioxidant properties of flours obtained from house cricket (Acheta domesticus). European Food Research and Technology, 245(7), 1451–1458.

Mai, H. C., Dao, N. D., Lam, T. D., Nguyen, B. V., Nguyen, D. C., & Bach, L. G. (2019). Purification process, physicochemical properties, and fatty acid composition of black soldier fly (Hermetia illucens Linnaeus) larvae oil. Journal of the American Oil Chemists' Society, 96(11), 1303–1311.

Makkar, H. P. S., Tran, G., Heuzé, V., & Ankers, P. (2014). State-of-the-art on use of insects as animal feed. Animal Feed Science and Technology, 197, 1–33.

Martins da Silva, R., Köhler, A., de Souza Schneider, R., Prado de Vargas, D., Köhler, A. L., da Costa e Silva, D., & Soares, J. (2024). Proximate and fatty acid profile analysis of Tenebrio molitor and Zophobas morio using different killing methods. Food Chemistry, 445, 138719.

Matthäus, B., Piofczyk, T., Katz, H., & Pudel, F. (2019). Renewable resources from insects: Exploitation, properties, and refining of fat obtained by cold-pressing from Hermetia illucens (black soldier fly) larvae. European Journal of Lipid Science and Technology, 121(7), 1800376.

Mattioli, S., Fratini, F., Cacchiarelli, C., Martinis, V., Tuccinardi, T., Paci, G., Dal Bosco, A., & Mancini, S. (2024). Chemical composition, fatty acid profile, antioxidant content and microbiological loads of lesser mealworm, mealworm, and superworm larvae. Italian Journal of Animal Science, 23, 125–137.

Mattioli, S., Paci, G., Fratini, F., Dal Bosco, A., Tuccinardi, T., & Mancini, S. (2021). Former foodstuff in mealworm farming: Effects on fatty acids profile, lipid metabolism and antioxidant molecules. LWT – Food Science and Technology, 147, 111644.

Mazı, I. B. (2024). Comparative analysis of nutritional quality and color properties of flours derived from Locusta migratoria at different developmental stages. Food Science and Technology International. Advance online publication.

Meyer-Rochow, V. B., Gahukar, R. T., Ghosh, S., & Jung, C. (2021). Chemical composition, nutrient quality and acceptability of edible insects are affected by species, developmental stage, gender, diet, and processing method. Foods, 10(5), 1036.

Miranda, C. D., Cammack, J. A., & Tomberlin, J. K. (2020). Life-history traits of house fly, Musca domestica L. (Diptera: Muscidae), reared on three manure types. Journal of Insects as Food and Feed, 6(1), 81–90.

Mlcek, J., Adamková, A., Adámek, M., Borkovcová, M., Bednárova, M., & Kourimská, L. (2018). Selected nutritional values of field cricket (Gryllus assimilis) and its possible use as a human food. Indian Journal of Traditional Knowledge, 17(3).

Mohamed, E. H. (2015). Determination of nutritive value of the edible migratory locust Locusta migratoria, Linnaeus, 1758 (Orthoptera: Acrididae). International Journal of Advances in Pharmacy, Biology and Chemistry, 4(1), 144–148.

Moon, R. D. (2019). Muscid flies (Muscidae). In G. R. Mullen & L. A. Durden (Eds.), Medical and veterinary entomology (3rd ed., pp. 345–368). Academic Press.

Moreki, J. C., Tiroesele, B., & Chiripasi, S. C. (2012). Prospects of utilizing insects as alternative sources of protein in poultry diets in Botswana: A review. Journal of Animal Science Advances, 2(8), 649–658.

Nitbani, F. O., Tjitda, P. J. P., Nitti, F., Jumina, J., & Detha, A. I. R. (2022). Antimicrobial properties of lauric acid and monolaurin in virgin coconut oil: A review. ChemBioEng Reviews, 9(5), 442–461.

Niu, Y., Zheng, D., Yao, B., Cai, Z., Zhao, Z., Wu, S., Cong, P., & Yang, D. (2017). A novel bioconversion for value-added products from food waste using Musca domestica. Waste Management, 61, 455–460.

Ochiai, M., & Komiya, Y. (2021). Detection of edible insect-derived phospholipids with polyunsaturated fatty acids by thin-layer chromatography, gas chromatography, and enzymatic methods. Journal of Food Composition and Analysis, 99.

Ogilvy, V., Fidgett, A. L., & Preziosi, R. F. (2012). Differences in carotenoid accumulation among three feeder-cricket species: Implications for carotenoid delivery to captive insectivores. Zoo Biology, 31, 470–478.

Oliveira, L. A., Pereira, S. M. S., Dias, K. A., Paes, S. D. S., Grancieri, M., Jimenez, L. G. S., de Carvalho, C. W. P., de Oliveira, E. E., Martino, H. S. D., & Della Lucia, C. M. (2024). Nutritional content, amino acid profile and protein properties of edible insects (Tenebrio molitor and Gryllus assimilis) powders at different stages of development. Journal of Food Composition and Analysis, 125, 105804.

Onsongo, V. O., Osuga, I. M., Gachuiri, C. K., Wachira, A. M., Miano, D. M., Tanga, C. M., Ekesi, S., Nakimbugwe, D., & Fiaboe, K. K. M. (2018). Insects for income generation through animal feed: Effect of dietary replacement of soybean and fish meal with black soldier fly meal on broiler growth and economic performance. Journal of Economic Entomology, 111(4), 1966–1973.

Oonincx, D. G. A. B., Laurent, S., Veenenbos, M. E., & Loon, J. J. A. (2020). Dietary enrichment of edible insects with omega-3 fatty acids. Insect Science, 27(3), 500–509.

Oonincx, D. G. A. B., & Van der Poel, A. F. B. (2011). Effects of diet on the chemical composition of migratory locusts (Locusta migratoria). Zoo Biology, 30(1), 9–16.

Orkusz, A., Dymińska, L., Banaś, K., & Harasym, J. (2023). Chemical and nutritional fat profile of Acheta domesticus, Gryllus bimaculatus, Tenebrio molitor and Rhynchophorus ferrugineus. Foods, 13(1), 32.

Osborn, H. T., & Akoh, C. C. (2002). Structured lipids – Novel fats with medical, nutraceutical, and food applications. Comprehensive Reviews in Food Science and Food Safety, 1(3), 110–120.

Otero, P., Gutierrez-Docio, A., Del Hierro, J. N., Reglero, G., & Martin, D. (2020). Extracts from the edible insects Acheta domesticus and Tenebrio molitor with improved fatty acid profile due to ultrasound-assisted or pressurized liquid extraction. Food Chemistry, 314, 126200.

Pastell, H., Mellberg, S., Ritvanen, T., Raatikainen, M., Mykkänen, S., Niemi, J., & Wirtanen, G. (2021). How does locally produced feed affect the chemical composition of reared house crickets (Acheta domesticus)? ACS Food Science & Technology, 1(4), 625–635.

Paul, A., Frederich, M., Megido, R. C., Alabi, T., Malik, P., Uyttenbroeck, R., Francis, F., Blecker, C., Haubruge, E., Lognay, G., & Danthine, S. (2017). Insect fatty acids: A comparison of lipids from three orthopterans and Tenebrio molitor L. larvae. Journal of Asia-Pacific Entomology, 20, 337–340.

Perez-Santaescolastica, C., de Pril, I., van de Voorde, I., & Fraeye, I. (2023). Fatty acid and amino acid profiles of seven edible insects: Focus on lipid class composition and protein conversion factors. Foods, 12(22), 4090.

Pieterse, E., & Pretorius, Q. (2014). Nutritional evaluation of dried larvae and pupae meal of the housefly (Musca domestica) using chemical- and broiler-based biological assays. Animal Production Science, 54(3), 347–355.

Pimentel, A. C., Montali, A., Bruno, D., & Tettamanti, G. (2017). Metabolic adjustment of the larval fat body in Hermetia illucens to dietary conditions. Journal of Asia-Pacific Entomology, 20(4), 1307–1313.

Pranil, T., Moontree, T., Moongngarm, A., & Loypimai, P. (2024). Ultrasound-assisted extraction of edible cricket oils from Gryllus bimaculatus and Acheta domesticus. Journal of Food Measurement and Characterization, 1–11.

Purschke, B., Tanzmeister, H., Meinlschmidt, P., Baumgartner, S., Lauter, K., & Jäger, H. (2018). Recovery of soluble proteins from migratory locust (Locusta migratoria) and characterisation of their compositional and techno-functional properties. Food Research International, 106, 271–279.

Rabani, V., Cheatsazan, H., & Davani, S. (2019). Proteomics and lipidomics of black soldier fly (Diptera: Stratiomyidae) and blow fly (Diptera: Calliphoridae) larvae. Journal of Insect Science, 19(3).

Ramos-Bueno, R. P., González-Fernández, M. J., Sánchez-Muros-Lozano, M. J., García-Barroso, F., & Guil-Guerrero, J. L. (2016). Fatty acid profiles and cholesterol content of seven insect species assessed by several extraction systems. European Food Research and Technology, 242, 1471–1477.

Rawski, M., Mazurkiewicz, J., Kierończyk, B., & Józefiak, D. (2020). Black soldier fly full-fat larvae meal as an alternative to fish meal and fish oil in Siberian sturgeon nutrition: The effects on physical properties of the feed, animal growth performance, and feed acceptance and utilization. Animals, 10(11), 2119.

Renna, M., Coppa, M., Lussiana, C., Le Morvan, A., Gasco, L., & Maxin, G. (2022). Full-fat insect meals in ruminant nutrition: In vitro rumen fermentation characteristics and lipid biohydrogenation. Journal of Animal Science and Biotechnology, 13, 138.

Renna, M., Gasco, L., Livorsi, L., Mele, M., Conte, G., Meneguz, M., & Lussiana, C. (2024). Growth performance, proximate composition and fatty acid profile of black soldier fly larvae reared on two grape pomace varieties. Animal, 18(8), 101240.

Ribeiro, J. C., Lima, R. C., Maia, M. R. G., Almeida, A. A., Fonseca, A. J. M., Cabrita, A. R. J., & Cunha, L. M. (2019). Impact of defatting freeze-dried edible crickets (Acheta domesticus and Gryllodes sigillatus) on the nutritive value, overall liking and sensory profile of cereal bars. LWT – Food Science and Technology, 113, 108335.

Riekkinen, K., Väkeväinen, K., & Korhonen, J. (2022). The effect of substrate on the nutrient content and fatty acid composition of edible insects. Insects, 13(7), 590.

Romero-Lorente, M., Fabrikov, D., Montes, J., Morote, E., Barroso, F. G., Vargas-García, M. C., Varga, T., & Sánchez-Muros, M. (2022). Pre-treatment of fish by-products to optimize feeding of Tenebrio molitor L. larvae. Insects, 13, 125.

Roncolini, A., Milanović, V., Aquilanti, L., Cardinali, F., Garofalo, C., Sabbatini, R., Clementi, F., Belleggia, L., Pasquini, M., Mozzon, M., Foligni, R., Trombetta, M. F., Haouet, M. N., Altissimi, M. S., Di Bella, S., Piersanti, A., Griffoni, F., Reale, A., Niro, S., & Osimani, A. (2020). Lesser mealworm Alphitobius diaperinus powder as a novel baking ingredient for manufacturing high-protein, mineral-dense snacks. Food Research International, 131, 109031.

Rose, A., Jaczynski, J., & Matak, K. (2021). Extraction of lipids from insect powders using a one-step organic solvent extraction process. Future Foods, 4, 100073.

Sánchez, M., Gómez, C., Avendaño, C., Harmsen, I., Ortiz, D., Ceballos, R., Villamizar-Sarmiento, M. G., Oyarzun-Ampuero, F., Wacyk, J., & Valenzuela, C. (2021). House fly (Musca domestica) larvae meal as an ingredient with high nutritional value: Microencapsulation and improvement of organoleptic characteristics. Food Research International, 145, 110423.

Shumo, M., Osuga, I. M., Khamis, F. M., Tanga, C. M., Fiaboe, K. K. M., Subramanian, S., Ekesi, S., van Huis, A., & Borgemeister, C. (2019). The nutritive value of black soldier fly larvae reared on common organic waste streams in Kenya. Scientific Reports, *9*(1), 10110.

Singh, Y., Cullere, M., Kovitvadhi, A., Chundang, P., & Dalle Zotte, A. (2020). Effect of different killing methods on physicochemical traits, nutritional characteristics, in vitro human digestibility and oxidative stability during storage of the house cricket (Acheta domesticus L.). Innovative Food Science & Emerging Technologies, 65, 102444.

Soares Araújo, R. R., dos Santos Benfica, T. A. R., Ferraz, V. P., & Moreira Santos, E. (2019). Nutritional composition of insects Gryllus assimilis and Zophobas morio: Potential foods harvested in Brazil. Journal of Food Composition and Analysis, 76, 22–26.

Soetemans, L., Gianotten, N., & Bastiaens, L. (2020). Agri-food side-stream inclusion in the diet of Alphitobius diaperinus. Part 2: Impact on larvae composition. Insects, 11, 190.

Solà Marsiñach, M., & Cuenca, A. P. (2019). The impact of sea buckthorn oil fatty acids on human health. Lipids in Health and Disease, 18(1), 145.

Son, Y., Choi, S., Hwang, I., Nho, C. W., & Kim, S. H. (2020). Could defatted mealworm (Tenebrio molitor) and mealworm oil be used as food ingredients? Foods, 9, 40.

Spranghers, T., Ottoboni, M., Klootwijk, C., Ovyn, A., Deboosere, S., De Meulenaer, B., Michiels, J., Eeckhout, M., De Clercq, P., & De Smet, S. (2017). Nutritional composition of black soldier fly (Hermetia illucens) prepupae reared on different organic waste substrates. Journal of the Science of Food and Agriculture, 97(8), 2594–2600.

Stefanov, K., Nechev, J., Lavchieva-Nacheva, G., Nikolova, N., Seizova, K., Kwartirnikov, M., Lavchiev, V., & Popov, S. (2002). Lipids and sterols in Musca domestica L. (Diptera: Muscidae): Changes after treatment with sucrose and lead. Comparative Biochemistry and Physiology Part B: Biochemistry and Molecular Biology, 131(3), 543–550.

Suryati, T., Julaeha, E., Farabi, K., Ambarsari, H., & Hidayat, A. T. (2023). Lauric acid from the black soldier fly (Hermetia illucens) and its potential applications. Sustainability, 15(13), 10383.

Susirirut, P., Thitipramote, N., & Chaiwut, P. (2023). Simultaneous extraction of oil and protein from silkworm (Bombyx mori L.) pupae (Lueng Parroj var.) and their in vitro skin moisturization. Molecules, 28(20), 7032.

Tangsanthatkun, J., Peanparkdee, M., Katekhong, W., Harnsilawat, T., Tan, C. P., & Klinkesorn, U. (2022). Application of aqueous saline process to extract silkworm pupae oil (Bombyx mori): Process optimization and composition analysis. Foods, 11(3), 291.

Tognocchi, M., Conte, G., Rossi, E., Perioli, R., Mantino, A., Serra, A., & Mele, M. (2023). Characterization of polar and non-polar lipids of Hermetia illucens and Tenebrio molitor meals as animal feed ingredients. Animal Feed Science and Technology, 295, 115524.

Toprak, U., Hegedus, D., Doğan, C., & Güney, G. (2020). A journey into the world of insect lipid metabolism. Archives of Insect Biochemistry and Physiology, 104(2), e21682.

Tran, H. Q., Nguyen, T. T., Prokesova, M. D., Matousek, J., Tomcala, A., Van Doan, H., Kiljunen, M., & Stejskal, V. (2023). Insight into bioavailability of various insect meals for European perch (Perca fluviatilis): A nutritional and stable isotopic evaluation. Aquaculture, 563, 738912.

Turck, D., Bohn, T., Castenmiller, J., De Henauw, S., Hirsch-Ernst, K. I., Maciuk, A., Mangelsdorf, I., McArdle, H. J., Naska, A., Peláez, C., . . . Knutsen, H. K. (2023). Safety of UV-treated powder of whole yellow mealworm (Tenebrio molitor larva) as a novel food pursuant to Regulation (EU) 2015/2283. EFSA Journal, 21(5), 8009. https://doi.org/10.2903/j.efsa.2023.8009

Tzompa-Sosa, D. A., Dewettinck, K., Provijn, P., Brouwers, J. F., De Meulenaer, B., & Oonincx, D. G. (2021). Lipidome of cricket species used as food. Food Chemistry, 349, 129077.

Tzompa-Sosa, D. A., Yi, L., van Valenberg, H. J., van Boekel, M. A., & Lakemond, C. M. (2014). Insect lipid profile: Aqueous versus organic solvent-based extraction methods. Food Research International, 62, 1087–1094.

Ugur, A. E., Bolat, B., Oztop, M. H., & Alpas, H. (2021). Effects of high hydrostatic pressure (HHP) processing and temperature on physicochemical characterization of insect oils extracted from Acheta domesticus (house cricket) and Tenebrio molitor (yellow mealworm). Waste and Biomass Valorization, 12, 4277–4286.

Ukanwoko, A. I., & Olalekan, O. A. (2015). Effects of source and time of harvest on the proximate composition of maggot (Musca domestica) larva meal. International Journal of Livestock Research, 5(8), 4.

Ushakova, N. A., Brodskii, E. S., Kovalenko, A. A., Bastrakov, A. I., Kozlova, A. A., & Pavlov, D. S. (2016). Characteristics of lipid fractions of larvae of the black soldier fly Hermetia illucens. Doklady Biochemistry and Biophysics, 468(1), 209–212.

Van Broekhoven, S., Oonincx, D. G. A. B., Van Huis, A., & Van Loon, J. J. A. (2015). Growth performance and feed conversion efficiency of three edible mealworm species (Coleoptera: Tenebrionidae) on diets composed of organic by-products. Journal of Insect Physiology, 73, 1–10.

Van Huis, A. (2015). Edible insects contributing to food security? Agriculture & Food Security, 4, 1–9.

Verheyen, G. R., Ooms, T., Vogels, L., Vreysen, S., Bovy, A., Van Miert, S., & Meersman, F. (2018). Insects as an alternative source for the production of fats for cosmetics. Journal of Cosmetic Science, 69(3), 187–202.

Vlahova-Vangelova, D., Balev, D., Kolev, D., Dragoev, S., Petkov, E., & Popova, T. (2024). Comparison of the effect of drying treatments on physicochemical parameters, oxidative stability, and microbiological status of yellow mealworm (Tenebrio molitor L.) flours as an alternative protein source. Agriculture, 14, 436.

Wang, W., Xu, L., Zou, Y., Pang, D., Shi, W., Mu, L., & Liao, S. (2020). Comprehensive identification of principal lipid classes and tocochromanols in silkworm (Antheraea pernyi and Bombyx mori) pupae oils. European Journal of Lipid Science and Technology, 122(2), 1900280.

Yeasmin, M. S., Ferdousi, L., Dey, S. S., Uddin, M. J., Chowdhury, T. A., & Rana, M., Muzahid, A. A. (2024). Sustainable production of liquid phenyl using oil extracted from discarded pupae of mulberry silkworms (Bombyx mori). Journal of the Indian Chemical Society, 101248.

Zafar, A., Shaheen, M., Tahir, A. B., Gomes da Silva, A. P., Manzoor, H. Y., & Zia, S. (2024). Unraveling the nutritional, biofunctional, and sustainable food application of edible crickets: A comprehensive review. Trends in Food Science & Technology, 143, 104254.

Zhai, M., Wang, W., Zou, Y., Liao, S., Wang, S., & Mu, L. (2024). Function, composition, digestion, and processing and utilization of silkworm pupae oil: A review. European Journal of Lipid Science and Technology, 126(2), 2300094.

Zhang, S., Tan, Z. Z., Zhang, J., Li, S. Z., Qiu, K. R., & Zhu, F. (2023). Comparison of vitamin E isomer composition among three animal feeds containing the insects Musca domestica (Diptera: Muscidae), Hermetia illucens (Diptera: Stratiomyidae), and Tenebrio molitor (Coleoptera: Tenebrionidae). Entomological News, 130(5), 416–431.

Zhao, L., Wu, B., Liang, S., Min, D., & Jiang, H. (2022). Insight of silkworm pupae oil regulating oxidative stress and lipid metabolism in Caenorhabditis elegans. Foods, 11(24), 4084.

Zielinska, E., Baraniak, B., Karas, M., Rybczynska, K., & Jakubczyk, A. (2015). Selected species of edible insects as a source of nutrient composition. Food Research International, 77, 460–466.

Zulkifli, N. F. N. M., Seok-Kian, A. Y., Seng, L. L., Mustafa, S., Kim, Y. S., & Shapawi, R. (2022). Nutritional value of black soldier fly (Hermetia illucens) larvae processed by different methods. PLoS One, 7(2), e0263924.

Nutritional and Health Benefits of Insects

5

José Eleazar Aguilar-Toalá and Andrea M. Liceaga

5.1 INTRODUCTION

The use of insects as food has been part of the regular diet in many cultures around the world, with about 119 countries mainly in Asia, Africa, and Latin America reported to regularly practice entomophagy (Alexander et al., 2017). In these regions, eating insects is a long-standing tradition, with insects valued not only as an accessible nutrition source but also for their distinctive flavours and culinary versatility. Across these diverse cultures, a wide variety of insect species are prepared in numerous ways, from frying and grilling to being incorporated into soups, sauces, stews, tacos, and other dishes in high-end restaurants or in street food stalls (Lesnik, 2019; Liceaga, 2021; Melgar-Lalanne, Hernández-Álvarez, & Salinas-Castro, 2019).

Edible insects are commonly obtained by farming operations in commercial facilities or directly from their natural habitat. However, regulations across several European and in North America countries prohibit the use of wild-harvested insects for human food/animal feed applications. In this context, many of the current insect biomass facilities can be produced organically, which eliminates environmental pollution factors and allows insects to convert agricultural side-streams (i.e., by-products) into high-nutrient feed that can serve as an alternative to the costly animal feed ingredients.

To date, it has been reported that close to 1,900 insect species are classified as edible, which suggests their biodiversity and their potential role in the human diet, offering a wide range of nutritional options, flavours, and textures, making edible insects a viable supplement or alternative to conventional protein sources (Liang, Zhu, Leonard, & Fang, 2024). In general, the most frequently consumed insects are beetles (order: Coleoptera) (31%), caterpillars (order: Lepidoptera) (18%), and bees, wasps, and ants (order: Hymenoptera) (14%). These are followed by grasshoppers, locusts, and crickets (order: Orthoptera) (13%),

DOI: 10.1201/9781003488156-5

cicadas, leafhoppers, planthoppers, scale insects, and true bugs (order: Hemiptera) (10%), termites (order: Blattodea) (3%), dragonflies (order: Odonata) (3%), flies (order: Diptera) (2%), and insects from other orders (5%) (Zielińska, Baraniak, Karaś, Rybczyńska, & Jakubczyk, 2015).

Despite the increasing interest in the great potential of insects as food, scientific knowledge about consumer behaviour in the field of entomophagy indicate that in some Western regions of the world, principally in Europe and the United States, the consumption of edible insects is largely unfamiliar and often perceived with a sense of disgust (Ardoin & Prinyawiwatkul, 2021). However, continuous efforts in marketing and educating the public about the nutritional advantages and health benefits of edible insects can gradually influence people's willingness to include them in their diet (Kim, Yong, Kim, Kim, & Choi, 2019).

In contrast, the Food and Agricultural Organization (FAO) of the United Nations has stated that the global population in the world will grow to nearly 10 billion people by 2050; thus, the demand for food will increase by 50%, with the increase of animal-based food by 70% (Searchinger, Waite, Hanson, Ranganathan, & Dumas, 2019). Current conventional food production systems are unable to supply adequate food for the global population, as such systems imply high environmental issues such as land, water, air pollution, and biodiversity loss (Liceaga, Aguilar-Toalá, Vallejo-Cordoba, González-Córdova, & Hernández-Mendoza, 2022). Therefore, the consumption of alternative protein sources such as plant-based foods and edible insects has been proposed as possible solutions. For a more detailed overview about plant-based foods as an alternative protein source, the reader is referred to Hidalgo-Fuentes et al. (2024). Overall, edible insects are a definite food source of nutrients for humans by providing high-quality proteins with noticeable digestibility, essential amino acids, unsaturated fatty acids, vitamins, fibre, and minerals, which can help address nutritional deficiencies and contribute to sustainable food systems. In addition to their nutritional benefits, insects are also recognized for their potential health benefits, as shown in Figure 5.1 (Liceaga et al., 2022; Rumpold & Schlüter, 2013).

5.2 NUTRITIONAL COMPOSITION OF EDIBLE INSECTS

5.2.1 Macronutrients

Insects are particularly rich in protein, which represent the main component of their nutrient composition, often containing levels comparable to or even higher than those found in traditional livestock such as beef, chicken, and pork (Rumpold & Schlüter, 2013; Zielińska et al., 2015). Additionally, the protein derived from insects is considered as complete, meaning it contains all the essential amino acids required by the human body, making them a valuable resource for maintaining muscle mass, repairing tissues, and supporting overall growth and development (Aguilar-Toalá, Cruz-Monterrosa, & Liceaga, 2022). Additionally, research has shown that the amino acid profile of edible insects such as

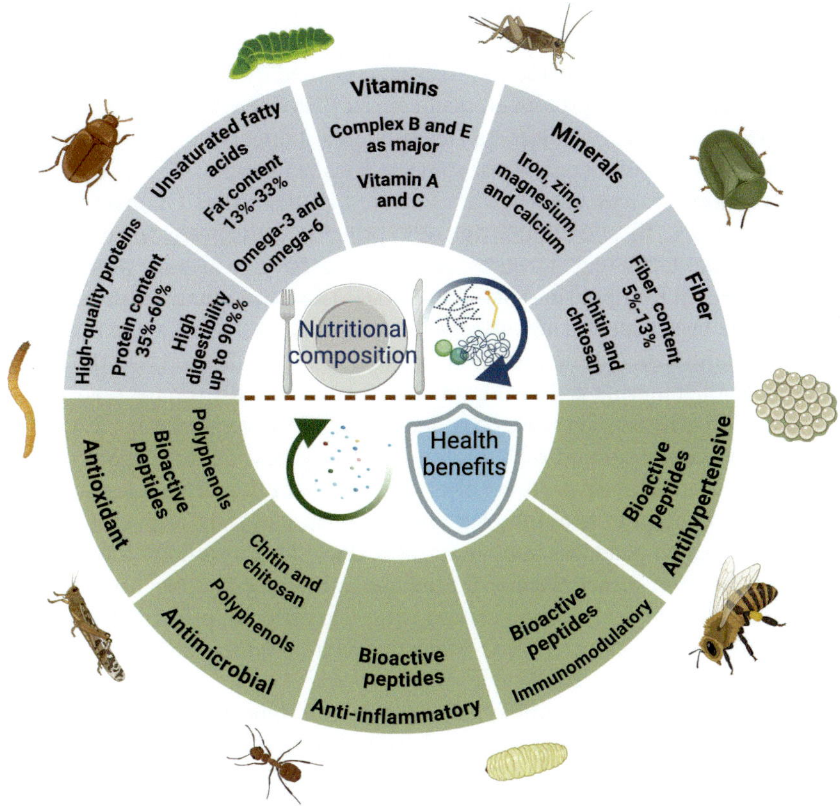

FIGURE 5.1 Nutritional and health benefits of farming edible insects. Created in BioRender. Liceaga, A. (2026).

silkworm larvae and crickets surpass the amino acid requirements set by FAO and the World Health Organization (Hall, Jones, O'Haire, & Liceaga, 2017; Wu, Jia, Tan, Xu, & Gui, 2011). Furthermore, biomass production is considered more environmentally sustainable than traditional livestock farming. This is primarily due to insects' requirement for fewer resources to produce an equivalent amount of protein and/or to achieve substantial body mass gain. It is well established in scientific literature that overall, insect rearing requires significantly less land, water, and feed resources while also emitting lower greenhouse gas emissions (Berggren, Jansson, & Low, 2019). For instance, farmed insects consume approximately 1.5, 2.5, and 5 times less water than poultry, pork, and beef, respectively, to produce the same amount (e.g., 1 kg) of protein. Likewise, in terms of agricultural land use, poultry and pork require nearly 3.5 times more land than insects, while for beef, this requirement is approximately 14 times greater (Ros-Baró et al., 2022). This makes insect biomass production a viable option for contributing to more sustainable food systems, particularly in the face of growing global population pressures and the need to reduce the environmental impact of food production. By integrating edible insects into human diets, we can improve nutritional intake and address deficiencies, as well as promote a more resilient and eco-friendly approach to food and feed production.

The overall nutrient composition of insects, classified by order, is listed in Table 5.1, where it can be seen that the protein content can vary between 35% up to 61%. For example, the species of the order Orthoptera (crickets, grasshoppers, locusts) have an average protein of 61%, with protein contents of some species being up to 77% (e.g., *Sphenarium histrio, Melanoplus mexicanus*). They are followed by the orders Odonata (e.g., dragonflies) and Diptera (e.g., flies) with 55% and 49% protein, respectively. In contrast, the species of the order Blattodea (e.g., termites, cockroaches) have an average protein content of 35% (Liceaga, 2022); Rumpold and Schlüter (2013). The variation in the protein content is influenced by the insect species and its life-cycle stage, as well as what the insects are fed, environment conditions, and the protein measurement methods used. On the other hand, other authors (Nongonierma & FitzGerald, 2017; van Huis, 2016) have reported that the protein content of insects, measured on a dry-matter basis, varies from 7% to 91%, with many species averaging around 60% protein. Recently, researchers reported that flours obtained from different body portions of crickets (*Acheta domesticus* L.) contained different protein contents. For example, leg and antennae flour had a much higher protein content (73% d.w.) than head and body flour (66% d.w.). This information will be valuable for the incorporation of insect protein in food and animal feed applications as separation of insects based on their body's composition will yield different protein contents aimed at different food/feed formulations (Brena-Melendez, Garcia-Amezquita, Liceaga, Pascacio-Villafán, & Tejada-Ortigoza, 2024).

The protein digestibility of edible insects is also an important factor that needs to be considered. Researchers have shown protein digestibility values ranging between 86% and 90%, measured by true faecal nitrogen levels in rat models. These levels are higher than many plant-based proteins but slightly lower than those observed for certain animal proteins, such as egg (95%), beef (98%), and casein (99%), which are considered the benchmark for protein quality and digestibility (Finke, 2013). This lower protein digestibility values have been attributed to some of the protein being complexed with the chitin polymerized matrix. Thus, efforts are made to separate the chitin-complexed protein by a variety of physico-chemical methods (e.g., extrusion, enzymatic hydrolysis), making insects a competitive, alternative to traditional animal proteins, with potential applications in developing balanced and sustainable dietary options. However, further studies on the bioaccessibility (i.e., protein released during gastrointestinal digestion) and bioavailability (i.e., absorbed protein, metabolized by the human body) are needed to fully appreciate the protein value from insects.

Lipids constitute the second largest component of the nutrient composition in edible insects varying from 13% in Orthoptera (such as grasshoppers, crickets, and locusts) to 33% in Coleoptera (including beetles and grubs) (Rumpold & Schlüter, 2013; van Huis, 2016). The lipid content of insects (Table 5.1) will vary depending on insect order, species, life-cycle, migratory flight stage, among others. For example, within the order Blattodea, African termites on average are reported to have 46% lipid content, whereas tree termites are reported with a lower 2% lipid content. Lipids in insects are primarily found in the form of triglycerides (about 80% of the total lipid content), which tend to serve as an energy source during periods of high energy requirements (e.g., during reproduction stages and long flights) (Sosa & Fogliano, 2017). After triglycerides, phospholipids represent the next

TABLE 5.1 Proximate Composition (%) of Some Examples of Edible Insects Classified by Order.

ORDER	*SCIENTIFIC NAME*	*COMMON NAME*	*MOISTURE*	*PROTEIN*	*LIPIDS*	*FIBRE*	*ASH*
Orthoptera	*Gryllodes sigillatus*	Tropical banded cricket	11.2	56.8	6.1	7.8	18.1
	Acheta domesticus	House cricket	69.2	66.6	22.1	10.2	3.6
	Brachytrupes sp.	Cricket	73.3	47.9	21.3	13.5	9.4
	Gryllotalpa africana	Mole cricket	71.2	53.5	21.9	9.7	9.4
	Sphenarium histro	Grasshoppers	77.0	4.0	ND	12.0	2.0
Blattodea	*Cortaritermes silvestri*	South American termites	77.8	48.6	6.9	ND	8.5
	Macrotermes bellicosus	African termites	6.0	34.8	46.1	ND	10.2
	Nasutitermes corniger	Tree termite	75.3	66.7	2.2	27.1	4.6
Lepidoptera	*Agrotis infusa*	Bogong moth	49.2	52.7	39.0	5.3	ND
	Anaphe panda	African moth	73.9	45.6	35.0	6.5	3.7
	Ascalapha odorata	Black witch moth	56.0	15.0	ND	12.0	6.0
	Bombyx mori	Silkworm	82.7	53.8	8.1	6.4	6.4
	Pseudaletia unipuncta	Army worm	2.0	55.5	15.2	5.1	7.0
Coleoptera	*Oileus rimator*	Beetle	26.0	36.0	ND	15.0	3.0
	Passalus punctiger	Beetle	26.0	44.0	ND	15.0	3.0
	Tenebrio molitor	Yellow mealworm	63.7	65.3	14.9	20.4	3.3
	Zophobas morio	Darkling beetle	57.9	46.8	42.0	6.3	2.4
Hymenoptera	*Apis mellifera*	European honeybee	76.8	40.5	20.3	1.3	3.4
	Atta mexicana	Leaf-cutter ant	46.0	39.0	ND	11.0	4.0
	Olecophylla smaragdina	Weaver ant	74.0	53.5	13.5	6.9	6.5
	Oecophylla virescens	Green tree ant	78.32	41.0	26.7	ND	6.0
	Polybia sp.	Wasp	63.0	13.0	ND	15.0	6.0
Diptera	*Drosophila melanogaster*	Common fruit fly	67.1	56.3	17.9	ND	5.2
	Hermetia illucens	Black soldier fly	3.8	47.0	32.6	6.7	8.6

ND, not determined/reported.

Source: Finke, 2002, 2004, 2007, 2013; Ramos-Elorduy, 2009.

major lipid group in insects, followed by steroids. Insects are also reported to be a source of cholesterol (Antunes et al., 2024; Tzompa-Sosa et al., 2021). Particularly, it is reported that the fatty acids found in insects are generally similar to those in poultry and fish in terms of unsaturation, but insects have a higher content of monounsaturated fatty acids (i.e., palmitoleic acid and oleic acid) and polyunsaturated fatty acids (i.e., linolenic acid and linoleic acid) (Antunes et al., 2024; Dobermann, Swift, & Field, 2017; Rumpold & Schlüter, 2013). The main saturated fatty acids found in insects are C16:0 (palmitic acid) and C 18:0 (stearic acid) (Finke & Oonincx, 2023). In this context, edible insects are a good source of essential fatty acids, including omega-3 and omega-6 fatty acids (e.g., eicosapentaenoic acid), which are important for cardiovascular health, brain function, and reducing inflammation (Riekkinen, Väkeväinen, & Korhonen, 2022). These essential fatty acids have been reported in insect larva meal at comparable values to those for fish and even higher than those reported for cattle and pigs (Antunes et al., 2024).

Overall, in addition to being a good protein source, based on the lipid content of insects, edible insects can also be considered as a good energy source and provide essential unsaturated fatty acids to meet human nutrition needs. Similar to the protein content of edible insects, the lipid content varies between species and developmental stages but also from the feedstuff, environmental conditions, and lipid/fatty acid determination methods (Antunes et al., 2024; Finke & Oonincx, 2023; Sosa & Fogliano, 2017).

Perhaps the major type of carbohydrate found in edible insects corresponds to the overall fibre content mainly present in the form of chitin. Chitin, the major compound in the exoskeleton of insects, is a polysaccharide consisting of β-(1–4) N-acetyl-D-glucosamine units that form a linear polymer through the covalent β-1,4-linkages (Figure 5.2). It is suggested that chitin represents the main fibre content in insects due to chitin's structural similarity to plant cellulose (linear polymer of β-(1→4)-D-glucopyranose units) (Finke & Oonincx, 2023). Chitin is an abundant polysaccharide, but its inability to dissolve in water due to this tight packing structure limits its applicability. It can somewhat be digested by humans, but it mostly constitutes as a fibre source in the human body (Yhoung-aree, 2008). As a result, for commercial applications, chitin is frequently converted to its deacetylated counterpart, chitosan, due to its advanced properties. The chitin deacetylation process involves defatting, demineralization, decolorization, deproteinization, and deacetylation.

Chitin and chitosan have been successfully extracted from crickets, grasshoppers, cicada sloughs, silkworms, beetles, and black soldier flies (Chae, Shin, & Shin, 2018; Ibitoye et al., 2018; Malm & Liceaga, 2021; Y.-S. Park et al., 2017); however, its content will vary depending on the insect type and species (Table 5.1). The synthesis and modification of chitin will depend on the species, growth, and development stage of insects. For example, it can range from 5% for Blattodea to 13% for Hemiptera (Rumpold & Schlüter, 2013). Compared to crustacean (shellfish) chitosan, insect chitosan is reported to be smaller. This is an important physicochemical feature to consider as it can have an effect on the chitosan's solubility and biological activity (e.g., antimicrobial and antioxidant properties) (Bumgardner et al., 2017; Hosseinnejad & Jafari, 2016; Vinsova & Vavrikova, 2011; Xing et al., 2005). For example, Park et al. (2017) reported a molecular weight of 308 kDa for African field cricket chitosan, whereas cicada sloughs, grasshoppers, mealworms, and silkworm chrysalis had chitosan with a smaller molecular weight at ca. 4 kDa (Luo et al., 2019). In addition, Malm and Liceaga (2021) identified differences in the molecular weight of chitosan obtained from two cricket species, where *Gryllodes sigillatus* chitosan was larger (524 kDa) than *Acheta domesticus* chitosan (344 kDa). The

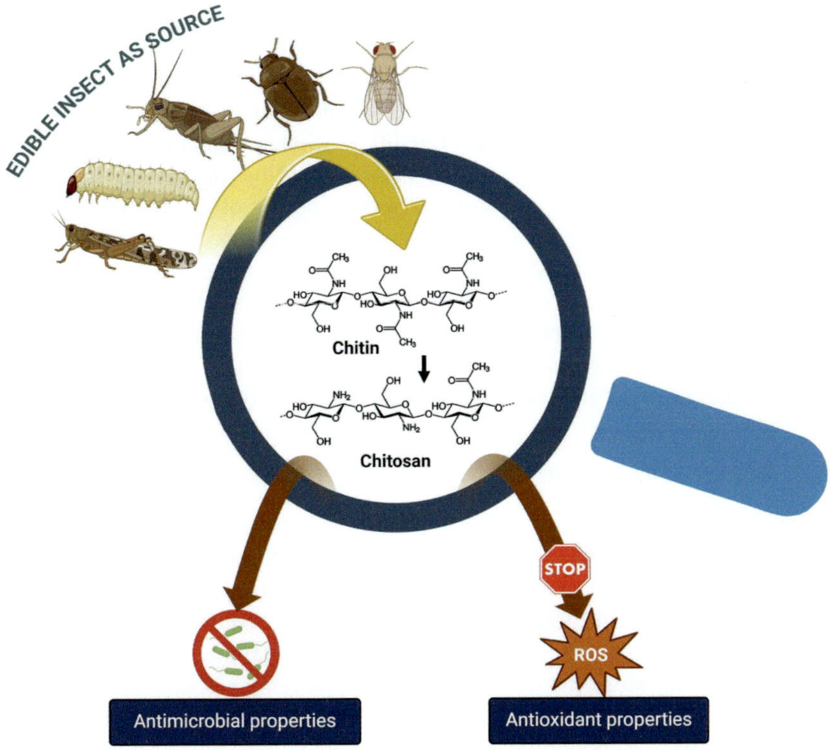

FIGURE 5.2 Insect chitin and chitosan structure. Created in BioRender. Liceaga, A. (2025).

differences in molecular weight are believed to be related to their living environments such as aerodynamics required for jumping and flying, thus their chitin cuticles tend to be arranged differently, influencing the molecular weight (Berezina & Hubert, 2019).

5.2.2 Micronutrients

5.2.2.1 Vitamin Content

Vitamin content of edible insects seems to differ in literature, and it is directly affected by the feedstuff used in the rearing of insects (e.g., using food waste materials versus commercial feed) (Baiano, 2020). A systematic review by (Weru, Chege, & Kinyuru, 2021) identified that vitamin E had the highest value overall (0.925 mg/100 g) for the termite (*Nasutitermes* spp.), whereas vitamin C had the lowest (0.0046 mg/100 g) for the termite (*Macrotermes bellicosus*). Insects are reported to have a variety of B vitamins (including B12, which is typically lacking in plant-based diets). Although thiamine is relatively unstable, it has been identified in several insects, such as palm weevil butterworms, superworms, silkworms, mealworm larvae, and waxworms. In contrast, riboflavin was reported at levels ranging from 17.6 to 306.3 mg/kg d.w. in caterpillars, palm weevil larvae, and termites. Schmidt, Call, Macheiner, and Mayer (2019) reported that levels of vitamin B12 were 1.08 μg/100 g for mealworm, 2.88 μg/100 g for cricket, and 0.84 μg/100 g for grasshopper. Likewise, Lenaerts, Van Der Borght, Callens, and Van Campenhout (2018) found values of vitamin B12 between 0.31 and 0.85 μg/100 g dry weight for mealworm (larvae). These values align closely with the range found in other animal-based foods, such as pork (1.0 μg/100 g) or fish-like mackerel (9.0 μg/100 g) (Souci, Fachmann, & Kraut, 2024). The majority of recorded vitamin A (retinol) levels in edible insects are low, with some reports indicating values as low as 300 μg of retinol per kg of dry matter (Finke, 2013). While edible insects generally exhibit low retinol content, insects harvested in the wild were found to have high concentrations of carotenoids, such as β-carotene, a precursor of vitamin A. In contrast, commercially raised insects tended to contain lower levels of carotenoids (Finke, 2013; Oonincx, van Broekhoven, van Huis, & van Loon, 2015). This information highlights the important role that feedstuff can have on nutrient composition and the role of a controlled diet versus plants and other resources in the wild.

5.2.2.2 Mineral Content

Insects are also abundant in minerals like iron, zinc, magnesium, and calcium, which are critical for oxygen transport, immune response, bone health, and enzyme activity (Nowakowski, Miller, Miller, Xiao, & Wu, 2022). Edible insects contain minerals, classified as macro-minerals and trace elements, in varying amounts. Macro-minerals (e.g., calcium, phosphorus, and magnesium) are crucial for maintaining the skeletal structure of vertebrates, while chloride, potassium, and sodium function as electrolytes and help regulate the body's acid-base balance. Essential trace elements (i.e., micro-minerals), including iron, zinc, copper, manganese, iodine, and selenium, are involved in various functions, from oxygen transport to serving as co-factors for several enzymes (Finke & Oonincx, 2023). Several edible insects can provide considerable levels of trace elements, such as copper, iron, manganese, and zinc, which are essential for humans. (Akhtar & Isman, 2018) found that crickets offer about 15 times more magnesium and three times more iron than beef, while soldier flies provide 71 times more calcium and twice as much magnesium and zinc as beef. According to (van Huis, 2016), mineral deficiencies (e.g., iron and zinc), which are common in developing countries, can have severe consequences for pregnant women and young children. Therefore, consuming edible insects, such as termites and crickets that are rich in these minerals, could play a significant role in alleviating these nutrient deficiencies.

5.3 POTENTIAL HEALTH BENEFITS OF EDIBLE INSECTS

The health benefits of edible insects go beyond basic nutrition, as they are also a rich source of bioactive compounds that can positively impact various aspects of health. Table 5.2 highlights some of the studies available in scientific literature related to the diverse health benefits associated with different compounds derived from edible insect species. Among the bioactive compounds that have been reported in different edible insects, these include peptides, chitin, chitosan, fatty acids, and polyphenols (Aguilar-Toalá et al., 2022). Also,

TABLE 5.2 Recent Scientific Literature Related to the Health Benefits of Different Compounds Present in Some Edible Insect Species.

BIOACTIVITY/ HEALTH BENEFIT	*INSECT SPECIES*	*SAMPLE USED OR BIOACTIVE COMPOUND IDENTIFIED*	*TYPE OF STUDY*	*REFERENCE*
Antioxidant	Mealworms (*Tenebrio molitor*) larvae	Protein	*In vitro* and cell line model	(Hoon Lee, Kim, Kim, Song, & Choi, 2024)
	Caterpillars (Gynanisa maja, Gonimbrasia belina)	Protein hydrolysates	*In vitro*	(Mokaya, Mudalungu, Tchouassi, & Tanga, 2024)
	Beetle (*Protaetia brevitarsis*)	Protein	Cell line model	(Lee, Kim, Kang et al., 2023)
	Crickets (*Gryllus bimaculatus*)	Protein	*In vitro*	(Fashakin et al., 2023)
	Mealworms (*Tenebrio molitor*), silkworms (*Bombyx mori*), giant worms (*Zophobas morio*)	Water extract	Cell line model	(Veronica D'Antonio et al., 2023)
	Crickets (*Gryllodes sigillatus*)	Bioactive peptides	*In vitro and in vivo*	(F. Hall et al., 2018; F. Hall et al., 2020; Mudd et al., 2022)
	Green beetle (*Mimela* sp.)	Aqueous extract	*In vivo*	(Tukshipa, Chakravorty, & Prveen, 2022)
	Mealworm (*Tenebrio molitor*) grasshopper (*Sphenarium purpurascens*)	Flour fermented with *Lactococcus lactis* strains	*In vitro*	(Mendoza-Salazar et al., 2021)
	House cricket *(Acheta domesticus*)	Polyphenolic methanolic extracts	*In vitro*	(Nino et al., 2021)
	House cricket *(Acheta domesticus*) and mealworms (*Tenebrio molitor*)	Polyphenolic ethanol and ethanol: water extracts	*In vitro*	(del Hierro et al., 2020)
	Mealworms (*Tenebrio molitor*) and Chinese beetle (*Ulomoides dermestoides*)	Extract with main components of saponins, carbohydrates, and proteins	*In vitro*	(Flores et al., 2020)
	Mealworms (*Tenebrio molitor*), buffalo worms (*Alphitobius diaperinus*), Palm worm larvae (*Rhynchophorus ferrugineus*), Evening cicada (*Tanna japonensis*), Black ants (*Lasius niger*), African caterpillars (*Imbrasia oyemensis*)	Water and liposoluble extracts	*In vitro*	(Di Mattia et al., 2019)

(*Continued*)

TABLE 5.2 (Continued)

BIOACTIVITY/ HEALTH BENEFIT	*INSECT SPECIES*	*SAMPLE USED OR BIOACTIVE COMPOUND IDENTIFIED*	*TYPE OF STUDY*	*REFERENCE*
	Silkworm (*Bombyx mori*), Grasshoppers (*Calliptamus italicus*), Crickets (*Acheta domesticus*), Mini crickets (*Acheta domesticus*), Giant water bugs (*Lethocerus indicus*), Scolopendra gigantea (*Scolopendra*)			
	Silkworm (*Bombyx mori*)	Oil	*In vivo*	(Nguyen et al., 2016)
Immunomodulatory	Beetle (*Protaetia brevitarsis*)	Acidic polysaccharide	*In vivo*	(Olawuyi et al., 2025)
	Bug (*Coridius nepalensis*)	Aqueous extract	*In vivo*	(Chakravorty et al., 2024)
	Beetle (*Protaetia brevitarsis*)	Protein	Cell line model	(Lee, Kim, Kim et al., 2023)
	Green beetle (*Mimela* sp.)	Aqueous extract	*In vivo*	(Tukshipa et al., 2022)
	Bee pupae	Polypeptide components	*In vitro* and *in vivo*	(Chen et al., 2022)
	Mealworms (*Tenebrio molitor*)	Supercritical fluid CO_2 extract	*In vivo*	(Q. F. Tang & Dai, 2016)
	Weaver ants (*Polyrhachis dives*)	13 non-peptide nitrogen compounds (most were identified as alkaloids)	*In vitro*	(J.-J. Tang et al., 2015)
Anti-inflammatory	Bug (*Coridius chinensis*)	Hydro-alcoholic extract	*In vitro*	(Bonysana et al., 2024)
	Mealworms (*Tenebrio molitor*)	Bioactive peptides	Cell line model	(Y. Yoon et al., 2023)
	Beetle (*Protaetia brevitarsis*)	Ethanolic extract	Cell line model	(Choi & Bae, 2023)
	Black soldier fly (*Hermetia illucens*)	Oil	*In vivo*	(Xu et al., 2021)
	Crickets (*Gryllodes sigillatus*)	Bioactive peptides	*In vitro*	(F. Hall et al., 2018; F. Hall et al., 2020)
	Weaver ants (*Polyrhachis dives*)	13 non-peptide nitrogen compounds (most were identified as alkaloids)	*In vitro*	(J.-J. Tang et al., 2015)

(Continued)

TABLE 5.2 (Continued)

BIOACTIVITY/ HEALTH BENEFIT	*INSECT SPECIES*	*SAMPLE USED OR BIOACTIVE COMPOUND IDENTIFIED*	*TYPE OF STUDY*	*REFERENCE*
Antimicrobial	Beetle (*Pachnoda sinuata*) and crickets (*Gryllus bimaculatus* and *Scapsipedus icipe*)	Sterols	*In vitro*	(Mudalungu et al., 2023)
	House cricket (*Acheta domesticus*) and tropical banded cricket (*Gryllodes sigillatus*)	Chitin and chitosan	*In vitro*	(Morgan Malm & Andrea M. Liceaga, 2021)
	Mealworms (*Tenebrio molitor*) and Chinese beetle (*Ulomoides dermestoides*)	Extract with main components of saponins, carbohydrates, and proteins	*In vitro*	(Flores et al., 2020)
Antiobesity	Mealworms (*Tenebrio molitor*)	Oil and fermented extract	*In vivo*	(Mun et al., 2024)
	Mealworms (*Tenebrio molitor*)	Powder	*In vivo*	(B. M. Park, Lim, & Lee, 2021)
	House cricket *(Acheta domesticus*) and -mealworms (*Tenebrio molitor*)	Polyphenolic ethanol and ethanol: water extracts	*In vitro*	(del Hierro et al., 2020)
	Mealworms (*Tenebrio molitor*)	Ethanol extract	*In vitro* and *in vivo*	(Seo et al., 2017)
Antihypertensive	Mealworm (*Tenebrio molitor*) and grasshopper (*Sphenarium purpurascens*)	Flour fermented with *Lactococcus lactis* strains	*In vitro*	(Mendoza-Salazar et al., 2021)
	Crickets (*Gryllodes sigillatus*)	Bioactive peptides	*In vitro*	(F. Hall et al., 2018; F. Hall et al., 2020)
Hypolipidemic	House cricket (*Acheta domesticus*) and tropical banded cricket (*Gryllodes sigillatus*)	Chitin and chitosan	*In vitro*	(Morgan Malm & Andrea M. Liceaga, 2021)
	Housefly larvae (*Musca domestica* vicina)	Protein-rich extract	*In vivo*	(Attia et al., 2020)
	Clanis bilineata	Chitooligosaccharides	*In vivo*	(Xia et al., 2013)
Anti-adipogenic	Mealworms (*Tenebrio molitor*)	Ethanol extract	*In vitro* and *in vivo*	(Seo et al., 2017)
Anti-Parkinson activity	Silkworm (*Bombyx mori*)	Powder	*In vivo*	(Nguyen et al., 2016)
Anti-dyslipidemia	Silkworm (*Bombyx mori*)	Oil	*In vivo*	(Nguyen et al., 2016)

there are reports that bioactive extracts from edible insects, containing a mixture of the above mentioned compounds, can also exhibit some bioactivities (Di Mattia, Battista, Sacchetti, & Serafini, 2019; Flores, Casados, Velasco, Ramírez, & Velázquez, 2020). These bioactive compounds or extracts are reported to have antioxidant, antihypertensive, anti-inflammatory, antimicrobial, and immunomodulatory bioactivities (del Hierro, Gutiérrez-Docio, Otero, Reglero, & Martin, 2020; Nino, Reddivari, Ferruzzi, & Liceaga, 2021). Other authors (D'Antonio, Battista, Sacchetti, Di Mattia, & Serafini, 2023; Zhou et al., 2022) have indicated that edible insects have other functional effects such as blood glucose and lipid regulation, regulation of intestinal microbiota, and cardiovascular protection, among others. From Table 5.2, the antioxidant activity seems to be one of the most commonly investigated bioactivities across studies, followed by antihypertensive, anti-inflammatory, antimicrobial, and immunomodulatory effects. Similarly, a significant portion of the research has concentrated on yellow mealworms (*Tenebrio molitor*) and house crickets (*Acheta domesticus*), since these are the two of the most widely used species in commercial insect biomass production (Liceaga, 2022).

The most used method for obtaining bioactive peptides from edible insects has been through enzymatic proteolysis. In this regard, Yoon, Wong, Chae, and Auh (2019) compared the bioactivities of hydrolysates from yellow mealworm, crickets, and silkworm pupae. They found that hydrolysis of the proteins in the three insect species increased the inhibitory activity towards angiotensin converting enzyme (ACE) and α-glucosidase. Among the protein hydrolysates, those obtained using the commercial enzyme Alcalase showed the greatest inhibition towards ACE. With respect to the inhibition of α-glucosidase, Alcalase-treated silkworm pupae and the mixture of Flavourzyme and Alcalase-treated mealworms were determined to be the most effective inhibitors. Finally, in this investigation into the anti-inflammatory effects of the insect proteins, only silkworm pupae showed anti-inflammatory activity. Likewise, Ahn et al. (2020) reported that field cricket glycosaminoglycan can have *in vivo* antioxidant effect in the diabetic state by lowering blood glucose, low-density lipoprotein (LDL)-cholesterol, and oxidative damage markers while significantly enhancing antioxidative enzymes. Similarly, it was observed that the administration of cricket (*G. bimaculatus*) extract attenuated the alcohol-induced steatosis and apoptotic responses in the liver and intestinal permeability to bacterial endotoxin. These protective effects were associated with suppression of reactive oxygen species (ROS)-mediated oxidative stress in both biological samples.

A significant number of studies have explored the antioxidant potential of edible insects. Extracts and protein fractions from species like *Tenebrio molitor* (mealworms), *Gryllus bimaculatus* (crickets), and *Bombyx mori* (silkworms) have demonstrated high antioxidant activity (Table 5.2). These were tested using *in vitro* models, cell line assays, and, in some cases, *in vivo* models. Cricket (*Gryllodes sigillatus*) bioactive peptides (Hall, Johnson, & Liceaga, 2018; Hall, Reddivari, & Liceaga, 2020; Mudd, Martin-Gonzalez, Ferruzzi, & Liceaga, 2022), cricket (*Acheta domesticus*) polyphenols (del Hierro et al., 2020; Nino et al., 2021), and fermented grasshopper (*Sphenarium purpurascens*) and mealworm (*Tenebrio molitor*) flours (Mendoza-Salazar et al., 2021) were among the key preparations evaluated, showcasing their capacity to provide antioxidant activity, improve cellular health (using caco-2 cells and RAW 264.7 macrophage cells), and mitigate oxidative stress *in vivo* (using a *Caenorhabditis elegans* model). (Kaya, Erdogan, Mol, & Baran, 2015) also demonstrated the scavenging potential of chitosan from seven Orthoptera species. Likewise, the immunomodulatory effects of edible insects have been documented in studies using acidic polysaccharides (Olawuyi et al., 2025), polypeptides (Chen et al., 2022), and aqueous extracts (Chakravorty, Tukshipa, Bhadra, & Muni, 2024). For instance, beetle *(Protaetia brevitarsis)* (Lee, Kim, Kang, Kim, & Choi, 2023) and bee pupae-derived extracts have shown immune-enhancing properties in both *in vitro* and *in vivo* models (Chen et al., 2022). These findings suggest their potential role in modulating immuno-responses and combating immune-related disorders. Likewise, anti-inflammatory properties were reported for species such as *Tenebrio molitor* (Yoon, Oh, Park, Kim, & Kim, 2023) and *Hermetia illucens* (Xu, Ji, Belghit, Liland, Wu, & Li, 2021), with bioactive compounds, like peptides and ethanolic extracts, able to decrease inflammatory markers in cell models and animal studies. These findings indicate their potential use in managing chronic inflammatory conditions.

The antimicrobial properties of sterols, chitin, and chitosan extracted from insects like *Acheta domesticus* and *Pachnoda sinuata* have been evaluated *in vitro* for their antimicrobial action against a range of microbial pathogens, underscoring their utility in food preservation and health applications. For example, grasshopper chitin was effective against Gram-negative bacteria and *S. enteritidis* and *V. alginolyticus* (Kaya et al., 2015; Malm & Liceaga, 2021; Mudalungu, Mokaya, & Tanga, 2023). Similarly, extracts from insects like *Tenebrio molitor* and *Sphenarium purpurascens* were tested for their role in weight management and blood pressure regulation. Fermented flours, ethanolic extracts, and peptides were found to inhibit adipogenesis and exhibit hypotensive effects, supporting their use as functional foods for metabolic health (Mendoza-Salazar et al., 2021; Seo et al., 2017), while (Xia, Chen, & Wu, 2013) determined that insect chitin and chitosan could help lower plasma LDL-cholesterol in the prevention of cardiovascular disease.

Other bioactivities such as anti-dyslipidemia, hypolipidemic, and anti-adipogenic effects were observed in studies on *Acheta domesticus*, *Gryllodes sigillatus,* and *Musca domestica* vicina (Attia, El-Samad, & Zaghloul, 2020; Malm & Liceaga, 2021). Notably, silkworm extracts were linked to improved lipid profiles and enhanced health span in animal models. Unique bioactivities like anti-Parkinson effects were also attributed to silkworm powder in *in vivo* studies (Nguyen et al., 2016). The findings summarized in Table 5.2 reveal the significant potential of edible insects as functional food ingredients and natural therapeutic agents. The extensive use of both in vitro and in vivo models provides robust evidence of their bioactivities, paving the way for further research into their mechanisms of action, safety, and efficacy in human applications. These studies also support the integration of insect-derived products into sustainable diets, addressing global challenges related to nutrition, health, and food security. The variety of bioactive compounds examined, ranging from proteins and peptides to polysaccharides and oils, demonstrates the versatility of edible insects as sources of health-promoting molecules.

5.4 FUTURE DIRECTIONS AND RESEARCH GAPS

While edible insects with high nutritional value are commonly consumed in many countries, food neophobia and biases against eating insects continue to restrict their acceptance as food sources worldwide (Sogari et al., 2023). To address this, the food industry has considered the processing of insects in order to alter their visual (i.e., recognizable) form by grinding them into powders or extracting their protein and other nutrients before incorporating them into conventional foods.

Edible insects are a good nutrition source, particularly of high-quality proteins with noticeable digestibility, essential amino acids, as well as a good source of unsaturated fatty acids, fibre, vitamins, and minerals. Their versatile nutritional compounds allow them to be used as ingredients or substitutes to enhance the nutritional value of other foods and feed. For instance, bread made with crickets (*Acheta domesticus*) showed higher levels of protein, fat, fibre, and dietary fibre compared to standard bread (Osimani et al., 2018). Similarly, the crude fibre and protein content of egg pasta was enhanced by supplementing it with mealworm (*Tenebrio molitor*) and grasshopper (*Locusta migratoria*) powder (Çabuk & Yılmaz, 2020). In contrast, incorporating silkworm (*Bombyx mori*) pupae into pork meat batter has been shown to enhance the protein content of the meat product (Park et al., 2017). Likewise, substituting 10% of lean meat with pre-treated mealworms and silkworms in an emulsified sausage was found to significantly boost its protein and mineral content (Kim, Setyabrata, Lee, Jones, & Kim, 2016). Therefore, more research is necessary to evaluate the incorporation of edible insects for the enhancement of the nutritional value of plant-based or animal-based foods.

The diversification in product offerings aligns with a growing market trend for sustainable and eco-friendly solutions across the food, feed, and agricultural sectors. These data suggest a positive consumer shift towards insects as alternative, sustainable protein sources, with companies adopting innovative strategies to modify insect appearances and forms such as powders, protein bars, and oils to improve consumer acceptance and utilization in various applications.

5.5 CONCLUSION

In conclusion, edible insects present a promising alternative as a sustainable and nutritionally rich food source. With a history of widespread consumption across many cultures, particularly in Asia, Africa, and Latin America, insects offer a range of essential nutrients, including high-quality proteins, fats, fibre, vitamins, and minerals that contribute to human health. Insects provide complete proteins with all essential amino acids, a significant level of polyunsaturated fatty acids, a food source of chitin, and an assortment of vitamins and minerals that could help address nutritional deficiencies and support overall bodily functions. Additionally, commercial insect biomass production is environmentally beneficial, requiring fewer resources and emitting less greenhouse gases compared to conventional livestock, making it a viable solution to the food demands of a growing global population. In addition to their high nutritional content, edible insects have shown to provide additional health benefits such as antioxidant, anti-inflammatory, and antihypertensive, among others. However, until human (clinical) trials are conducted, it is still too early to assert the full extent of these health benefits.

Finally, the food industry has innovated processing of insects into flours, powders, protein hydrolysates, oils, and other forms that make them less visual (recognizable) and more acceptable to consumers, particularly in Western regions where food neophobia remains a major barrier. The availability of insect-based products, from snacks and energy bars to insect-derived oils and fertilizers, reflects a growing consumer shift towards sustainable food options. As awareness increases around the environmental and health benefits of edible insects, these products have the potential to play a critical role in achieving global food security, improving human nutrition, and contributing to more sustainable food systems. The insect-based product market is actively evolving, with companies adapting to meet sustainability goals across food, feed, and agriculture sectors; as companies continue to innovate, insect-based products are likely to become a more common solution in sustainable food systems.

REFERENCES

Aguilar-Toalá, J. E., Cruz-Monterrosa, R. G., & Liceaga, A. M. (2022). Beyond human nutrition of edible insects: Health benefits and safety aspects. *Insects*, *13*(11). https://doi.org/10.3390/insects13111007

Ahn, M. Y., Kim, B. J., Kim, H. J., Jin, J. M., Yoon, H. J., Hwang, J. S., & Lee, B. M. (2020). Anti-diabetic activity of field cricket glycosaminoglycan by ameliorating oxidative stress. *BMC Complementary Medicine and Therapies*, *20*(1), 232. https://doi.org/10.1186/s12906-020-03027-x

Akhtar, Y., & Isman, M. B. (2018). Insects as an alternative protein source. In *Proteins in food processing* (pp. 263–288). Elsevier.

Alexander, P., Brown, C., Arneth, A., Dias, C., Finnigan, J., Moran, D., & Rounsevell, M. D. A. (2017). Could consumption of insects, cultured meat or imitation meat reduce global agricultural land use? *Global Food Security*, *15*, 22–32. https://doi.org/10.1016/j.gfs.2017.04.001

Antunes, A. L. M., Mesquita, B. M. A. D. C., Fonseca, F. S. A. D., Carvalho, L. M. D., Brandi, I. V., Carvalho, G. G. P. D., & Coimbra, J. S. D. R. (2024). Extraction and application of lipids from edible insects. *Critical Reviews in Food Science and Nutrition*, 1–9.

Ardoin, R., & Prinyawiwatkul, W. (2021). Consumer perceptions of insect consumption: a review of western research since 2015. *International Journal of Food Science & Technology*, *56*(10), 4942–4958. https://doi.org/10.1111/ijfs.15167

Attia, A. A., El-Samad, L. M., & Zaghloul, N. (2020). Effects of protein extract from the housefly-larvae (*Musca domestica vicina*) on thehyperlipidemic mice induced by Triton WR–1339. *The Swedish Journal of Scientific Research*, *1*, 16–27.

Baiano, A. (2020). Edible insects: An overview on nutritional characteristics, safety, farming, production technologies, regulatory framework, and socio-economic and ethical implications. *Trends in Food Science & Technology*, *100*, 35–50.

Berezina, N., & Hubert, A. (2019). Marketing and regulations of chitin and chitosan from insects. *Chitin and Chitosan: Properties and Applications*, 477–489.

Berggren, A., Jansson, A., & Low, M. (2019). Approaching ecological sustainability in the emerging insects-as-food industry. *Trends in Ecology and Evolution*, *34*(2), 132–138. https://doi.org/10.1016/j.tree.2018.11.005

Bonysana, R., Singh, K. D., Devi, W. D., koijam, A. S., Kapesa, K., Kalita, J.,. . . . Rajashekar, Y. (2024). Ethno-entomotherapeutic and metabolite profiling of *Coridius chinensis* (Dallas), a traditional edible insect species of North-East India. *Scientific Reports*, *14*(1), 6545. https://doi.org/10.1038/s41598-024-57202-y

Brena-Melendez, A., Garcia-Amezquita, L. E., Liceaga, A., Pascacio-Villafán, C., & Tejada-Ortigoza, V. (2024). Novel food ingredients: Evaluation of commercial processing conditions on nutritional and technological properties of edible cricket (*Acheta domesticus*) and its derived parts. *Innovative Food Science & Emerging Technologies*, *92*, 103589.

Bumgardner, J. D., Murali, V. P., Su, H., Jenkins, O. D., Velasquez-Pulgarin, D., Jennings, J. A.,. . . . Jayakumar, R. (2017). 4 – Characterization of chitosan matters. In J. A. Jennings & J. D. Bumgardner (Eds.), *Chitosan based biomaterials* (Vol. 1, pp. 81–114). Woodhead Publishing.

Çabuk, B., & Yılmaz, B. (2020). Fortification of traditional egg pasta (erişte) with edible insects: Nutritional quality, cooking properties and sensory characteristics evaluation. *Journal of Food Science and Technology*, *57*(7), 2750–2757. https://doi.org/10.1007/s13197-020-04315-7

Chae, K. S., Shin, C. S., & Shin, W. S. (2018). Characteristics of cricket (*Gryllus bimaculatus*) chitosan and chitosan-based nanoparticles. *Food Science and Biotechnology*, *27*(3), 631–639. https://doi.org/10.1007/s10068-018-0314-4

Chakravorty, J., Tukshipa, S. D., Bhadra, P., & Muni, N. (2024). Immunomodulatory potential of *Coridius nepalensis* on cyclophosphamide induced mice: Most desired edible insect among ethnic people of arunachal Pradesh, India. *Food Science and Engineering*, 241–254.

Chen, Y., Zhao, J., Zhang, W., Zhao, T., Zhang, Q., Mao, G.,. . . . Wu, X. (2022). Purification of novel polypeptides from bee pupae and their immunomodulatory activity *in vivo* and *in vitro*. *Journal of Insects as Food and Feed*, *8*(10), 1117–1132. https://doi.org/10.3920/JIFF2021.0190

Choi, Y. J., & Bae, I. Y. (2023). White-spotted flower chafer (*Protaetia brevitarsis)* ameliorates inflammatory responses in LPS-stimulated RAW 264.7 macrophages. *Journal of Insects as Food and Feed*, *9*(8), 1037–1046. https://doi.org/10.3920/JIFF2022.0015

D'Antonio, V., Battista, N., Di Mattia, C. D., Sacchetti, G., Ramal-Sanchez, M., Prete, R.,. . . . Serafini, M. (2023). Edible insects and legumes exert an antioxidant effect on human colon mucosal cells stressed with 2,2′-azobis (2-amidinopropane)-dihydrochloride. *Frontiers in Nutrition*, *10*. https://www.frontiersin.org/journals/nutrition/articles/10.3389/fnut.2023.1219837

D'Antonio, V., Battista, N., Sacchetti, G., Di Mattia, C., & Serafini, M. (2023). Functional properties of edible insects: A systematic review. *Nutrition Research Reviews*, *36*(1), 98–119. https://doi.org/10.1017/S0954422421000366

del Hierro, J. N., Gutiérrez-Docio, A., Otero, P., Reglero, G., & Martin, D. (2020). Characterization, antioxidant activity, and inhibitory effect on pancreatic lipase of extracts from the edible insects *Acheta domesticus* and *Tenebrio molitor*. *Food Chemistry*, *309*, 125742. https://doi.org/10.1016/j.foodchem.2019.125742

Di Mattia, C., Battista, N., Sacchetti, G., & Serafini, M. (2019). Antioxidant activities in vitro of water and liposoluble extracts obtained by different species of edible insects and invertebrates. *6*. https://doi.org/10.3389/fnut.2019.00106

Dobermann, D., Swift, J., & Field, L. (2017). Opportunities and hurdles of edible insects for food and feed. *Nutrition Bulletin*, *42*(4), 293–308.

Fashakin, O. O., Tangjaidee, P., Unban, K., Klangpetch, W., Khumsap, T., Sringarm, K.,. . . . Phongthai, S. (2023). Isolation and identification of antioxidant peptides derived from cricket (*Gryllus bimaculatus*) protein fractions. *Insects*, *14*(8). https://doi.org/10.3390/insects14080674

Finke, M. D. (2002). Complete nutrient composition of commercially raised invertebrates used as food for insectivores. *Zoo Biology*, *21*(3), 269–285.

Finke, M. D. (2004). *Nutrient content of insects*. Springer.

Finke, M. D. (2007). Estimate of chitin in raw whole insects. *Zoo Biology*, *26*(2), 105–115.

Finke, M. D. (2013). Complete nutrient content of four species of feeder insects. *Zoo Biology*, *32*(1), 27–36. https://doi.org/10.1002/zoo.21012

Finke, M. D., & Oonincx, D. (2023). Insects as food for insectivores. In *Mass production of beneficial organisms* (pp. 511–540). Elsevier.

Flores, D. R., Casados, L. E., Velasco, S. F., Ramírez, A. C., & Velázquez, G. (2020). Comparative study of composition, antioxidant and antimicrobial activity of two adult edible insects from Tenebrionidae family. *BMC Chemistry*, *14*(1), 55. https://doi.org/10.1186/s13065-020-00707-0

Hall, F., Johnson, P. E., & Liceaga, A. (2018). Effect of enzymatic hydrolysis on bioactive properties and allergenicity of cricket (Gryllodes sigillatus) protein. *Food Chemistry*, *262*, 39–47. https://doi.org/10.1016/j.foodchem.2018.04.058

Hall, F. G., Jones, O. G., O'Haire, M. E., & Liceaga, A. M. (2017). Functional properties of tropical banded cricket (Gryllodes sigillatus) protein hydrolysates. *Food Chemistry*, *224*, 414–422. https://doi.org/10.1016/j.foodchem.2016.11.138

Hall, F., Reddivari, L., & Liceaga, A. M. (2020). Identification and characterization of edible cricket peptides on hypertensive and glycemic *in vitro* inhibition and their anti-inflammatory activity on RAW 264.7 macrophage cells. *Nutrients*, *12*(11). https://doi.org/10.3390/nu12113588

Hidalgo-Fuentes, B., de Jesús-José, E., Cabrera-Hidalgo, A. d. J., Sandoval-Castilla, O., Espinosa-Solares, T., González-Reza, R. M.,. . . . Aguilar-Toalá, J. E. (2024). Plant-based fermented beverages: Nutritional composition, sensory properties, and health benefits. *Foods*, *13*(6). https://doi.org/10.3390/foods13060844

Hoon Lee, J., Kim, Y.-J., Kim, T.-K., Song, K.-M., & Choi, Y.-S. (2024). Effect of ethanol treatment on the structural, techno-functional, and antioxidant properties of edible insect protein obtained from Tenebrio molitor larvae. *Food Chemistry*, *437*, 137852. https://doi.org/10.1016/j.foodchem.2023.137852

Hosseinnejad, M., & Jafari, S. M. (2016). Evaluation of different factors affecting antimicrobial properties of chitosan. *International Journal of Biological Macromolecules*, *85*, 467–475.

Ibitoye, E. B., Lokman, I. H., Hezmee, M. N. M., Goh, Y. M., Zuki, A. B. Z., & Jimoh, A. A. (2018). Extraction and physicochemical characterization of chitin and chitosan isolated from house cricket. *Biomed Mater*, *13*(2), 025009. https://doi.org/10.1088/1748-605X/aa9dde

Kaya, M., Erdogan, S., Mol, A., & Baran, T. (2015). Comparison of chitin structures isolated from seven Orthoptera species. *International Journal of Biological Macromolecules*, *72*, 797–805.

Kim, H.-W., Setyabrata, D., Lee, Y. J., Jones, O. G., & Kim, Y. H. B. (2016). Pre-treated mealworm larvae and silkworm pupae as a novel protein ingredient in emulsion sausages. *Innovative Food Science & Emerging Technologies*, *38*, 116–123. doi:https://doi.org/10.1016/j.ifset.2016.09.023

Kim, T. K., Yong, H. I., Kim, Y. B., Kim, H. W., & Choi, Y. S. (2019). Edible insects as a protein source: A review of public perception, processing technology, and research trends. *Food Science of Animal Resources*, *39*(4), 521–540. https://doi.org/10.5851/kosfa.2019.e53

Lee, J. H., Kim, T.-K., Kang, M.-C., Kim, B.-K., & Choi, Y.-S. (2023). Protective effects of edible insect protein extracts from Protaetia brevitarsis against H2O2-induced oxidative stress in mouse C2C12 myoblast cells. *Food Bioscience*, *52*, 102396. https://doi.org/10.1016/j.fbio.2023.102396

Lee, J. H., Kim, T.-K., Kim, Y. J., Kang, M.-C., Song, K.-M., Kim, B.-K., & Choi, Y.-S. (2023). Structural, physicochemical, and immune-enhancing properties of edible insect protein isolates from Protaetia brevitarsis larvae. *Food Chemistry: X*, *18*, 100722. https://doi.org/10.1016/j.fochx.2023.100722

Lenaerts, S., Van Der Borght, M., Callens, A., & Van Campenhout, L. (2018). Suitability of microwave drying for mealworms (Tenebrio molitor) as alternative to freeze drying: Impact on nutritional quality and colour. *Food Chemistry*, *254*, 129–136. https://doi.org/10.1016/j.foodchem.2018.02.006

Lesnik, J. J. (2019). *Edible insects and human evolution*. University Press of Florida.

Liang, Z., Zhu, Y., Leonard, W., & Fang, Z. (2024). Recent advances in edible insect processing technologies. *Food Research International*, *182*, 114137. https://doi.org/10.1016/j.foodres.2024.114137

Liceaga, A. M. (2021). Processing insects for use in the food and feed industry. *Current Opinion in Insect Science*, *48*, 32–36. https://doi.org/10.1016/j.cois.2021.08.002

Liceaga, A. M. (2022). Chapter Four – Edible insects, a valuable protein source from ancient to modern times. In J. Wu (Ed.), *Advances in food and nutrition research* (Vol. 101, pp. 129–152). Academic Press.

Liceaga, A. M., Aguilar-Toalá, J. E., Vallejo-Cordoba, B., González-Córdova, A. F., & Hernández-Mendoza, A. (2022). Insects as an alternative protein source. *Annual Review of Food Science and Technology*, *13*(1), 19–34. https://doi.org/10.1146/annurev-food-052720-112443

Luo, Q., Wang, Y., Han, Q., Ji, L., Zhang, H., Fei, Z., & Wang, Y. (2019). Comparison of the physicochemical, rheological, and morphologic properties of chitosan from four insects. *Carbohydrate Polymers*, *209*, 266–275. https://doi.org/10.1016/j.carbpol.2019.01.030

Malm, M., & Liceaga, A. M. (2021). Physicochemical properties of chitosan from two commonly reared edible cricket species, and its application as a hypolipidemic and antimicrobial agent. *Polysaccharides*, *2*(2), 339–353. https://doi.org/10.3390/polysaccharides2020022

Melgar-Lalanne, G., Hernández-Álvarez, A.-J., & Salinas-Castro, A. (2019). Edible insects processing: Traditional and innovative technologies. *Comprehensive Reviews in Food Science and Food Safety*, *18*(4), 1166–1191. https://doi.org/10.1111/1541-4337.12463

Mendoza-Salazar, A., Santiago-López, L., Torres-Llanez, M. J., Hernández-Mendoza, A., Vallejo-Cordoba, B., Liceaga, A. M., & González-Córdova, A. F. (2021). *In vitro* antioxidant and antihypertensive activity of edible insects flours (mealworm and grasshopper) fermented with *Lactococcus lactis* strains. *Fermentation*, *7*(3). https://doi.org/10.3390/fermentation7030153

Mokaya, H. O., Mudalungu, C. M., Tchouassi, D. P., & Tanga, C. M. (2024). Techno-functional and antioxidant properties of extracted protein from edible insects. *ACS Food Science & Technology*, *4*(5), 1130–1141. https://doi.org/10.1021/acsfoodscitech.3c00706

Mudalungu, C. M., Mokaya, H. O., & Tanga, C. M. (2023). Beneficial sterols in selected edible insects and their associated antibacterial activities. *Scientific Reports*, *13*(1), 10786. https://doi.org/10.1038/s41598-023-37905-4

Mudd, N., Martin-Gonzalez, F. S., Ferruzzi, M., & Liceaga, A. M. (2022). *In vivo* antioxidant effect of edible cricket (*Gryllodes sigillatus*) peptides using a *Caenorhabditis elegans* model. *Food Hydrocolloids for Health*, *2*, 100083. https://doi.org/10.1016/j.fhfh.2022.100083

Mun, S.-K., Jang, C. J., Jo, S., Park, S.-H., Sim, H. B., Ramos, S. C., ... Kim, J.-J. (2024). Anti-obesity and immunomodulatory effects of oil and fermented extract dried from *Tenebrio molitor* larvae on aged obese mice. *Animal Cells and Systems*, *28*(1), 340–352. https://doi.org/10.1080/19768354.2024.2374547

Nguyen, P., Kim, K.-Y., Kim, A. Y., Kim, N.-S., Kweon, H., Ji, S.-D., & Koh, Y. H. (2016). Increased healthspan and resistance to Parkinson's disease in *Drosophila* by boiled and freeze-dried mature silk worm larval powder. *Journal of Asia-Pacific Entomology*, *19*(2), 551–561. https://doi.org/10.1016/j.aspen.2016.05.003

Nino, M. C., Reddivari, L., Ferruzzi, M. G., & Liceaga, A. M. (2021). Targeted phenolic characterization and antioxidant bioactivity of extracts from edible *Acheta Domesticus*. *Foods*, *10*(10). https://doi.org/10.3390/foods10102295

Nongonierma, A. B., & FitzGerald, R. J. (2017). Unlocking the biological potential of proteins from edible insects through enzymatic hydrolysis: A review. *Innovative Food Science & Emerging Technologies*, *43*, 239–252. https://doi.org/10.1016/j.ifset.2017.08.014

Nowakowski, A. C., Miller, A. C., Miller, M. E., Xiao, H., & Wu, X. (2022). Potential health benefits of edible insects. *Critical Reviews in Food Science and Nutrition*, *62*(13), 3499–3508. https://doi.org/10.1080/10408398.2020.1867053

Olawuyi, I. F., Heo, E., Jeong, M., Kim, J. H., Park, J.-J., Chae, J.,. . . . Byun, S. (2025). Acidic polysaccharide from the edible insect *Protaetia brevitarsis* seulensis activates antiviral immunity to suppress norovirus infection. *Carbohydrate Polymers*, *347*, 122587. https://doi.org/10.1016/j.carbpol.2024.122587

Oonincx, D. G. A. B., van Broekhoven, S., van Huis, A., & van Loon, J. J. A. (2015). Feed conversion, survival and development, and composition of four insect species on diets composed of food by-products. *PLoS One*, *10*(12), e0144601. https://doi.org/10.1371/journal.pone.0144601

Osimani, A., Milanović, V., Cardinali, F., Roncolini, A., Garofalo, C., Clementi, F.,. . . . Aquilanti, L. (2018). Bread enriched with cricket powder (*Acheta domesticus*): A technological, microbiological and nutritional evaluation. *Innovative Food Science & Emerging Technologies*, *48*, 150–163. https://doi.org/10.1016/j.ifset.2018.06.007

Park, B. M., Lim, H. J., & Lee, B. J. (2021). Anti-obesity effects of *Tenebrio molitor* larvae powder in high-fat diet-induced obese mice. *Journal of Nutrition and Health*, *54*(4), 342–354. Retrieved from https://doi.org/10.4163/jnh.2021.54.4.342

Park, Y.-S., Choi, Y.-S., Hwang, K.-E., Kim, T.-K., Lee, C.-W., Shin, D.-M., & Han, S. G. (2017). Physicochemical properties of meat batter added with edible silkworm pupae (*Bombyx mori*) and transglutaminase. *Korean Journal for Food Science of Animal Resources*, *37*(3), 351–359.

Ramos-Elorduy, J. (2009). Anthropo-entomophagy: Cultures, evolution and sustainability. *Entomological Research*, *39*(5), 271–288.

Riekkinen, K., Väkeväinen, K., & Korhonen, J. (2022). The effect of substrate on the nutrient content and fatty acid composition of edible insects. *Insects*, *13*(7). https://doi.org/10.3390/insects13070590

Ros-Baró, M., Casas-Agustench, P., Díaz-Rizzolo, D. A., Batlle-Bayer, L., Adrià-Acosta, F., Aguilar-Martínez, A.,. . . . Bach-Faig, A. (2022). Edible insect consumption for human and planetary health: A systematic review. *International Journal of Environmental Research and Public Health*, *19*(18). https://doi.org/10.3390/ijerph191811653

Rumpold, B. A., & Schlüter, O. K. (2013). Nutritional composition and safety aspects of edible insects. *Molecular Nutrition & Food Research*, *57*(5), 802–823. https://doi.org/10.1002/mnfr.201200735

Schmidt, A., Call, L.-M., Macheiner, L., & Mayer, H. K. (2019). Determination of vitamin B12 in four edible insect species by immunoaffinity and ultra-high performance liquid chromatography. *Food Chemistry*, *281*, 124–129. https://doi.org/10.1016/j.foodchem.2018.12.039

Searchinger, T., Waite, R., Hanson, C., Ranganathan, J., & Dumas, P. (2019). *Creating a sustainable food future – A menu of solutions to feed nearly 10 Billion people by 2050*. World Resources Institute.

Seo, M., Goo, T.-W., Chung, M. Y., Baek, M., Hwang, J.-S., Kim, M.-A., & Yun, E.-Y. (2017). *Tenebrio molitor* larvae inhibit adipogenesis through AMPK and MAPKs signaling in 3T3-L1 adipocytes and obesity in high-fat diet-induced obese mice. *International Journal of Molecular Sciences*, *18*(3). https://doi.org/10.3390/ijms18030518

Sogari, G., Riccioli, F., Moruzzo, R., Menozzi, D., Tzompa Sosa, D. A., Li, J.,. . . . Mancini, S. (2023). Engaging in entomophagy: The role of food neophobia and disgust between insect and non-insect eaters. *Food Quality and Preference*, *104*, 104764. https://doi.org/10.1016/j.foodqual.2022.104764

Sosa, D. A. T., & Fogliano, V. (2017). Potential of insect-derived ingredients for food applications. *Insect Physiology and Ecology*, *2017*, 215–231.

Souci, S. W., Fachmann, W., & Kraut, H. (2024). *Food composition and nutrition tables: Die Zusammensetzung der Lebensmittel – Nährwert-Tabellen I La composition des aliments – Tableaux des valeurs nutritives* (p. 1276). https://doi.org/10.52778/9783804751019

Tang, J.-J., Fang, P., Xia, H.-L., Tu, Z.-C., Hou, B.-Y., Yan, Y.-M.,. . . . Cheng, Y.-X. (2015). Constituents from the edible Chinese black ants (*Polyrhachis dives*) showing protective effect on rat mesangial cells and anti-inflammatory activity. *Food Research International*, *67*, 163–168. https://doi.org/10.1016/j.foodres.2014.11.022

Tang, Q. F., & Dai, Y. (2016). Immunomodulatory effects of supercritical fluid CO2 extracts from freeze-dried powder of *Tenebrio molitor* larvae (yellow mealworm). *Food Science and Technology*, *36*, 493–498.

Tukshipa, S. D., Chakravorty, J., & Prveen. (2022). *In vivo* antioxidant and immunomodulatory effect of the aqueous extract of *Mimela* sp. on cyclophosphamide induced im-munocompromised mice. *International Journal of Pharmacognosy and Life Science*, *3*(2), 13–19.

Tzompa-Sosa, D. A., Dewettinck, K., Provijn, P., Brouwers, J. F., De Meulenaer, B., & Oonincx, D. G. (2021). Lipidome of cricket species used as food. *Food Chemistry*, *349*, 129077.

van Huis, A. (2016). Edible insects are the future? *Proceedings of the Nutrition Society*, *75*(3), 294–305. https://doi.org/10.1017/S0029665116000069

Vinsova, J., & Vavrikova, E. (2011). Chitosan derivatives with antimicrobial, antitumour and antioxidant activities-a review. *Current Pharmaceutical Design*, *17*(32), 3596–3607.

Weru, J., Chege, P., & Kinyuru, J. (2021). Nutritional potential of edible insects: A systematic review of published data. *International Journal of Tropical Insect Science*, *41*, 2015–2037.

Wu, Q. Y., Jia, J. Q., Tan, G. X., Xu, J. L., & Gui, Z. Z. (2011). Physicochemical properties of silkworm larvae protein isolate and gastrointestinal hydrolysate bioactivities. *African Journal ofBiotechnology*, *10*(32), 6145–6153.

Xia, Z., Chen, J., & Wu, S. (2013). Hypolipidemic activity of the chitooligosaccharides from *Clanis bilineata* (Lepidoptera), an edible insect. *International Journal of Biological Macromolecules*, *59*, 96–98.

Xing, R., Liu, S., Guo, Z., Yu, H., Wang, P., Li, C.,. . . . Li, P. (2005). Relevance of molecular weight of chitosan and its derivatives and their antioxidant activities in vitro. *Bioorganic & Medicinal Chemistry*, *13*(5), 1573–1577. https://doi.org/10.1016/j.bmc.2004.12.022

Xu, X., Ji, H., Belghit, I., Liland, N. S., Wu, W., & Li, X. (2021). Effects of black soldier fly oil rich in n-3 HUFA on growth performance, metabolism and health response of juvenile mirror carp (*Cyprinus carpio* var. specularis). *Aquaculture*, *533*, 736144. https://doi.org/10.1016/j.aquaculture.2020.736144

Yhoung-aree, J. (2008). Edible insects in Thailand: Nutritional values and health concerns. In *Proceedings of a workshop on Asia-Pacific resources and their potential for development. Food and Agriculture Organization of the United Nations. Regional Office of Asia and the Pacific*. Bangkok, Thailand.

Yoon, S., Wong, N. A. K., Chae, M., & Auh, J.-H. (2019). Comparative characterization of protein hydrolysates from three edible insects: Mealworm larvae, adult crickets, and silkworm pupae. *Foods*, *8*(11). https://doi.org/10.3390/foods8110563

Yoon, Y., Oh, E., Park, W.-J., Kim, H. S., & Kim, Y. (2023). Effects of enzymatic hydrolysis and size fractionation of mealworm (*Tenebrio molitor* larvae) protein on antioxidant and anti-inflammatory activities and inhibition of muscle loss. *ACS Food Science & Technology*, *3*(5), 850–857. https://doi.org/10.1021/acsfoodscitech.3c00034

Zhou, Y., Wang, D., Zhou, S., Duan, H., Guo, J., & Yan, W. (2022). Nutritional composition, health benefits, and application value of edible insects: A review. *Foods*, *11*(24). https://doi.org/10.3390/foods11243961

Zielińska, E., Baraniak, B., Karaś, M., Rybczyńska, K., & Jakubczyk, A. (2015). Selected species of edible insects as a source of nutrient composition. *Food Research International*, *77*, 460–466. https://doi.org/10.1016/j.foodres.2015.09.008

6 Use of Insects and/or Insect-Derived Compounds as Ingredients in Food Formulation

Yun-Sang Choi, Jae Hoon Lee, Ji Yoon Cha, and Yea-Ji Kim

6.1 INTRODUCTION TO INSECT-BASED FOODS: EXPLORING A SUSTAINABLE PROTEIN SOURCE

In recent years, the quest for sustainable and environmentally friendly sources of protein has intensified concerns regarding the ecological footprint of traditional livestock farming. As global populations continue to increase and food security has become an increasingly pressing issue, alternative protein sources have emerged as viable solutions to address these challenges. Among these alternatives, insect-based foods have gained significant attention because of their potential to provide a sustainable and nutritious protein source while mitigating the negative environmental impacts associated with conventional livestock production. The consumption of insects, known as entomophagy, is not a new concept and has been practiced in various cultures worldwide for centuries. However, in recent years, insects have been seriously considered a mainstream protein source in Western societies. This shift in perception can be attributed to several factors, including

DOI: 10.1201/9781003488156-6

a growing awareness of the environmental impact of meat production, advancements in food technology, and changing consumer attitudes toward food sustainability and ethical consumption.

Insects offer several advantages over traditional livestock as protein sources. Insects are highly efficient converters of feed to protein, requiring significantly less land, water, and feed resources to produce the same amount of protein as conventional livestock. This efficiency is particularly important as pressure mounts to feed a growing global population while minimizing environmental degradation. Additionally, insects can be reared using organic waste materials, which further reduces their environmental footprint and contributes to waste reduction.

Furthermore, insects are nutritionally dense and contain high levels of proteins, healthy fats, vitamins, and minerals. They are also a rich source of essential amino acids, making them a complete protein source, comparable to meat, fish, and poultry. Thus, incorporating insects into the diet can contribute to addressing malnutrition and food insecurity, particularly in regions with limited access to animal proteins.

Despite these advantages, the widespread adoption of insect-based foods faces several challenges. One primary barrier is the cultural stigma associated with insect consumption in Western societies. Overcoming this stigma will require efforts to educate consumers about the nutritional and environmental benefits of insect-based foods, as well as innovative marketing strategies to make them more appealing and accessible. Additionally, there are regulatory hurdles to overcome as many countries lack clear guidelines and regulations governing the production and sale of insect-based foods. Establishing robust regulatory frameworks is essential to ensure food safety and consumer confidence in these products.

In conclusion, insect-based foods represent a promising and sustainable solution to the challenges of feeding a growing global population, while reducing the environmental impact of food production. By leveraging their nutritional benefits, environmental advantages, and innovative food technologies, insect-based foods have the potential to revolutionize the way we think about protein sources and contribute to a more sustainable and resilient food system.

6.2 INSECT DIVERSITY: EXPLORING POTENTIAL SPECIES FOR FOOD FORMULATION

Insects represent a remarkably diverse group of organisms, with estimates suggesting as many as 10 million species worldwide, although only a fraction has been formally described. This vast diversity offers a wealth of potential options for the development of insect-based foods, each with unique nutritional, flavor, and culinary characteristics. In this chapter, we explore some of the most promising insect species for food formulation, highlighting their nutritional benefits, culinary versatility, and potential for sustainable production. Crickets are among the most widely studied and utilized insect species in food. Crickets are rich in proteins, comprising up to 70% of protein by dry weight, as well as healthy fats, vitamins, and minerals. They are also relatively easy to rear, requiring minimal space, water, and feed compared with traditional livestock. Crickets have a mild nutty flavor that makes them suitable for a variety of culinary applications, including protein bars, snacks, and baked goods. In addition to their nutritional value, crickets have been shown to have a low environmental impact, making them an attractive option for sustainable food production.

Another promising insect species for food formulation is the mealworm. Mealworms have high protein and fat content, with the protein content ranging from 50% to 65% by dry weight. They are also rich in essential amino acids, which makes them a complete protein source. Mealworms have a slightly earthy flavor with a hint of sweetness, making them suitable for both savory and sweet dishes. Similar to crickets, mealworms are relatively easy to rear and require minimal resources, making them an environmentally friendly protein source. In addition to crickets and mealworms, several other insect species show promise for use in food formulation. Black soldier flies are rich in protein and fat and can be reared on organic waste materials, making them excellent candidates for sustainable food production. Their larvae have a mild nutty flavor and can be used in a variety of dishes, including protein powders, pasta, and animal feed.

Other insect species explored for food purposes include grasshoppers, locusts, and waxworms. Grasshoppers and locusts are rich in proteins and have a slightly tangy flavor that pairs well with spices and herbs. They are commonly consumed in many parts of the world and are gaining popularity in Western countries. Waxworms, the larvae of wax moths, are high in fat and have a buttery flavor, which makes them well-suited for use in sauces, dressings, and spreads. Although the nutritional and culinary potential of insect-based foods is clear, challenges remain in terms of production, regulation, and consumer acceptance. Scaling up insect production to meet the growing demand for insect-based foods will require investment in research and development, as well as infrastructure for mass rearing and processing. Additionally, the regulatory frameworks governing the production and sale of insect-based foods vary widely across countries, creating barriers to producers' market entry.

Despite these challenges, growing interest in sustainable and alternative protein sources is driving innovation in the insect-based food industry. With continued research and investment, insect-based foods have the potential to play a significant role in addressing global food security and sustainability challenges while also offering consumers delicious and nutritious dietary options for their diets.

6.3 NUTRITIONAL COMPOSITION OF INSECTS: PROTEIN, LIPIDS, AND MICRONUTRIENTS

Given the growing recognition of insects as future protein sources, edible insect proteins are regarded as crucial nutritional components. Previous studies have indicated that the protein content of edible insects ranges from approximately 18–77% and varies depending on the species, sex, feed, life cycle, and rearing environment (Table 6.1). Some insects have been recognized as livestock food in several countries. Like other livestock, insects contain muscle proteins from the skeletal and visceral muscles. However, when proteins are extracted from insects for use as food ingredients, there are many types of proteins

TABLE 6.1 Nutrition Composition of Edible Insects.

SPECIES	*CRUDE PROTEIN (%)*[a]	*CRUDE FAT (%)*	*CRUDE FIBER (%)*	*CRUDE ASH (%)*	*MICRONUTRIENTS*	*REFERENCES*
Coleoptera (beetles)						
Allomyrina dichotoma (L)[b]	39.3–54.2	20.2–25.2	0.3–4.0	3.9–5.3	Thiamine 2.5 ppm, Ca 0.1%, P 0.9%, K 1.2%, Mg 0.3%, Fe 142.6 ppm, Zn 102.6 ppm	Baek et al. (2017), Ghosh et al. (2017), Jang et al. (2022)
Alphitobius diaperinus (L)	58.8–63.2	23.0–27.3	4.1–7.3	3.4–3.7	–	EFSA Panel on Nutrition et al., (2022)
Copris nevinsoni Waterhouse (A)	54.4	13.6	15.2	9.2	–	Raksakantong et al. (2010)
Protaetia brevitarsis (L)	44.2–57.9	15.4	11.1–16.6	6.9–8.4	Ca 0.3%, P 1.1%, K 2.0%, Mg 0.3%, Fe 162.0 ppm, Zn 118.9 ppm	Baek et al. (2017), Ghosh et al. (2017)
Tenebrio molitor (L)	46.4–59.2	25.9–35.9	4.6–11.8	2.9–3.7	Cobalamin 10 ppb, Ca 2.0–2.9%, P 1.1–1.3%	Baek et al. (2019), Ravzanaadii et al. (2012), Schmidt et al. (2019), Yu et al. (2021)
Rhynchophorus ferrugineus (L)	18.0–28.5	40.8–60.1	3.8–4.5	2.4–2.9	P 0.2–0.3%, K 0.5–0.8%, Mg 0.1–0.2%, Zn 80–90 ppm	Chinarak et al. (2020), Kavle et al. (2022), Orkusz et al. (2023)
Zophobas morio (L)	48.9–52.2	36.3–38.9	5.0–5.2	2.8–3.6	P 0.9%, K 0.9%, Mg 0.2%	Kim et al. (2019)
Diptera (flies)						
Hermetia illucens (L)	29.9–56.2	4.8–49.0	7.4–10.0	5.0–15.4	Ca 1.0–3.1%, P 0.8–1.7%, K 0.8–1.7%	Chia et al. (2020), Liu et al. (2017), Zulkifli et al. (2022)
Musca domestica (L)	42.3–55.4	20.8–29.7	6.0–7.1	6.2–7.6	–	Aniebo and Owen (2010), Sánchez et al. (2021)
Hemiptera (stink bug)						
Tessaratoma papillosa (A)	50.5	23.6	13.8	5.4	–	Raksakantong et al. (2010)

(*Continued*)

TABLE 6.1 (Continued)

Hymenoptera (bees)						
Apis mellifera (L)	35.3–45.7	14.5–28.1	2.7	2.8–4.1	K 1.9%, Mg 0.2%	Ghosh et al. (2016), Haber et al. (2019), Mekuria et al. (2021)
Apis mellifera (P)	24.6–51.8	16.0–26.2	2.5–2.7	3.2–4.2	K 2.2%, Mg 0.2%	Choi et al. (2019), Ghosh et al. (2016), Haber et al. (2019)
Isoptera (termites)						
Macrotermes bellicosus (A)	25.4–68.2	1.8–47.0	1.6–6.2	0.9–4.7	Vitamin A 1.6–26.5 ppm, Fe 0.1%, Zn 5.9–107.6 ppm	Atowa et al. (2021), Kinyuru et al. (2013), Ntukuyoh et al. (2012)
Macrotermes nigeriensis (A)	59.0	18.9	7.1	5.3	–	Muinat et al. (2021)
Lepidoptera (moths)						
Bombyx mori (P)	53.1–60 7	23.5–33.3	1.1	0.9–4.5	Thiamine 31.8 ppm, riboflavin 91.3 ppm, Ca 0.9%, P 1.1%, K 2.2%, Mg 0.4%, Zn 127.5 ppm	Akande et al. (2020), Brogan et al. (2021), Jang et al. (2022), Yeruva et al. (2023)
Gonimbrasia belina (L)	69.8	9.0	14.5	5.3	Ca 0.2%, Fe 312.3 ppm, Zn 143.1 ppm	Nantanga and Amakali (2020)
Orthoptera (crickets, grasshoppers, and locusts)						
Acheta domesticus (A)	56.8–72 0	10.4–22.8	3.3–6.2	3.5–5.4	Ca 0.1%, P 0.9%, K 0.4%, Mg 0.1%	Brogan et al. (2021), Kulma et al. (2019), Orkusz et al. (2023), Udomsil et al. (2019)
Gryllus bimaculatus (A)	53.4–62.2	20.5–26.4	6.4–10.9	2.8–4.0	Riboflavin 15.5 ppm, Ca 0.1%, P 0.7%, K 0.3%, Mg 0.1%	Jang et al. (2022), Loypimai et al. (2024), Orkusz et al. (2023), Udomsil et al. (2019)
Kraussaria angulifera (A)	49.7	8.8	2.6	0.7	–	Muinat et al. (2021)

(Continued)

TABLE 6.1 (Continued)

SPECIES	*CRUDE PROTEIN (%)*[a]	*CRUDE FAT (%)*	*CRUDE FIBER (%)*	*CRUDE ASH (%)*	*MICRONUTRIENTS*	*REFERENCES*
Locusta migratoria (A)	50.8–77.3	6.5–34.9	9.4–13.4	2.4–4.1	Cobalamin 8.4 ppb, P 0.6%, K 0.3%, Mg 0.1%, Zn 162–298 ppm	Brogan et al. (2021), Clarkson et al. (2018), Jang et al. (2022), Kim et al. (2020), Schmidt et al. (2019)
Oxya chinensis sinuosa (A)	74.2	3.0	–	4.4	Riboflavin 34.5 ppm	Jang et al. (2022), Kim et al. (2017)
Zonocerus variegatus (A)	41.1	32.2	–	4.0	Vitamin A 117.0 ppm, Fe 381.0 ppm, Zn 181.1 ppm	Atowa et al. (2021)

[a] All nutritional compositions are presented on a dry basis.
[b] Life stage of insect. (L) larvae; (P) pupae; (A) adults.

because insect proteins are extracted from the entire insect and not from a specific body part. For instance, hemocyanin, which functions as an oxygen transporter in insects, and hexamerin, which is involved in amino acid storage and energy production, have been detected in some insect extracts. The cuticle, the main component of the insect integument, is composed of proteins. However, when utilizing protein extracts from insects, the proteins in the cuticle are bound to chitin, making it difficult to extract and present their nutritional and functional properties (Cha et al., 2025). The protein composition of the extract can vary depending on the extraction conditions such as pH, ionic strength of the solvent, extraction temperature, and other physical treatments. Furthermore, protein fragmentation and aggregation can be induced during processing, such as drying and defatting.

Lipid content and fatty acid profiles can differ according to factors similar to those of proteins (Choi et al., 2023). With respect to lipid content, locusts and termites among insect species and larval stages among rearing stages are known to have high lipid content. Regarding fatty acids, unsaturated fatty acids (UFAs), including oleic acid (C18:1), linoleic acid (C18:2), and α-linoleic acid (C18:3 n-3), are abundant in edible insects; hence, the lipid from the insects generally is in liquid state. Several studies have demonstrated the health benefits of fatty acids. In addition, differences in fat content and fatty acid profiles obviously affect the nutritional value of insect fats, as well as the availability of food processing. The fats present in edible insects are composed of phospholipids, triglycerides, sterols, and waxes, with the main components being phospholipids and triglycerides, similar to those in other animals. Phospholipids are amphiphilic molecules consisting of hydrophilic phosphate and glycerol groups, and hydrophobic fatty acid chains. Triglycerides are composed of glycerol and fatty acids without phosphates; thus, they are hydrophobic and uncharged. Phospholipids from some foods are well-known emulsifiers, and the charge of phospholipids, such as phosphatidylcholine, phosphatidylethanolamine, and phosphatidylinositol, differs depending on their type. Phospholipid content and composition are also influenced by feed, species, and metamorphosis. Therefore, the type and content of lipids and fatty acids are important for the use of edible insect lipids as food ingredients.

Chitin is an N-acetylglucosamine polymer, a monomer of a glucose derivative, and a major carbohydrate in edible insects. Chitin is located in insect cuticles; thus, the chitin content of insects is affected by the species and metamorphosis. *Hermetia illucens, Tenebrio molitor, Zophobas morio, Gryllus bimaculatus*, and *Bombyx mori* have been identified as sources of chitin. In *Hermetia illucens* larvae, chitin accounts for 9.5% of the biomass, whereas in *Hermetia illucens* shedding, chitin accounts for 31.1% of the biomass (Soetemans et al., 2020). Chitin is considered a dietary fiber and has numerous health benefits, including weight loss, gastrointestinal health, anticancer, and antiaging (Lv et al., 2023). Stull et al. (2018) demonstrated that the consumption of *Gryllodes sigillatus* powder has a positive effect on the gut microbiota, increasing *Bifidobacterium animalis* and reducing systemic inflammation. The results of this study were assumed to be due to the presence of chitin and other fibers.

The types and contents of vitamins and minerals in edible insects vary considerably depending on the species, diet, environmental conditions, and metamorphosis. Nevertheless, when used as an alternative protein source, edible insects can be a valuable source of vitamin B, which is also present in meat. Among the B vitamins, thiamine (vitamin B1) and riboflavin (vitamin B2) are present in several edible insects. However, cobalamin (vitamin B12) is only found in some insects, such as *Tenebrio molitor* and *Orthoptera*.

According to the Food and Agriculture Organization (FAO) of the United Nations, most edible insects are rich sources of iron and zinc and could be a solution to the deficiency of these nutrients in developing countries (Van Huis et al., 2013). Copper and manganese have also been detected in some insects, although they can vary considerably depending on the factors mentioned above.

6.4 SAFETY ASSESSMENT OF INSECT-DERIVED COMPOUNDS IN FOOD PRODUCTION

When using compounds obtained from edible insects in food, it is important to ensure the safety of the materials. Potential hazard risks of edible insects include allergenic, microbial, and chemical risks. Therefore, the effective management of these risks should be a priority before using insect-derived compounds as food.

Many consumers worry about allergen risk when consuming edible insects. Allergies caused by edible insects include primary allergies caused by allergens present in insects, such as tropomyosin, arginine kinase, hexamerin 1 B, and paramyosin, and allergies in which IgE is confused with allergens in insects that are similar to specific allergens, resulting in an allergic reaction. Efforts are being made to reduce allergen risk in edible insects by decomposing these antigen proteins. It has been reported that the allergen risk can be reduced by reducing binding activity to IgE through food processing processes such as enzymatic hydrolysis or heat treatment, and it has also been reported that the antigenicity of some allergen proteins is reduced through microwave vacuum drying or tray drying processes (Khammeethong et al., 2024). Despite these efforts, some allergens (tropomyosin from crickets) have been reported to be so stable that their antigenicity is maintained even after digestion (De Marchi et al., 2021). Therefore, research is needed on appropriate processing methods to reduce the antigenicity of each antigenic protein.

Second, microbial risk is a potential risk factor for the consumption of edible insects. Microbiological contamination of edible insects can be divided into contamination of the outside and inside of the insects. Microbiological contamination of insects can be treated to some extent through food processing. However, it is difficult to treat the contamination of the gut microbiota of edible insects with pathogenic microorganisms. The biomass of gut microorganisms in edible insects is so high that it accounts for up to 10% of total body weight. Additionally, owing to the small size of edible insects, it is difficult to separate the gut microbiota from other parts. Therefore, it is important to prevent the contamination of edible insects with pathogenic microorganisms. To achieve this, microbiological management must be implemented starting with a breeding environment in which edible insects are raised. The traditional consumption of edible insects may have been difficult to manage because wild insects were consumed, but currently, edible insects are raised for consumption as food; therefore, contamination with pathogenic microorganisms can be sufficiently prevented. Therefore, microbial risk can be eliminated by thoroughly managing microbial contamination in the cages, feed, and water supply systems used to raise edible insects.

Finally, pesticide or heavy metal contamination of edible insects is a problematic chemical risk. Wild insects may consume crops contaminated with pesticides, and the accumulation of chemical materials (such as cadmium (Cd), arsenic (As), lead (Pb), and mercury (Hg)) in insects may occur through the consumption of unsafe food. However, chemical and microbiological risks can be sufficiently resolved through a systematic breeding environment.

6.5 FUNCTIONAL PROPERTIES OF INSECT INGREDIENTS IN FOOD APPLICATIONS

It is possible to use edible insects as alternative protein and lipid sources; however, the possibility of using edible insects as functional food materials is also increasing because of the various bioactive compounds they contain. A variety of bioactive compounds, ranging from macromolecules such as proteins, lipids, and carbohydrates to micromolecules such as peptides, fatty acids, polysaccharides, and phenolic compounds, are known to be the major causes of functional activity in edible insects (Table 6.2).

6.5.1 Protein and Peptides Isolated from Edible Insects

Proteins account for the largest proportion of edible insect ingredients, and various functional activities of proteins and peptides have been reported through many studies. The antioxidant activities of edible insect proteins and peptides have been reported by several researchers. *Tenebrio molitor* larvae, also known as mealworms, show significant free-radical scavenging activity (Lee et al., 2024). Additionally, *T. molitor* protein effectively restored the viability of Vero cells following H_2O_2 treatment. It was also confirmed that the antioxidant activity of *T. molitor* increased when it was pretreated with various concentrations of ethanol. Treatment with 20% ethanol resulted in the highest increase in antioxidant activity and was confirmed to effectively regulate the expression of apoptosis-related proteins following H_2O_2 treatment. Proteins obtained from *Protaetia brevitarsis* have been reported to induce a significant increase in the expression of proinflammatory mediators in mouse macrophages (Lee, Kim, Kim et al., 2023). It was found to induce the activation of macrophages through the mitogen-activated protein kinase (MAPK) and NF-κB cell signaling pathways, and through this, it was confirmed that the edible insect protein has excellent immune-enhancing activity. In addition, research results confirming the anticancer activity of *Helophorus aquaticus* and *Helophorus syriacus* protein extracts using PC-3 cells, a prostate cancer cell line, have also been reported (Elhazar et al., 2023). The proteins isolated from seven edible insects (*Omphisa fuscidentalis, Gryllodes sigillatus, Musca domestica, Schistocerca gregaria, T. molitor, Bombyx mori, and Oecophylla smaragdina*) showed strong free-radical scavenging activity and inhibitory activity on LOX and COX-2 (Matheswaran et al., 2019).

Various studies are being attempted to produce peptides with improved functional activities from edible insect-derived proteins with the above functional activities. The most

TABLE 6.2 Functional Properties of Edible Insect Ingredients in Food Application.

EDIBLE INSECT INGREDIENT	*EDIBLE INSECT TYPE*	*FUNCTIONAL PROPERTIES*	*REFERENCES*
Protein and peptide	*Tenebrio molitor*	Antioxidant activity • scavenges 2,2′-azino-bis(3-ethylbenzothiazoline-6-sulfonic acid) (ABTS) radicals • regulates the expression of apoptosis-related protein	Lee et al. (2024)
	Protaetia brevitarsis	Immune-enhancing activity • increases the expression of pro-inflammatory cytokines • activates MAPK and NF-κB cell signaling pathways	Lee, Kim, Kim et al. (2023)
	Helophorus aquaticus and *Helophorus syriacus*	Anticancer activity • inhibits PC-3 cell growth	Elhazar et al. (2023)
	Omphisa fuscidentalis, Gryllodes sigillatus, Musca domestica, Schistocerca gregaria, Tenebrio molitor, Bombyx mori, and *Oecophylla smaragdina*	Antioxidant activity • scavenges DPPH and ABTS radicals Anti-inflammatory activity • inhibits lipoxygenase (LOX) and cyclooxygenase-2 (COX)	Matheswaran et al. (2019)
	Bombyx mori	Antioxidant activity • scavenges DPPH radicals • chelates ferrous ion	Chukiatsiri et al. (2024)
	Hermetia illucens	Antioxidant activity • scavenges hydroxyl, 2,2-diphenyl-1-picrylhydrazyl (DPPH), and ABTS radicals	Zhu et al. (2020)
	Tenebrio molitor	Antihypertensive activity • inhibits ACE	Dai et al. (2013)
	Protaetia brevitarsis	Antihypertensive activity • inhibits ACE	Lee, Kim, Yong et al. (2023)

(*Continued*)

TABLE 6.2 (Continued)

	Gryllodes sigillatus, Tenebrio molitor, and *Schistocerca gregaria*	Anti-inflammatory activity • inhibits LOX and COX-2	Zielińska et al. (2017)
	Gryllodes sigillatus	Antihypertensive activity • inhibits ACE Anti-inflammatory activity • inhibits the expression of NF-κB	Hall et al. (2020)
	Bombyx mori	Anticancer activity • induces cell cycle arrest in S phase • activates the expression of caspase-3 • down-regulates the expression of Bcl-2 protein	Kaewkorn et al. (2012)
Lipid and fatty acids	*Tenebrio molitor*	Anticancer activity • inhibit HepG2 and Caco-2 cancer cell growth • increases early and late apoptosis	Wu et al. (2020)
	Tenebrio molitor	Anti-inflammatory activity • reduces the production of nitric oxide	Son et al. (2020)
	Hermertia illucens	Positive effect on immunity • regulates lipid metabolism • improve antioxidant capacity	Xu et al. (2021)
Carbohydrates, chitin, and chitosan	*Gryllus bimaculatus*	Anti-inflammatory activity • reduces pro-inflammatory cytokines (VEGF, IL-6, PGE2, and TNF-α) • down-regulates NF-κB activity	Ahn et al. (2014), Ahn et al. (2020)
	Gryllus bimaculatus	Anti-diabetic activity • reduces blood glucose level • inhibits LDL-cholesterol and alkaline phosphatase level	Ahn et al. (2020)
	Antheraea yamamai	Immune-enhancing activity • promotes the synthesis of pro-inflammatory cytokines (TNF-α, IFN-β, and IL-1β)	Ohta et al. (2016)

(Continued)

TABLE 6.2 (Continued)

EDIBLE INSECT INGREDIENT	*EDIBLE INSECT TYPE*	*FUNCTIONAL PROPERTIES*	*REFERENCES*
	Acheta domesticus and *Gryllodes sigillatus*	Antimicrobial activity • inhibits the growth of *Listeria innocua* and *Escherichia coli*	(Malm et al., 2021)
	Scapsipedus icipe	Prebiotic activity • boost the antimicrobial activity of probiotics	Kipkoech et al. (2021)
Extracts of edible insects	*Musca domestica*	Antioxidant activity • scavenges DPPH radicals Antihypertensive activity • inhibits ACE	Li et al. (2017)
	Bombyx mori	Anticancer activity • up-regulates pro-apoptotic proteins • down-regulates anti-apoptotic proteins	Cho et al. (2019)
	Allomyrina dichotoma	Anti-obesity activity • inhibits intracellular lipid accumulation • inhibits the expression of adipocyte-specific marker genes	Chung et al., (2014), Seo et al. (2017)
	Tenebrio molitor	Anti-obesity activity • inhibits intracellular lipid accumulation • inhibits the expression of adipocyte-specific marker genes	Seo et al. (2017)
	Lycorma delicatula	Anti-inflammatory activity • inhibits the production of nitric oxide and IL-13	Baek et al. (2018)
	Allomyrina dichotoma, *Protaetia brevitarsis*, and *Tenebrio molitor*	Antioxidant activity • increase storage stability of meat products by suppressing lipid oxidation	Yong et al. (2023)

widely used method utilizes enzymes, and there is a report in which three types of peptides, ENIILFR, LNKDLMR, and MLIIIMR, were obtained from a fraction obtained by hydrolyzing *Bombyx mori* pupae with pepsin (Chukiatsiri et al., 2024). The three types of peptides obtained in this study were confirmed to have higher DPPH radical scavenging and ferrous ion chelating activities than butylated hydroxyanisole (BHA) and butylated hydroxytoluene (BHT), which are used as synthetic antioxidants. The alcalase hydrolysate of *Hermetia illucens* L., a well-known black soldier fly, has excellent antioxidant activity (Zhu et al., 2020). After fractionation of the hydrolysate of *H. illucens* L. by molecular weight, excellent radical scavenging activity was confirmed in the fraction below 3 kDa, and 17 novel antioxidant peptides were identified in this fraction through liquid chromatography–tandem mass spectrometry (LC-MS/MS) analysis. In addition to their antioxidant activities, these functional peptides have been reported to exhibit various functional activities. YAN (tyrosine – alanine – asparagine), a tripeptide obtained from *T. molitor* larvae through alcalase enzyme hydrolysis, has been reported to exhibit excellent ACE inhibitory activity (Dai et al., 2013). YAN has been shown to effectively regulate blood pressure in a spontaneously hypertensive rat model. The SY (serine – tyrosine), PF (proline – phenylalanine), YPY (tyrosine – proline – tyrosine), and WI (tryptophan – isoleucine) peptides obtained from the hydrolysis of *Protaetia brevitarsis* using Flavourzyme have excellent antihypertensive activities (Lee, Kim, Yong et al., 2023). All four peptides showed excellent ACE inhibitory activity, and molecular docking analysis confirmed that multiple hydrogen bonds were formed between the active site of ACE and the peptides. The anti-inflammatory activity of enzymatic hydrolysates obtained from *Gryllodes sigillatus*, *T. molitor*, and *Schistocerca gregaria* using an in vitro digestion model has been reported (Zielińska et al., 2017). In this study, we confirmed the inhibitory activity of insect protein-derived hydrolysates against LOX and COX-2, which are associated with the production of leukotrienes and prostaglandins. The overproduction of these substances is closely related to the dysregulation of inflammatory responses. In addition, a study was reported in which hydrolysate was obtained from tropical crickets (*Gryllodes sigillatus*) through a simulated gastrointestinal digestion model, and 28 types of peptides were identified after undergoing separation/purification processes through size-exclusion and reverse-phase chromatography (Hall et al., 2020). These peptides exhibit antihypertensive, anti-glycemic, and anti-inflammatory activities. In particular, two peptides, PHGAP (proline – histidine – glycine – alanine – proline) and VGPPQ (valine – glycine – proline – proline – glutamine), have been reported to show antihypertensive activity similar to that of captopril, a well-known antihypertensive drug. Additionally, sericin (antimicrobial peptides, AMPs, obtained from *Bombyx mori*) was found to reduce cell viability of SW 480 cells (human colorectal cancer cells) (Kaewkorn et al., 2012). Sericin induces cell cycle arrest in the S phase of SW480 cells. This induced the activation of caspase-3 and downregulation of Bcl-2 protein expression. It was confirmed that sericin showed an anticancer effect through cell apoptosis in SW480 cells.

6.5.2 Lipids and Fatty Acids Isolated from Edible Insects

Lipids are the second most abundant ingredient in edible insects after proteins, and it is generally known that the content of UFAs is higher than that of saturated fatty acids (SFAs) (dos Santos Aguilar, 2021).

The oil extracts of *Tenebrio molitor* larvae showed antiproliferative effects against human hepatocellular carcinoma (HepG2) and colorectal adenocarcinoma (Caco-2) cells (Wu et al., 2020). *T. molitor* larval oil exhibited caspase 3-mediated cytotoxicity, and the high omega-3 fatty acid content of the oil was identified as the main cytotoxic compound. In addition, the unsaponifiable lipid of *T. molitor* oil has been reported to effectively reduce the amount of nitric oxide produced by lipopolysaccharides (LPS) treatment of mouse macrophage RAW 264.7 cells (Son et al., 2020). Thus, the possibility of using *T. molitor* oil to alleviate inflammation was confirmed. And also, black soldier fly (*Hermetia illucens*) oil was found to have a positive effect on improving immunity by effectively regulating lipid metabolism and enhancing antioxidant activity (X. Xu et al., 2021).

6.5.3 Carbohydrates, Chitin, and Chitosan Isolated from Edible Insects

Although they are not major compounds in edible insects, such as proteins and lipids, functional activities of carbohydrates and related derivatives have been reported.

Glycosaminoglycan obtained from *Gryllus bimaculatus* has been reported to exhibit anti-inflammatory, antilipidemic, and antidiabetic activities (Ahn et al., 2014, 2016, 2020). It has been confirmed that glycosaminoglycan acts as an anti-inflammatory substance in a rat model of chronic arthritis, and supplementation with glycosaminoglycan was effective in lowering blood glucose, low-density lipoprotein cholesterol, and alkaline phosphatase levels in a diabetic mouse model. In particular, the activities of antioxidant enzymes such as catalase, superoxide dismutase, and glutathione peroxidase are increased by glycosaminoglycans (Ahn et al., 2020). Furthermore, an acidic polysaccharide isolated from *Antheraea yamamai* pupae induced the activation of NF-κB in RAW 264 cells (Ohta et al., 2016). As a result, the synthesis of pro-inflammatory cytokines (TNF-α, IFN-β, and IL-1β) was promoted, ultimately resulting in a positive effect on innate immunity.

Chitin, a representative polysaccharide obtained from edible crickets (*Acheta domesticus* and *Gryllodes sigillatus*), and chitosan converted from it have been reported to have antimicrobial activity, inhibiting the growth of *Listeria innocua* and *Escherichia coli* by more than 90% (Malm & Liceaga, 2021). Chitosan obtained from *Scapsipedus icipe* has been reported to boost the antimicrobial activity of probiotics as a prebiotic (Kipkoech et al., 2021). Chitosan has been found to increase the growth of *Lactobacillus fermentum*, *Lactobacillus acidophilus*, and *Bifidobacterium adolescentis*, while reducing the growth of *Salmonella typhi*, a pathogenic bacterium. Additionally, packaging films manufactured using chitosan obtained from *Periplaneta americana* were reported to be effective in inhibiting the growth of *Serratia marcescens* and *Escherichia coli* owing to their antioxidant activity (Chen et al., 2021).

6.5.4 Extracts Isolated from Edible Insects

Although specific ingredients such as proteins, lipids, and carbohydrates have not been identified, studies have reported that extracts obtained from edible insects have various biofunctional properties.

Water extracts prepared from housefly larvae (*Musca domestica*) exhibit multifunctional biological activities (Li et al., 2017). This water extract exhibited strong DPPH radical scavenging activity and inhibitory activity of ACE. Fermented ethanol extract obtained from mulberry silkworm larvae (*Bombyx mori*) effectively regulated apoptosis of human hepatocellular carcinoma cells (HepG2) (Cho et al., 2019). The ethanolic extract upregulated pro-apoptotic proteins and downregulated antiapoptotic proteins in HepG2 cells, resulting in cell cycle arrest and apoptosis. The ethanol extracts of *A. dichotoma* and *T. molitor* were also exhibited antiobesity activity using 3T3-L1 adipocytes (Chung et al., 2014; Seo et al., 2017). These extracts exhibited antiobesity activity by inhibiting intracellular lipid accumulation and the expression of adipocyte-specific marker genes. Additionally, water extract from *Lycorma delicatula* was reported to exhibit anti-inflammatory activity in LPS-induced macrophage RAW 264.7 cells (Baek et al., 2018). It was confirmed to effectively inhibit the production of nitric oxide and interleukin-13, known as inflammatory mediators, and as a result, it was found to have high potential for use as a biomaterial to prevent inflammatory damage. A study has also been reported to increase storage stability by applying ethanol extracts of three types of edible insects (*A. dichotoma*, *P. brevitarsis*, and *T. molitor*) to tteokgalbi, a meat product (Yong et al., 2023). It has been reported that because of the strong antioxidant activity of insect ethanol extract, it can be expected to increase storage stability by effectively suppressing lipid oxidation even on the 14th day of storage.

6.6 CULINARY INNOVATIONS: RECIPES AND TECHNIQUES WITH INSECT-BASED INGREDIENTS

6.6.1 Recipes for Insect-Based Ingredients

Insects have been consumed as food and medicines in many parts of the world, including Africa, Asia, and Latin America, since ancient times. In Western societies, consumption has traditionally been low; however, the FAO promotes consumption by sharing knowledge and raising awareness about the use of insects as food and feed. Research on the use of edible insects, which are nutritious and function as sustainable food resources (Table 6.3).

The incorporation of edible insects into grain-based foods such as snacks, bakeries, and pasta directly enhances the nutritional profile of these products. Recently, the use of crickets (*Acheta domesticus*) and mealworms (*Tenebrio molitor)* as grain-based food has become increasingly common. The incorporation of edible insects into pasta products has been shown to significantly increase the nutritional profile of the finished products, including proteins, essential amino acids, lipids, minerals, and crude fiber. The incorporation of insects into pasta formulations had a positive impact on consumer acceptance, as evidenced by changes in firmness and adhesion, which were perceived as improvements in the sense of organization. Furthermore, the replacement of durum wheat, which is used in pasta, with 5% cricket powder resulted in a color similar to that of whole-grain pasta. The addition of insect flour to bread or cookies enriches proteins, essential amino acids, lipids, and minerals. The protein digestibility corrected amino acid score increased in

TABLE 6.3 Culinary Innovation Recipes with Insect-Based Ingredients.

PRODUCT TYPE	*EDIBLE INSECT TYPE*	*ALTERNATIVE QUANTITY*	*MAIN RESULTS*	*REFERENCES*
Pasta	*Locusta migratoria Tenebrio molitor*	15% of flour	• A significant enhance in nutritional profile of protein, ash, and crude fiber content	Pasini et al. (2022)
Pasta	*Acheta domesticus, Tenebrio molitor*	14% of durum wheat semolina	• Enhancing the quantity and quality of protein • Achieving desired texture in terms of firmness and adhesion • Conducting sensory assessments to evaluate consumer acceptability in relation to insect odor	Çabuk and Yılmaz (2020)
Pasta	*Bombyx mori*	5 and 10% of buckwheat flour	• Improved nutritional parameters • The pasta with the highest overall acceptance contains 10% insect content.	Biró et al. (2019)
Pasta	*Acheta domesticus*	5% of durum meal semolina	• Carbohydrate content decreased, while protein and fat increased • Color similar to whole-grain pasta • Among the textural properties, no difference was found in robustness, cohesion, and elasticity	Ho et al. (2022)
Bread	*Acheta domesticus, Tenebrio molitor*	10% of refined wheat flour	• Pretreated according to the degree of defatting of the insect powder. • *Tenebrio molitor* (TM) fat contributes to the quality of bread production • Defatted *Acheta domesticus* (AD) increases protein digestibility-corrected amino acid score value, making it a nutritious alternative to partially replace wheat	Bottle et al. (2024)
Bread	*Tenebrio molitor, Alphitobius diaperinus, Acheta domesticus*	10, 20, and 30% of wheat flour	• 10% insect powder content contributes to a significant increase in protein content compared to wheat bread • According to sensory analysis, the addition of edible insect powder to bakery products is generally acceptable up to 10%	Kowalski et al. (2022)

(Continued)

TABLE 6.3 (Continued)

Bread	*Acheta domesticus*	5, 10, and 15% of wheat flour	• The addition of insect powder increases the protein and fat content of the product while reducing its carbohydrate content • The optimal substitution level that does not significantly affect the physicochemical and sensory properties in bread making is 5% insect powder	Gantner et al. (2024)
Bread	*Acheta domesticus*, *Tenebrio molitor*	5, 10, and 15% of wheat flour	• Improving the nutritional value of insect bread in terms of protein, lipid, fiber, and mineral content • Insect bread showed a significant increase in the essential amino acids, valine and tyrosine • In vitro digestion experiments revealed that insect bread has higher bioaccessibility of Na, K, Ca, Mg, P, Fe, Zn, Mn, and Li compared to white bread	Mihaly Cozmuta et al. (2022)
Cracker	*Acheta domesticus*, *Locusta migratoria*, *Tenebrio molitor*	5% of wheat flour	• Increased content of crude protein, fat, and all amino acids in insect crackers • Insect crackers have been rated as acceptable, but some evaluators perceive an unfamiliar taste, aroma • Cricket crackers score better in sensory evaluation than other insect crackers	Ivanišová et al. (2023)
Cracker	*Tenebrio molitor*	2, 4, 6, 10, 15, and 20% of wheat flour	• Increased the protein content by 15% and enhanced the levels of minerals (potassium, phosphorus, copper, and zinc), thereby improving the nutritional profile of the snack • Sensory evaluation revealed that crackers with 6% addition received the highest scores • The addition of insect powder reduced firmness and increased crispness	Djouadi et al. (2022)

(Continued)

TABLE 6.3 (Continued)

PRODUCT TYPE	*EDIBLE INSECT TYPE*	*ALTERNATIVE QUANTITY*	*MAIN RESULTS*	*REFERENCES*
Cookie	*Ruspolia differens*	10% of self-raising wheat	• The insect cookie contains sufficient omega-3 fatty acids • Increased protein, amino acids, and minerals, enhancing its nutritional value	Ochieng et al. (2023)
Beef patties	*Acheta domesticus*	5, 7.5, and 10% of beef lean meat	• Higher in protein content than beef patties and has a firm, brown appearance • Including insect powder up to 5.0% in beef patties is optimal in terms of composition, technology, and culinary characteristics	Pasqualin Cavalheiro et al. (2024)
Beef patties	*Gryllus assimilis*	10, 15, and 20% of beef	• The 10% replacement patties demonstrated no effect on lipid oxidation and exhibited reduced liquid loss, contraction, and sensory changes. • While beef patties received high sensory ratings, the 10% replacement patties were also well accepted by consumers.	Gomes Martins et al. (2024)
Beef burger	*Acheta domesticus* *Locusta migratoria* *Tenebrio molitor*	5% of beef meat	• The addition of insect powder enhances the bioactive compound content in cooked burgers without affecting their texture. • In cricket burgers, it reduced the accumulation of heterocyclic amines and decreased lipid accumulation, while increasing the content of glutaminylarginine in in vitro gastrointestinal digestion experiments. • Mealworm burgers demonstrated higher sensory acceptance.	Rocchetti et al. (2024)

(*Continued*)

TABLE 6.3 (Continued)

Meat analogues	*Acheta domestica*	5, 6, and 7% of soy protein	• The soluble protein content (5.9%) is higher than that of soy protein powder (4.7%), forming a fibrous structure that enhances texture, increases firmness and adhesion, and reduces chewiness • The optimal blend for texture, fibrous structure, and nutritional value was found to be insect powder and glutinous rice flour in a 6:1 ratio	Nakagawa et al. (2024)
Frankfurters	*Acheta domesticus*	2.5, 5, and 7.5% of pork meat	• Increased the protein content without affecting the fat content, and their structural properties remained stable in lipid–protein interactions • The levels of zinc, calcium, and manganese increased, while the sodium content remained unchanged • Incorporating insects up to 5% is a viable alternative to replacing pork	Cavalheiro et al. (2023)
Frankfurters	*Tenebrio molitor L*	5, 10, and 15% of lean meat	• Microwave-dried insect powder exhibited excellent textural properties and demonstrated low cooking loss. • Although the sensory score decreased with higher replacement rates compared to the standard Frankfurt, the microwave-dried powder performed better than other dried powders.	Zhang et al. (2022)
Sausages	*Sphenarium purpurascens*	3, 5, 7, and 10% of potato starch	• Replacement sausages, protein content, hardness, springiness, gumminess, and chewiness all increased. • Addition of 10% insect flour was less acceptable to consumers.	Cruz-López et al. (2022)

(*Continued*)

TABLE 6.3 (Continued)

PRODUCT TYPE	*EDIBLE INSECT TYPE*	*ALTERNATIVE QUANTITY*	*MAIN RESULTS*	*REFERENCES*
Sausages	*Zophobas morio L.*	5, 10, 25, and 50% of pork shoulder	• Hybrid sausages with 5 and 10% insect content exhibited viscoelastic properties comparable to standard sausages when heated to 90°C • Cooking loss was minimized due to the moisture and fat stabilization in the hybrid sausages; however, they became unstable under external pressure.	Scholliers et al. (2020)
Tteokgalbi	*Tenebrio molitor L.* *Protaetia brevitarsis L.* *Allomyrina dichotoma L.*	1 and 2% of ground meat	• Freeze-drying ethanol-extracted residue to produce insect extract • A 2% concentration of *A. dichotoma* extract demonstrated high cooking water retention, the highest radical scavenging and metal chelating activities, and the lowest TBARS values. • No negative effects on microbial counts were observed during the storage period.	Yong et al. (2023)
Goat meat emulsion	*Acheta domesticus*	2% of minced meat	• Evaluating the potential of cricket protein hydrolysates as novel food biopreservatives. • Their antioxidant, antimicrobial properties, lipid stability, protein oxidation prevention, and sensory quality positively affect storage stability.	Lone et al. (2023)

bread made with 10% cricket flour, indicating the potential for it to be used as a nutritionally viable replacement for wheat flour. Overall, consumers demonstrated acceptance of insect-based bread when incorporated at concentrations below 10%. This raw material can be enhanced by improving its nutritional profile compared to that of primarily carbohydrate-based flours. However, they must be capable of meeting consumers' flavor and texture expectations.

The rationale behind the incorporation of insect proteins into meat products is that they exhibit a meat-like nutritional profile comprising essential amino acids, protein content, and essential fatty acids. Additionally, it is a sustainable source of protein. The substitution of insect powder in beef patties resulted in a high protein content, although the resulting product had a firm brown appearance. A preference for a product is indicated when a low content is used as a replacement. The consumption of cricket-based burgers has been demonstrated to result in a reduction in the accumulation of heterocyclic amines and a concomitant reduction in lipid accumulation. In the meat analog study, the ratio of 6% cricket powder to 1% glutinous rice powder was identified as the optimal blend because it contained a high concentration of water-soluble protein and formed fiber structures. The substitution of pork in sausages resulted in increased protein content while maintaining fat content. This allows the lipid-protein structural properties to remain stable, thereby minimizing cooking loss. However, external pressures can destabilize this process. Alternative sausages exhibited increased tissue profiles such as firmness, springiness, and stickiness. Consumer preferences are generally positive toward the addition of insects to products, with a preference for up to a 10% replacement. Research is underway to identify strategies for reducing the structural properties of these products and enhancing consumer acceptance.

6.6.2 Techniques for Insect-Based Ingredients

Conventional techniques employed for the preparation of insects typically include roasting, frying, and boiling. These processing methods are characterized by the inactivation of microorganisms and enzymes through heat treatment, which extends the shelf life of the product and improves its color, flavor, and texture through the Maillard reaction. However, improper heat treatment can result in the degradation of protein quality due to protein denaturation, fatty acid oxidation, and nutrient loss. With the expansion of the edible insect industry, the need for improved processing and extraction technologies has increased. The use of industrial processing technologies can improve quality, maintain consistency, enable mass production, reduce production costs, and develop various forms of products. To address these challenges, various studies have been conducted to develop insect industries (Table 6.4).

PEFs are applied at a high intensity to induce electrical perforation of the cell membrane, resulting in deformation or destruction if the strength is sufficiently high. This process facilitates the transport of substances between intracellular and extracellular environments, leading to the outflow of physiologically active substances and nutrients. The nonthermal treatment method has the potential to minimize denaturation while simultaneously increasing efficiency, owing to its relatively low energy requirements. The application of this technology to insects has resulted in an increase in the yield of proteins and

TABLE 6.4 Processing Techniques for Application of Insect-Based Ingredients.

PROCESSING TECHNOLOGY	*PROCESSING CONDITION*	*EDIBLE INSECT TYPE*	*MAIN RESULTS*	*REFERENCES*
Pulsed electric field (PEF)	Rectangular, 4.4 kV/cm, 25 μs, 20 Hz,	*Acheta domesticus*	• PEF treatment had a positive effect on oven drying at 60°C by reducing oven drying time and energy consumption by 14.22%. • It also increased protein solubility by 53.07%, antioxidant activity by 24.05%, and reduced histamine content in EHD dried samples by 25.87%.	Psarianos et al. (2024)
	Monopolar, 1 kV/cm, 7 μs, 2 Hz, 5, 20, and 40 kJ/kg	*Hermetia illucens* L.	• In convection dried PEF-treated samples, fat extraction yield increased with increasing treatment intensity. Protein extraction yield was highest with the 5 kJ/kg treatment • In convection and infrared-dried samples, antioxidant activity increased with increasing PEF intensity.	Bogusz et al. (2023)
	Monopolar and near rectangular, 1.07 kV/cm, 40 μs, 2 Hz, 5 and 20 kJ/kg (500, 2000 pulses)	*Hermetia illucens* L. *Tenebrio molitor* L.	• PEF treatment has been demonstrated to enhance the rate of moisture evaporation during the drying process. • While the 20 kJ/kg treatment was effective in reducing certain microbial counts, further treatment was necessary for the safe production of food.	Bogusz et al. (2022)
	Bipolar and near rectangular, 1.5 kV/cm, 15 μs, 20 Hz, 4.9–49.1 kJ/kg (100–1000 pulses)	*Acheta domesticus*	• The application of treatment at 4.90 kJ/kg resulted in a notable enhancement in the oil binding capacity (19.53%), emulsifying capacity (22.06%), antioxidant activity (45.79%), fat yield (41.75%), and protein yield (18.62%) compared to the control group.	Psarianos et al. (2022)
	Monopolar, 2 kV/cm, 0.5 s, 2 Hz, 10, 15, and 20 kJ/kg	*Hermetia illucens* L.	• Drying time was reduced by 11–24.6% depending on the level of PEF treatment. • The response surface methodology shows that the optimum drying conditions were PEF energy of 11.2–13.1 kJ/kg and 81–84 C.	Shorstkii et al. (2022)

(*Continued*)

TABLE 6.4 (Continued)

Ultrasound-assisted extraction (UAE)	20 kHz, 500 W, 20–30 min, ambient temp	*Gryllus bimaculatus, Acheta domesticus*	• Oil extraction yield was 70–75% compared to n-hexane extraction, with a shorter extraction time. • UAE treatment resulted in an increase in total phenol content and antioxidant activity. There was also a gradual increase in free fatty acids, peroxidation value, and thiobarbituric acid.	Pranil et al. (2024)
	28 kHz, 4 W/cm^3, 10–50 min, 25 °C, pH 10	*Tenebrio molitor* L.	• Protein yield through UAE treatment was 60.04%, which is higher than that of traditional extraction methods. • Antioxidant activity, total phenol content, protein solubility, water-holding capacity, oil-holding capacity, and emulsifying property treated by UAE have all been enhanced.	Zhang et al. (2023)
	4.57 W/cm^3, 1 h, powder + NaOH (2.5 g/100ml) 1 g:56 ml	*Hermetia illucens* L.	• UAE treatment increased protein yield from 55.40% to 80.37%, reducing extraction time and energy consumption. • Protein secondary structure was characterized by an orderly and porous arrangement, with high solubility and emulsification capacity.	Xu et al. (2023)
	35 kHz, 240 W, 9.89–35.11 min, 16.36–83.64 °C, solvent of ethanol-isopropanol	*Alphitobius diaperinus* L.	• UAE treatment increases oil extraction yield from 60.09% to 88.08%, reducing energy consumption by approximately 18.5 times. • UAE extraction oil has increased physiological and antioxidant activity and improves oxidation stability, low peroxide prices, and maintains free fatty acid levels.	Gharibzahedi and Altintas (2022)
	20 kHz, 750 W, 10–30 min, 25–35 °C	*Patanga succincta* L.	• UAE treatment for 20 min at 60% amplitude had minimal effect on protein structure. • Protein yield increased with UAE treatment, improved solubility, foam formation capacity, and stability were maintained, and antioxidant activity was enhanced.	Kingwascharapong et al. (2021)

(*Continued*)

TABLE 6.4 (Continued)

PROCESSING TECHNOLOGY	*PROCESSING CONDITION*	*EDIBLE INSECT TYPE*	*MAIN RESULTS*	*REFERENCES*
	20 kHz, 15 min, 89.9 μm, room temp (final temp 70 °C)	*Acheta domesticus*, *Tenebrio molitor* L.	• Oil extraction yield increased and cholesterol content decreased during UAE treatment. • Fatty acid profiles were improved by increased polyunsaturated fatty acid (PUFA), decreased saturated fatty acid (SFA), and monounsaturated fatty acid (MUFA).	Otero et al. (2020)
High hydrostatic pressure (HHP)	200–600 MPa, 20 °C, 10 min	*Acheta domesticus*, *Alphitobius diaperinus*	• Treatment of HHP in mealworm samples resulted in less modification of the secondary protein structure. • The solubility improved to about 14–22%, with no effect on foam formation and emulsification properties.	Sweers et al. (2024)
	70–600 MPa, 20 °C, 5 min	*Tenebrio molitor* L.	• Modified protein structure in 600 MPa treatment and formed high molecular weight protein aggregates. Porous protein networks formed.	Boukil et al. (2022)
	100–500 MPa, 35 °C	*Protaetia brevitarsis* L.	• Protein solubility was highest when treated with 200 MPa and tended to decrease when over-treated. • The pH increased during high-pressure treatment, and the foam performance also increased.	Kim et al. (2021)
	500 MPa, 30–40 °C, 15 min	*Acheta domesticus*, *Tenebrio molitor* L.	• Functional properties and antioxidant activity are affected by HPP treatment and temperature.	Bolat et al. (2021)
	200, 380 MPa, 30–40 °C, 1min, (hydrolysis with Alcalase®)	*Gryllodes sigillatus*, *Tenebrio molitor*	• HPP treatment of mealworms and crickets further improved solubility and oil binding capacity in mealworms, but water binding capacity, foaming, and gelling properties were unaffected in both insects.	Dion-Poulin et al. (2020)

(Continued)

TABLE 6.4 (Continued)

Supercritical CO_2 extraction	25–30 MPa, 60°C, 90 min, 10 g/min	*Hermetia illucens* L.	• The yield was higher under the pressure of 30 MPa, but when ethanol was used as a cosolvent, the extraction efficiency was higher than that of the sole sc-CO_2 extraction. • The ratio of PUFAs/SFAs is shown to be relatively high.	Cruz et al. (2024)
	450 bar, 60°C, 85–130g/min	*Hermetia illucens* L.	• Samples including 16% w/w of moisture did not affect the recovery rate during supercritical extraction, but the moisture content was detected up to 1.7%. • There is no difference in oil fatty acid profile, and the peroxide index is lower than that of oil extracted with hexane.	Fornari et al. (2023)
	20–30 MPa, 40–65°C, 10 g/min	*Hermetia illucens* L.	• SC-CO_2 extraction can obtain oil without using organic solvents. • Intermittent extraction processes can reduce solvent consumption. • The extracted oil is high in lauric acid and rich in carotenoids.	Cruz et al. (2023)
	300 bar, 45°C, 3 h, 18 kg/h	*Hermetia illucens* L. *Tenebrio molitor* L.	• Powder from insects degreased with SC-CO_2 had lower fat content and higher protein content compared to centrifugal defatting. • The powder extracted with SC-CO_2 is in the state of a powder with a bright color and a larger average particle size.	Laurent et al. (2022)

lipids that can be used as raw food materials as well as a reduction in the time required for drying, thereby lowering energy consumption. Furthermore, it enhances solubility and antioxidant activity, which in turn influences protein processing. For large-scale production, it is essential that the high-strength electric field of the PEF equipment is maintained with a high degree of stability. Furthermore, the application of such technology to industrial production is contingent on a reduction in the cost of equipment design. The following describes the US-assisted extraction method. This process generates and collapses bubbles, causing ultrasonic cavitation and destroying the cell walls and membranes. The force exerted by this physical phenomenon allows the leakage of substances, which then diffuse more rapidly, thereby increasing the extraction speed. This process increases the yield of extracted protein and oil, as well as the total phenol content, which consequently enhances the antioxidant activity. Additionally, it increases the solubility of proteins, water-holding capacity, oil-holding capacity, and emulsification properties, enhancing the processing capability of insect extracts. However, it is essential to exercise caution when extracting specific components because ultrasound treatment may potentially affect heat-sensitive compounds owing to the generation of high temperatures. HHP, a traditional preservation technology, can also be employed to enhance the yield of insect extracts through the disruption of cell membranes, microbial inactivation, and the liberation of extracellular components, as a result of the application of high pressure. This research indicates that the application of HHP treatment results in increased protein solubility, reduced protein denaturation compared to heat treatment, and alteration of processing characteristics contingent upon the applied pressure. A comparable pressure extraction technique, supercritical carbon dioxide (SC-CO_2) extraction, has been used for the acquisition or removal of oils. This method exhibits the benefits of a low critical temperature and pressure, selective extraction, and the capacity to extract heat-sensitive substances. Additionally, it leaves no residual organic solvents and exhibits low toxicity. When applied to insects, the use of ethanol as a cosolvent can enhance the extraction yield and reduce peroxide formation compared to organic solvents, thereby improving oxidative stability.

The advent of these technologies is of significant consequence to the edible insect industry because they facilitate sustainability, efficiency, quality, safety, and economic benefits. They enable the use of insects as food sources and promote the development of insect-based foods and sustainable food markets.

6.7 CONSUMER PERCEPTION AND ACCEPTANCE OF INSECT-BASED FOODS

Although some consumers are willing to purchase insect foods because of their perceived nutritional value and sustainability benefits, most have no interest in them and remain unconvinced. Factors influencing this perception include psychological barriers such as aversion to insects and neophobia, cultural differences, negative media exposure, and views on environmental benefits. In Western cultures, where there is a paucity of traditions for consuming insects and there is a general preference for meat, protein intake is primarily derived from livestock. This has led to a prevailing sentiment of apprehension and revulsion regarding the prospect of consuming insects. The association of insects with uncleanliness and disease has led to the development of preconceived notions that prevent individuals from accepting insects as food sources. A lack of familiarity with the concept of consuming insects can result in feelings of anxiety or the rejection of new foods. Furthermore, depictions of insects in the media, whether negative or unconventional, can contribute to the formation of unfavorable public perceptions.

A substantial body of research has been conducted to develop strategies to alter consumer perceptions of insect-based foods. The principal objective of these strategies is to disseminate information regarding the environmental and nutritional benefits of insects to inform and educate consumers, with the ultimate goal of influencing their perceptions. Younger consumers generally display greater openness and curiosity regarding new foods, making them more receptive to the idea of consuming edible insects. Therefore, increasing exposure to insect foods from an early age may assist in reducing aversion. It is also vital to educate older generations about the benefits and uses of edible insects to prevent the transmission of negative perceptions.

It is crucial to select the most preferred and familiar insect ingredients among consumers and develop a variety of products that satisfy their taste preferences. It is essential that these products be continuously provided and marketed with innovative imagery to cultivate curiosity and enhance positive perceptions, thereby influencing consumption decisions. Furthermore, the creation of products fortified with nutrients preferred by consumers and the provision of tasting opportunities can facilitate increased acceptance. Emphasis on the safety and quality of these products in familiar forms can assist with purchase decisions. Through these methods, consumer perceptions can be positively altered, and as consumption frequency increases, the consumption of insects can become a normalized practice, thereby allowing the insect food market to establish itself as a significant food sector.

6.8 SUSTAINABLE SOURCING: FARMING AND HARVESTING PRACTICES FOR INSECTS

The consumption of insects as a food source is gaining attention as a potential solution to the food security challenges posed by global population growth as described earlier. A key factor in enhancing the utility of edible insects in diverse cuisines and food products is establishing a sustainable supply chain that is environmentally, economically, and socially stable. However, collecting wild insects for human consumption presents significant challenges in ensuring consistent quality and yield. Furthermore, there is a risk of heavy metal or pesticide accumulation in humans when wild insects are consumed. Therefore, the production of edible insects by farming in controlled environments is required. In addition, to ensure the sustainability of the environment, economy, and society during the domestication and industrialization of edible insects, several requirements should be considered when farming.

As edible insects have emerged as alternatives to meat, their environmental value as livestock is increasingly recognized. Indeed, production of the same weight of livestock requires less land, feed, and time when crickets are used as a substitute for cattle, and less methane, carbon dioxide, and ammonia are emitted from mealworms and locusts than from cattle. Nevertheless, prior to commencing farming, there are a few things to consider regarding environmentally and economically sustainable farming systems. The utilization of inedible by-products as feed substrates for the production of edible insects represents an eco-friendly approach for the generation of food from waste. Insects can digest food waste, plastic, livestock manure, and by-products discarded during food processing, such as brewer's spent grain and corn stover. However, feeding livestock manure to edible insects during breeding is prohibited in some countries. In addition, the energy efficiency of facilities for managing light and temperature, sterilization equipment, pest control, and feed for rearing insects is important. Native insects from certain climates in Africa and Southeast Asia or insects that have been traditionally eaten in certain countries should be carefully monitored to ensure similar husbandry conditions if they are reared in other countries. In the meantime, the domestication of exotic edible insects should be done cautiously because they may have the potential to become invasive species that destroy existing ecosystems. For example, *Clanis bilineata tsingtauica*, a traditional edible insect in China, feeds mainly on leguminous crops. This can result in severe economic losses if they deviate from the farm sites.

6.9 REGULATORY CHALLENGES AND OPPORTUNITIES IN INSECT FOOD PRODUCTION

Food regulation ensures safety against the hazards inherent in the raw material and those that may arise during processing and storage. Therefore, edible insect production requires several regulations. Notably, there are discrepancies in the regulations between countries that traditionally consume insects and those that are newly interested in edible insects.

In China, where insects are traditionally consumed, they are typically sourced from the wild rather than reared. However, this is not regarded as a significant concern that must be addressed. Thus, there is a paucity of legislation and regulations pertaining to the consumption of edible insects as food. Only one standard for fresh and frozen silkworm pupae was established in the Guangxi Zhuang Autonomous Region in January 2016. This standard addresses hygiene production, labeling, transportation, storage, and expiration dates. Thailand is one of the major sources of edible insects globally; thus, it distributes various reared and wild-harvested insects to local markets and other countries. Unlike China, insect farming is prevalent in small-scale household operations. However, there is still a need to establish systematic regulations regarding safety and shelf life. Furthermore, in Mexico, where insect biodiversity is extremely rich and a diverse range of edible insects is found, there are no regulations related to farming and killing.

In Europe, the European Union has approved whole insects and their parts as novel foods since 2018, according to Regulation (EU) 2015/2283. Subsequently, the safety of this novel food must be evaluated by the European Food Safety Authority (EFSA) for approval. To obtain approval, the application must provide information on the production process, compositional data, specifications, proposed uses, use levels, anticipated intake, toxicity, and allergenicity. In other countries, traditional edible insects without any safety problems can be approved using a simplified procedure (EFSA Panel on Dietetic Products et al., 2016). As of July 2024, eight insect species have been evaluated by the EFSA for their safety, and four species have been assessed as food sources in Europe: *Alphitobius diaperinus* larva, *Tenebrio molitor* larva, *Acheta domesticus*, and *Locusta migratoria*. In the United Kingdom (UK), seven insects are permitted as food according to transitional measures in the EU and the UK until 2023. From January 2024, Great Britain (GB), part of the UK, allowed only *Tenebrio molitor* larva, *Acheta domesticus*, *Gryllodes sigillatus*, and *Hermetia illucens* larva, which are applicated as novel food in GB before December 2023, to be in food products.

In the United States (US), the Food and Drug Administration (FDA) has stated that insects can be used as food if they meet the regulatory requirements and are used for their intended purpose, similar to any other food. In addition, the FDA indicated that it should be raised for human food separately from animal feed and that it must be clean and labeled properly in accordance with Good Manufacturing Practices, although there are no specific regulations regarding this. In July 2024, the Singapore Food Agency (SFA) approved 16 insect species by dividing the stages of growth for human consumption. This approval does not extend to insects harvested in the wild and is only meaningful for insects raised in a regulated facility, as well as in the US.

Several insect species, including *Bombyx mori* pupae and *Oxya japonica* Thunberg, are traditionally consumed in South Korea. However, certified edible insects with clear standards and regulations have gradually emerged after paying attention to the FAO. In September 2021, *Locusta migratoria* was acknowledged as the tenth most edible insect with standardized rearing and processing conditions. To ensure the safety of edible insects, microbial and heavy metal standards have been established for processed insect foods that are dried, powdered, and/or processed using food or food additives. Regarding the microbial count, the total microbiome, *Escherichia coli*, or coliforms should be evaluated according to the sterilization conditions and intake method. The levels of three types of heavy metals, lead, cadmium, and inorganic arsenic, should be lower than the respective standards for all edible insects. Moreover, the process of validating the discovery of additional edible insects is ongoing.

The regulation of novel foods can promote consumption, as it ensures the safety of foodstuffs and enables the production of large quantities in qualified facilities. Consequently, insect certification could encourage consumption by people who are unfamiliar with edible insects. In countries where insects have traditionally been consumed, the necessity for regulation has been underestimated; however, in light of global trends, clear regulations could contribute to the expansion of this industry.

6.10 FUTURE DIRECTIONS: EXPANDING THE ROLE OF INSECTS IN GLOBAL FOOD SECURITY

Global food security is a self-evident reality that will confront our future owing to the rapidly increasing global population around the world. To address this problem of food security, the food science community has been continuously researching alternative foods.

In the protein field, research on cell-cultured foods as a sustainable alternative protein resource is actively underway, but it is also true that there are many problems that need to be solved, such as securing mass production technology, ensuring the safety of cell-cultured foods, and securing economic feasibility. Therefore, until the problems with cell-cultured foods are resolved, it can be said that serving as an alternative protein source for edible insects is of utmost importance. Edible insect proteins have the advantage that the amount of protein produced is outstanding, whereas the land area, water, and feed used to raise edible insects are significantly lower than those used in traditional livestock farming. Additionally, from a nutritional perspective, edible insects have the advantage of high essential amino acid and protein contents. These advantages of edible insects increase the possibility of using edible insects as a sustainable protein resource.

Since ancient times, the consumption of edible insects as food has proven their safety. However, in some cultures, the refusal to consume edible insects due to their repulsive appearance is considered a problem that must be solved to utilize edible insects as a sustainable protein resource. Therefore, processing research is actively underway to reduce the disgust caused by appearance. In fact, looking at the current commercialization of edible insects, products that are difficult to recognize as edible insects by their appearance, such as bread, cookies, tortillas, energy bars, sauces, snacks, and chocolates, are being sold using edible insects (Melgar-Lalanne et al., 2019). In addition, by applying edible insects to the food culture of Western countries (e.g., hamburgers, meatballs, spreads, and dressings), products are being developed to be well-accepted by consumers (González et al., 2019). Therefore, research into the processing of these proteins is expected to continue. Additionally, as explained in the previous section, studies verifying the various biofunctional activities of edible insects can positively affect the intake of edible insect proteins. To lead a healthy life, consumer preference for functional health foods has been continuously increasing in recent years. Therefore, if the various biofunctional activities (antioxidant, anticancer, anti-inflammatory, antihypertensive, etc.) of edible insect proteins can be verified and presented, it is expected that the aversion to eating edible insect proteins can be alleviated to some extent. In addition to the disgust caused by appearance, there are cases where consumers have an aversion to some off-flavor ingredients, so it is expected that research will be needed to remove the off-flavor unique to edible insects. Through these studies, edible insects can fulfill their role in global food security.

6.11 CONCLUSION

The integration of insects and insect-derived compounds into food formulations is a promising advancement for addressing global food security challenges. As a sustainable and nutrient-rich protein source, insects offer substantial benefits over conventional livestock, including reduced environmental impacts and efficient resource utilization. This exploration of insect-based foods highlights several key aspects that are necessary for the successful adoption and integration of these novel ingredients.

First, the diversity of edible insect species provides a vast array of options for food formulation, each with a unique nutritional profile and functional properties. The nutritional composition of insects, which is rich in proteins, lipids, and essential micronutrients, is a valuable addition to the human diet. Ensuring the safety of these ingredients through rigorous assessments is crucial for gaining consumer trust and regulatory approval. The functional properties of insect ingredients such as their emulsifying, gelling, and water-holding capacities enhance their versatility in various culinary applications. Culinary innovations continue to evolve, demonstrating the potential of insect-based recipes and techniques to appeal to a broad audience. Consumer perception and acceptance play pivotal roles in the mainstream adoption of insect-based foods. Education and awareness campaigns, along with positive culinary experiences, can shift public attitudes toward embracing sustainable alternatives. Sustainable sourcing practices including ethical farming and harvesting have reinforced the environmental benefits of insect-based food production.

The role of insects in global food security is also expected to expand. Continued research and innovation will uncover new applications and benefits, thereby solidifying the position of insect-based foods in the global food system. By leveraging the advantages of insects as a sustainable protein source, we can make significant strides toward a more resilient and equitable food future.

REFERENCES

Ahn, M. Y., Han, J. W., Hwang, J. S., Yun, E. Y., & Lee, B. M. (2014). Anti-inflammatory effect of glycosaminoglycan derived from *Gryllus bimaculatus* (a type of cricket, insect) on adjuvant-treated chronic arthritis rat model. *Journal of Toxicology and Environmental Health, Part A*, 77(22–24), 1332–1345. https://doi.org/10.1080/15287394.2014.951591.

Ahn, M. Y., Hwang, J. S., Kim, M. J., & Park, K. K. (2016). Antilipidemic effects and gene expression profiling of the glycosaminoglycans from cricket in rats on a high fat diet. *Archives of Pharmacal Research*, 39, 926–936. https://doi.org/10.1007/s12272-016-0749-1.

Ahn, M. Y., Kim, B. J., Kim, H. J., Jin, J. M., Yoon, H. J., Hwang, J. S., & Lee, B. M. (2020). Anti-diabetic activity of field cricket glycosaminoglycan by ameliorating oxidative stress. *BMC Complementary Medicine and Therapies*, 20, 1–10. https://doi.org/10.1186/s12906-020-03027-x.

Akande, A. O., Jolayemi, O. S., Adelugba, V. A., & Akande, S. T. (2020). Silkworm pupae (*Bombyx mori*) and locusts as alternative protein sources for high-energy biscuits. *Journal of Asia-Pacific Entomology*, *23*(1), 234–241. https://doi.org/10.1016/j.aspen.2020.01.003.

Aniebo, A. O., & Owen, O. J. (2010). Effects of age and method of drying on the proximate composition of housefly larvae (*Musca domestica* Linnaeus) meal (HFLM). *Pakistan Journal of Nutrition*, 9(5), 485–487. https://doi.org/10.3923/pjn.2010.485.487.

Atowa, C. O., Okoro, B. C., Umego, E. C., Atowa, A. O., Emmanuel, O., Ude, V. C., & Ugbogu, E. A. (2021). Nutritional values of *Zonocerus variegatus*, *Macrotermes bellicosus* and *Cirina forda* insects: Mineral composition, fatty acids and amino acid profiles. *Scientific African*, 12, e00798. https://doi.org/10.1016/j.sciaf.2021.e00798.

Baek, M., Hwang, J. S., Kim, M., Kim, S. H., Goo, T. W., & Yun, E. Y. (2017). Comparative analysis of nutritional components of edible insects registered as novel foods. *Journal of Life Science*, 27(3), 334–338. https://doi.org/10.5352/JLS.2017.27.3.334.

Baek, M., Kim, M. A., Kwon, Y. S., Hwang, J. S., Goo, T. W., Jun, M., & Yun, E. Y. (2019). Effects of processing methods on nutritional composition and antioxidant activity of mealworm (*Tenebrio molitor*) larvae. *Entomological Research*, 49(6), 284–293. https://doi.org/10.1111/1748-5967.12363.

Baek, S. H., Joung, O., Lee, H. Y., Shin, J. C., Choi, W. S., Lee, T. H., Hwang, J,. S., Nam, S. H., Son, H. U., & Lee, S. H. (2018). Anti-oxidative fraction of *Lycorma delicatula* alleviates inflammatory indicators. *Natural Product Communications*, 13(4). https://doi.org/10.1177/1934578X1801300413.

Biró, B., Fodor, R., Szedljak, I., Pásztor-Huszár, K., & Gere, A. (2019). Buckwheat-pasta enriched with silkworm powder: Technological analysis and sensory evaluation. *LWT*, 116, 108542. https://doi.org/10.1016/j.lwt.2019.108542.

Bogusz, R., Bryś, J., Onopiuk, A., Rybak, K., Witrowa-Rajchert, D., & Nowacka, M. (2023). Effect of pulsed electric field technology on the composition and bioactive compounds of black soldier fly larvae dried with convective and infrared–convective methods. *Molecules*, 28(24), 8121. https://doi.org/10.3390/molecules28248121.

Bogusz, R., Smetana, S., Wiktor, A., Parniakov, O., Pobiega, K., Rybak, K., & Nowacka, M. (2022). The selected quality aspects of infrared-dried black soldier fly (*Hermetia illucens*) and yellow mealworm (*Tenebrio molitor*) larvae pre-treated by pulsed electric field. *Innovative Food Science & Emerging Technologies*, 80, 103085. https://doi.org/10.1016/j.ifset.2022.103085.

Bolat, B., Ugur, A. E., Oztop, M. H., & Alpas, H. (2021). Effects of high hydrostatic pressure assisted degreasing on the technological properties of insect powders obtained from *Acheta domesticus* & *Tenebrio molitor*. *Journal of Food Engineering*, 292, 110359. https://doi.org/10.1016/j.jfoodeng.2020.110359.

Bottle, E., Espinosa-Ramírez, J., Serna-Saldívar, S. O., & Tejada-Ortigoza, V. (2024). Effect of full fat and defatted insect meals in breadmaking quality. *LWT*, 191, 115602. https://doi.org/10.1016/j.lwt.2023.115602.

Boukil, A., Marciniak, A., Mezdour, S., Pouliot, Y., & Doyen, A. (2022). Effect of high hydrostatic pressure intensity on structural modifications in mealworm (*Tenebrio molitor*) proteins. *Foods*, 11(7), 956. https://doi.org/10.3390/foods11070956.

Brogan, E. N., Park, Y. L., Matak, K. E., & Jaczynski, J. (2021). Characterization of protein in cricket (*Acheta domesticus*), locust (*Locusta migratoria*), and silk worm pupae (*Bombyx mori*) insect powders. *LWT*, 152, 112314. https://doi.org/10.1016/j.lwt.2021.112314.

Çabuk, B., & Yılmaz, B. (2020). Fortification of traditional egg pasta (erişte) with edible insects: nutritional quality, cooking properties and sensory characteristics evaluation. *Journal of Food Science and Technology*, 57(7), 2750–2757. https://doi.org/10.1007/s13197-020-04315-7.

Cavalheiro, C. P., Ruiz-Capillas, C., Herrero, A. M., Pintado, T., Cruz, T. d. M. P., & da Silva, M. C. A. (2023). Cricket (*Acheta domesticus*) flour as meat replacer in frankfurters: Nutritional, technological, structural, and sensory characteristics. *Innovative Food Science & Emerging Technologies*, 83, 103245. https://doi.org/10.1016/j.ifset.2022.103245.

Cha, J. Y., Kim, Y. J., Kim, J. H., Keum, D. H., Han, J. & Choi, Y. S. (2025). Edible insect larvae as sustainable sources of chitosan: Influence of species and drying method on functional and structural properties. *Food Chemistry: X*, 31, 103133. https://doi.org/10.1016/j.fochx.2025.103133.

Chen, S., Wei, X., Sui, Z., Guo, M., Geng, J., Xiao, J., & Huang, D. (2021). Preparation of antioxidant and antibacterial chitosan film from *Periplaneta americana*. *Insects*, 12(1), 53. https://doi.org/10.3390/insects12010053.

Chia, S. Y., Tanga, C. M., Osuga, I. M., Cheseto, X., Ekesi, S., Dicke, M., & van Loon, J. J. (2020). Nutritional composition of black soldier fly larvae feeding on agro-industrial by-products. *Entomologia Experimentalis et Applicata*, 168(6–7), 472–481. https://doi.org/10.1111/eea.12940.

Chinarak, K., Chaijan, M., & Panpipat, W. (2020). Farm-raised sago palm weevil (*Rhynchophorus ferrugineus*) larvae: Potential and challenges for promising source of nutrients. *Journal of Food Composition and Analysis*, 92, 103542. https://doi.org/10.1016/j.jfca.2020.103542.

Cho, H. D., Min, H. J., Won, Y. S., Ahn, H. Y., Cho, Y. S., & Seo, K.-I. (2019). Solid state fermentation process with *Aspergillus kawachii* enhances the cancer-suppressive potential of silkworm larva in hepatocellular carcinoma cells. *BMC Complementary and Alternative Medicine*, 19, 1–15. https://doi.org/10.1186/s12906-019-2649-7.

Choi, H. M., Kim, H. Y., Woo, S. O., Kim, S. G., Bang, K. W., Moon, H. J., & Han, S. M. (2019). Drying techniques and nutritional composition of drone pupae (*Apis mellifera* L.) as edible food. *Journal of Apiculture*, 34(2), 161–167. https://doi.org/10.17519/apiculture.2019.06.34.2.161.

Choi, Y. S., Lee, J. H., Kim, T. K., & Shin, D. M. (2023). Chapter Six – Edible insects in food. *Advances in Food and Nutrition Research*, 108, 223–264. https://doi.org/10.1016/bs.afnr.2023.10.005.

Chukiatsiri, S., Wongsrangsap, N., Kiatwuthinon, P., & Phonphoem, W. (2024). Purification and identification of novel antioxidant peptides derived from *Bombyx mori* pupae hydrolysates. *Biochemistry and Biophysics Reports*, 38, 101707. https://doi.org/10.1016/j.bbrep.2024.101707.

Chung, M. Y., Yoon, Y. I., Hwang, J. S., Goo, T. W., & Yun, E. Y. (2014). Anti-obesity effect of *Allomyrina dichotoma* (Arthropoda: Insecta) larvae ethanol extract on 3T3-L1 adipocyte differentiation. *Entomological Research*, 44(1), 9–16. https://doi.org/10.1111/1748-5967.12044.

Clarkson, C., Mirosa, M., & Birch, J. (2018). Potential of extracted Locusta migratoria protein fractions as value-added ingredients. *Insects*, 9(1), 20. https://doi.org/10.3390/insects9010020.

Cruz, V. A., Ferreira, N. J., Cornelio-Santiago, H. P., Santos, G. M. T., & Oliveira, A. L. (2023). Oil extraction from black soldier fly (*Hermetia illucens* L.) larvae meal by dynamic and intermittent processes of supercritical CO_2 – Global yield, oil characterization, and solvent consumption. *The Journal of Supercritical Fluids*, 195, 105861. https://doi.org/10.1016/j.supflu.2023.105861.

Cruz, V. A., Ferreira, N. J., Le Roux, E., Destandau, E., & de Oliveira, A. L. (2024). Intensification of the SFE using ethanol as a cosolvent and integration of the SFE process with sc-CO2 followed by PLE using pressurized ethanol of black soldier fly (*Hermetia illucens* L.) larvae meal – extract yields and characterization. *Foods*, 13(11), 1620. https://doi.org/10.3390/foods13111620.

Cruz-López, S. O., Álvarez-Cisneros, Y. M., Domínguez-Soberanes, J., Escalona-Buendía, H. B., & Sánchez, C. N. (2022). Physicochemical and sensory characteristics of sausages made with grasshopper (*Sphenarium purpurascens*) flour. *Foods*, 11(5), 704. https://doi.org/10.3390/foods11050704.

Dai, C., Ma, H., Luo, L., & Yin, X. (2013). Angiotensin I-converting enzyme (ACE) inhibitory peptide derived from *Tenebrio molitor* (L.) larva protein hydrolysate. *European Food Research and Technology*, 236, 681–689. https://doi.org/10.1007/s00217-013-1923-z.

De Marchi, L., Mainente, F., Leonardi, M., Scheurer, S., Wangorsch, A., Mahler, V., Pilolli, R., Sorio, D., & Zoccatelli, G. (2021). Allergenicity assessment of the edible cricket *Acheta domesticus* in terms of thermal and gastrointestinal processing and IgE cross-reactivity with shrimp. *Food Chemistry*, 359, 129878. https://doi.org/10.1016/j.foodchem.2021.129878.

Dion-Poulin, A., Laroche, M., Doyen, A., & Turgeon, S. L. (2020). Functionality of cricket and mealworm hydrolysates generated after pretreatment of meals with high hydrostatic pressures. *Molecules*, 25(22), 5366. https://doi.org/10.3390/molecules25225366.

Djouadi, A., Sales, J. R., Carvalho, M. O., & Raymundo, A. (2022). Development of healthy protein-rich crackers using *Tenebrio molitor* flour. *Foods*, 11(5), 702. https://doi.org/10.3390/foods11050702.

dos Santos Aguilar, J. G. (2021). An overview of lipids from insects. *Biocatalysis and Agricultural Biotechnology*, 33, 101967. https://doi.org/10.1016/j.bcab.2021.101967.

EFSA Panel on Nutrition, Novel Foods and Food Allergens (NDA), Turck, D., Bohn, T., Castenmiller, J., De Henauw, S., Hirsch-Ernst, K. I., Maciuk, A., Mangelsdorf, I., McArdle, H. J., Naska, A., Pelaez, C., Pentieva, K., Siani, A., Thies, F., Tsabouri, S., Vinceti, M., Cubadda, F., Frenzel, T., Heinonen, M., Marchelli, R., Neuhäuser-Berthold, M., Poulsen, M., Maradona, M. P., Schlatter, J. R., van Loveren, H., Ververis, E., & Knutsen, H. K. (2022). Safety of frozen and freeze-dried formulations of the lesser mealworm (*Alphitobius diaperinus* larva) as a Novel food pursuant to Regulation (EU) 2015/2283. *EFSA Journal*, 20(7), e07325. https://doi.org/10.2903/j.efsa.2022.7325.

EFSA Panel on Dietetic Products, Nutrition and Allergies (NDA), Turck, D., Bresson, J. L., Burlingame, B., Dean, T., Fairweather-Tait, S., Heinonen, M., Hirsch-Ernst, K. I., Mangelsdorf, I., McArdle, H., Naska, A., Neuhäuser-Berthold, M., Nowicka, G., Pentieva, K., Sanz, Y., Siani, A., Sjödin, A., Stern, M., Tomé, D., Vinceti, M., Willatts, P., Engel, K. H., Marchelli, R., Pöting, A., Poulsen, M., Salminen, S., Schlatter, J., Arcella, D., Gelbmann, W., Sesmaisons-Lecarré, A., Verhagen, H., & van Loveren, H. (2016). Guidance on the preparation and presentation of an application for authorisation of a novel food in the context of Regulation (EU) 2015/2283. *EFSA Journal*, 14(11), e04594. https://doi.org/10.2903/j.efsa.2016.4594.

Elhazar, T., Kaya, B., & Caf, F. (2023). Determination of anti-cancer and antioxidant properties of protein extracts obtained from aquatic *Helophorus* (Coleoptera: Helophoridae) insects. *Ege Journal of Fisheries & Aquatic Sciences (EgeJFAS)/Su Ürünleri Dergisi*, 40(1). https://doi.org/10.12714/egejfas.40.1.05.

Fornari, T., Vázquez, L., Villanueva-Bermejo, D., Hurtado-Ribeira, R., Martín Hernández, D., & Martin, D. (2023). Effect of moisture and oil content in the supercritical CO2 defatting of *Hermetia illucens* Larvae. *Foods*, 12(3), 490. https://doi.org/10.3390/foods12030490.

Gantner, M., Sadowska, A., Piotrowska, A., Kulik, K., Sionek, B., & Kostyra, E. (2024). Wheat bread enriched with house cricket powder (*Acheta domesticus* L.) as an alternative protein source. *Molecules*, 29(3), 711. https://doi.org/10.3390/molecules29030711.

Gharibzahedi, S. M. T., & Altintas, Z. (2022). Ultrasound-assisted alcoholic extraction of lesser mealworm larvae oil: Process optimization, physicochemical characteristics, and energy consumption. *Antioxidants*, 11(10), 1943. https://doi.org/10.3390/antiox11101943.

Ghosh, S., Jung, C., & Meyer-Rochow, V. B. (2016). Nutritional value and chemical composition of larvae, pupae, and adults of worker honey bee, *Apis mellifera* ligustica as a sustainable food source. *Journal of Asia-Pacific Entomology*, 19(2), 487–495. https://doi.org/10.1016/j.aspen.2016.03.008.

Ghosh, S., Lee, S. M., Jung, C., & Meyer-Rochow, V. B. (2017). Nutritional composition of five commercial edible insects in South Korea. *Journal of Asia-Pacific Entomology*, 20(2), 686–694. https://doi.org/10.1016/j.aspen.2017.04.003.

Gomes Martins, V. M., Milano, P., Rodrigues Pollonio, M. A., dos Santos, M., de Oliveira, A. P., Savay-da-Silva, L. K., Câmara, A. K. F. I., & de Souza Paglarini, C. (2024). Adding cricket (*Gryllus assimilis*) flour in hybrid beef patties: physicochemical, technological and sensory challenges. *International Journal of Food Science & Technology*, 59(1), 450–459. https://doi.org/10.1111/ijfs.16826.

González, C. M., Garzón, R., & Rosell, C. M. (2019). Insects as ingredients for bakery goods. A comparison study of *H. illucens*, *A. domestica* and *T. molitor* flours. *Innovative Food Science & Emerging Technologies*, 51, 205–210. https://doi.org/10.1016/j.ifset.2018.03.021.

Haber, M., Mishyna, M., Martinez, J. I., & Benjamin, O. (2019). Edible larvae and pupae of honey bee (*Apis mellifera*): Odor and nutritional characterization as a function of diet. *Food Chemistry*, 292, 197–203. https://doi.org/10.1016/j.foodchem.2019.04.041.

Hall, F., Reddivari, L., & Liceaga, A. M. (2020). Identification and characterization of edible cricket peptides on hypertensive and glycemic in vitro inhibition and their anti-inflammatory activity on RAW 264.7 macrophage cells. *Nutrients*, 12(11), 3588. https://doi.org/10.3390/nu12113588.

Ho, I., Peterson, A., Madden, J., Huang, E., Amin, S., & Lammert, A. (2022). Will It cricket? Product development and evaluation of cricket (*Acheta domesticus*) powder replacement in sausage, pasta, and brownies. *Foods*, 11(19), 3128. https://doi.org/10.3390/foods11193128.

Ivanišová, E., Rajnoha, M., Harangozo, Ľ., Kunecová, D., Čech, M., Gabríny, L., Gálik, B., Banach, J. K., Kowalczewski, P. Ł., & Pietrzak-Fiećko, R. (2023). Physicochemical, nutritional, antioxidant, and sensory properties of crackers supplemented with edible insects. *Applied Sciences*, 13(21), 11911. https://doi.org/10.3390/app132111911.

Jang, Y., Jeon, J., Lee, S. H., Choi, Y. M., & Choung, M. G. (2022). Evaluation of the vitamin B content of edible insects. *Journal of the Korean Society of Food Science and Nutrition*, 51(11), 1223–1231. https://doi.org/10.3746/jkfn.2022.51.11.1223.

Kaewkorn, W., Limpeanchob, N., Tiyaboonchai, W., Pongcharoen, S., & Sutheerawattananonda, M. (2012). Effects of silk sericin on the proliferation and apoptosis of colon cancer cells. *Biological Research*, 45(1), 45–50. http://dx.doi.org/10.4067/S0716-97602012000100006.

Kavle, R. R., Pritchard, E. T. M., Carne, A., Bekhit, A. E. D. A., & Agyei, D. (2022). Fatty acid profile, mineral composition, and health implications of consuming dried sago grubs (*Rhynchophorus ferrugineus*). *Applied Sciences*, 13(1), 363. https://doi.org/10.3390/app13010363.

Khammeethong, T., Phiriyangkul, P., Inson, C., & Sinthuvanich, C. (2024). Effect of microwave vacuum drying and tray drying on the allergenicity of protein allergens in edible cricket, *Gryllus bimaculatus*. *Food Control*, 160, 110328. https://doi.org/10.1016/j.foodcont.2024.110328.

Kim, S. Y., Kim, H. G., Ko, H. J., Kim, M. A., Kim, I. W., Seo, M., Lee J. H., Lee, H. J., Baek, M., Hwang, J. S., & Yoon, H. J. (2019). Comparative analysis of nutrients and hazardous substances in *Zophobas atratus* larvae. *Journal of Life Science*, 29(12), 1378–1385. https://doi.org/10.5352/JLS.2019.29.12.1378.

Kim, S. Y., Kwak, K. W., Kim, E., Park, K., Kim, N. H., Song, M. H., Kim, Y. S., & Yoon, H. J. (2020). Comparative analysis of nutrients and hazardous substances in *Locusta Migratoria* from host plants. *Korean Journal of Environmental Agriculture*, 39(3), 253–262. https://doi.org/10.5338/KJEA.2020.39.3.30.

Kim, S. K., Weaver, C. M., & Choi, M. K. (2017). Proximate composition and mineral content of five edible insects consumed in Korea. *CyTA-Journal of Food*, 15(1), 143–146. https://doi.org/10.1080/19476337.2016.1223172.

Kim, T. K., Yong, H. I., Kang, M. C., Jung, S., Jang, H. W., & Choi, Y. S. (2021). Effects of high hydrostatic pressure on technical functional properties of edible insect protein. *Food Science of Animal Resources*, 41(2), 185–195. https://doi.org/10.5851/kosfa.2020.e85.

Kingwascharapong, P., Chaijan, M., & Karnjanapratum, S. (2021). Ultrasound-assisted extraction of protein from Bombay locusts and its impact on functional and antioxidative properties. *Scientific Reports*, 11(1), 17320. https://doi.org/10.1038/s41598-021-96694-w.

Kinyuru, J. N., Konyole, S. O., Roos, N., Onyango, C. A., Owino, V. O., Owuor, B. O., Estambale B. B., Friis, H., Aagaard-Hansen J., & Kenji, G. M. (2013). Nutrient composition of four species of winged termites consumed in western Kenya. *Journal of Food Composition and Analysis*, 30(2), 120–124. https://doi.org/10.1016/j.jfca.2013.02.008.

Kipkoech, C., Kinyuru, J. N., Imathiu, S., Meyer-Rochow, V. B., & Roos, N. (2021). In vitro study of cricket chitosan's potential as a prebiotic and a promoter of probiotic microorganisms to control pathogenic bacteria in the human gut. *Foods*, 10(10), 2310. https://doi.org/10.3390/foods10102310.

Kowalski, S., Mikulec, A., Mickowska, B., Skotnicka, M., & Mazurek, A. (2022). Wheat bread supplementation with various edible insect flours. Influence of chemical composition on nutritional and technological aspects. *LWT*, 159, 113220. https://doi.org/10.1016/j.lwt.2022.113220.

Kulma, M., Kouřimská, L., Plachý, V., Božik, M., Adámková, A., & Vrabec, V. (2019). Effect of sex on the nutritional value of house cricket, *Acheta domestica* L. *Food Chemistry*, 272, 267–272. https://doi.org/10.1016/j.foodchem.2018.08.049.

Laurent, S., Jury, V., de Lamballerie, M., & Fayolle, F. (2022). Effect of two defatting processes on the physicochemical and flow properties of *Hermetia illucens* and *Tenebrio molitor* larvae powders. *Journal of Food Processing and Preservation*, 46(10), e16853. https://doi.org/10.1111/jfpp.16853.

Lee, J. H., Kim, T. K., Kim, Y. J., Kang, M. C., Song, K. M., Kim, B. K., & Choi, Y. S. (2023). Structural, physicochemical, and immune-enhancing properties of edible insect protein isolates from *Protaetia brevitarsis* larvae. *Food Chemistry: X*, 18, 100722. https://doi.org/10.1016/j.fochx.2023.100722.

Lee, J. H., Kim, T. K., Yong, H. I., Cha, J. Y., Song, K. M., Lee, H. G., Je, J. G., Kang, M. C., & Choi, Y. S. (2023). Peptides inhibiting angiotensin-I-converting enzyme: Isolation from flavourzyme hydrolysate of *Protaetia brevitarsis* larva protein and identification. *Food Chemistry*, 399, 133897. https://doi.org/10.1016/j.foodchem.2022.133897.

Lee, J. H., Kim, Y. J., Kim, T. K., Song, K. M., & Choi, Y. S. (2024). Effect of ethanol treatment on the structural, techno-functional, and antioxidant properties of edible insect protein obtained from *Tenebrio molitor* larvae. *Food Chemistry*, 437, 137852. https://doi.org/10.1016/j.foodchem.2023.137852.

Li, H., Inoue, A., Taniguchi, S., Yukutake, T., Suyama, K., Nose, T., & Maeda, I. (2017). Multifunctional biological activities of water extract of housefly larvae (*Musca domestica*). *PharmaNutrition*, 5(4), 119–126. https://doi.org/10.1016/j.phanu.2017.09.001.

Liu, X., Chen, X., Wang, H., Yang, Q., ur Rehman, K., Li, W., Cai, M., Mazza, L., Zhang, J., Yu, Z., & Zheng, L. (2017). Dynamic changes of nutrient composition throughout the entire life cycle of black soldier fly. *PLoS One*, 12(8), e0182601. https://doi.org/10.1371/journal.pone.0182601.

Lone, A. B., Bhat, H. F., Aït-Kaddour, A., Hassoun, A., Aadil, R. M., Dar, B. N., & Bhat, Z. F. (2023). Cricket protein hydrolysates pre-processed with ultrasonication and microwave improved storage stability of goat meat emulsion. *Innovative Food Science & Emerging Technologies*, 86, 103364. https://doi.org/10.1016/j.ifset.2023.103364.

Loypimai, P., Moontree, T., Pranil, T., & Moongngarm, A. (2024). A comparative study of nutritional components of *Gryllus bimaculatus* and *Acheta domesticus* cricket powder prepared using different drying methods. *Journal of Food Measurement and Characterization*, 18(5), 3974–3983. https://doi.org/10.1007/s11694-024-02469-y.

Lv, J., Lv, X., Ma, M., Oh, D. H., Jiang, Z., & Fu, X. (2023). Chitin and chitin-based biomaterials: A review of advances in processing and food applications. *Carbohydrate Polymers*, 299, 120142. https://doi.org/10.1016/j.carbpol.2022.120142.

Malm, M., & Liceaga, A. M. (2021). Physicochemical properties of chitosan from two commonly reared edible cricket species, and its application as a hypolipidemic and antimicrobial agent. *Polysaccharides*, 2(2), 339–353. https://doi.org/10.3390/polysaccharides2020022.

Matheswaran, P., Raja, L., & Gani, S. (2019). Antioxidant and anti-inflammatory efficacy of functional proteins obtained from seven edible insects. *International Journal of Entomology Research*, 4(2), 24–31.

Mekuria, S. A., Kinyuru, J. N., Mokua, B. K., & Tenagashaw, M. W. (2021). Nutritional quality and safety of complementary foods developed from blends of staple grains and honey bee larvae (*Apis mellifera*). *International Journal of Food Science*, 2021(1), 5581585. https://doi.org/10.1155/2021/5581585.

Melgar-Lalanne, G., Hernández-Álvarez, A. J., & Salinas-Castro, A. (2019). Edible insects processing: Traditional and innovative technologies. *Comprehensive Reviews in Food Science and Food Safety*, 18(4), 1166–1191. https://doi.org/10.1111/1541-4337.12463.

Mihaly Cozmuta, A., Nicula, C., Peter, A., Mihaly Cozmuta, L., Nartea, A., Kuhalskaya, A., Pacetti, D., Silvi, S., Fiorini, D., & Pruteanu, L. (2022). Cricket and yellow mealworm powders promote higher bioaccessible fractions of mineral elements in functional bread. *Journal of Functional Foods*, 99, 105310. https://doi.org/https://doi.org/10.1016/j.jff.2022.105310.

Muinat, M. I., & Tariq, A. M. (2021). Evaluation of proximate compositions of some edible insects in Zaria Kaduna State Nigeria. *International Journal of Science for Global Sustainability*, 7(1), 44–52. https://doi.org/10.57233/ijsgs.v7i1.45.

Nakagawa, K., Chantanuson, R., Boonarsa, P., Seephua, N., & Siriamornpun, S. (2024). Meat analogue preparation from cricket and rice powder mixtures with controlled textural and nutritional quality by freeze alignment technique. *Food Chemistry: X*, 22, 101402. https://doi.org/10.1016/j.fochx.2024.101402.

Nantanga, K. K., & Amakali, T. (2020). Diversification of mopane caterpillars (*Gonimbrasia belina*) edible forms for improved livelihoods and food security. *Journal of Arid Environments*, 177, 104148. https://doi.org/10.1016/j.jaridenv.2020.104148.

Ntukuyoh, A. I., Udiong, D. S., Ikpe, E., & Akpakpan, A. E. (2012). Evaluation of nutritional value of termites (*Macrotermes bellicosus*): soldiers, workers, and queen in the Niger Delta region of Nigeria. *International Journal of Food Nutrition and Safety*, 1(2), 60–65.

Ochieng, B. O., Anyango, J. O., Khamis, F. M., Ekesi, S., Egonyu, J. P., Subramanian, S., Nduko, J. M., Nakimbugwe, D., Cheseto, X., Tanga, C. M. (2023). Nutritional characteristics, microbial loads and consumer acceptability of cookies enriched with insect (*Ruspolia differens*) meal. *LWT*, 184, 115012. https://doi.org/10.1016/j.lwt.2023.115012.

Ohta, T., Kusano, K., Ido, A., Miura, C., & Miura, T. (2016). Silkrose: A novel acidic polysaccharide from the silkmoth that can stimulate the innate immune response. *Carbohydrate Polymers*, 136, 995–1001. https://doi.org/10.1016/j.carbpol.2015.09.070.

Orkusz, A., Dymińska, L., Banaś, K., & Harasym, J. (2023). Chemical and nutritional fat profile of *Acheta domesticus*, *Gryllus bimaculatus*, *Tenebrio molitor* and *Rhynchophorus ferrugineus*. *Foods*, 13(1), 32. https://doi.org/10.3390/foods13010032.

Otero, P., Gutierrez-Docio, A., del Hierro, J. N., Reglero, G., & Martin, D. (2020). Extracts from the edible insects *Acheta domesticus* and *Tenebrio molitor* with improved fatty acid profile due to ultrasound assisted or pressurized liquid extraction. *Food Chemistry*, 314, 126200. https://doi.org/10.1016/j.foodchem.2020.126200.

Pasini, G., Cullere, M., Vegro, M., Simonato, B., & Dalle Zotte, A. (2022). Potentiality of protein fractions from the house cricket (*Acheta domesticus*) and yellow mealworm (*Tenebrio molitor*) for pasta formulation. *LWT*, 164, 113638. https://doi.org/10.1016/j.lwt.2022.113638.

Pasqualin Cavalheiro, C., Ruiz-Capillas, C., Herrero, A. M., Pintado, T., Avelar de Sousa, C. C., Sant'Ana Falcão Leite, J., & Costa Alves da Silva, M. (2024). Potential of cricket (*Acheta domesticus*) flour as a lean meat replacer in the development of beef patties. *Foods*, 13(2), 286. https://doi.org/10.3390/foods13020286.

Pranil, T., Moontree, T., Moongngarm, A., & Loypimai, P. (2024). Ultrasound-assisted extraction of edible cricket oils from *Gryllus bimaculatus* and *Acheta domesticus*. *Journal of Food Measurement and Characterization*. https://doi.org/10.1007/s11694-024-02634-3.

Psarianos, M., Dimopoulos, G., Ojha, S., Cavini, A. C. M., Bußler, S., Taoukis, P., & Schlüter, O. K. (2022). Effect of pulsed electric fields on cricket (*Acheta domesticus*) flour: Extraction yield (protein, fat and chitin) and techno-functional properties. *Innovative Food Science & Emerging Technologies*, 76, 102908. https://doi.org/10.1016/j.ifset.2021.102908.

Psarianos, M., Iranshahi, K., Rossi, S., Gottardi, D., & Schlüter, O. (2024). Quality evaluation of house cricket flour processed by electrohydrodynamic drying and pulsed electric fields treatment. *Food Chemistry*, 441, 138276. https://doi.org/10.1016/j.foodchem.2023.138276.

Raksakantong, P., Meeso, N., Kubola, J., & Siriamornpun, S. (2010). Fatty acids and proximate composition of eight Thai edible terricolous insects. *Food Research International*, 43(1), 350–355. https://doi.org/10.1016/j.foodres.2009.10.014.

Ravzanaadii, N., Kim, S. H., Choi, W. H., Hong, S. J., & Kim, N. J. (2012). Nutritional value of mealworm, *Tenebrio molitor* as food source. *International Journal of Industrial Entomology*, 25(1), 93–98. https://doi.org/10.7852/ijie.2012.25.1.093.

Rocchetti, G., Leni, G., Rebecchi, A., Dordoni, R., Giuberti, G., & Lucini, L. (2024). The distinctive effect of different insect powders as meat extenders in beef burgers subjected to cooking and in vitro gastrointestinal digestion. *Food Chemistry*, 442, 138422. https://doi.org/10.1016/j.foodchem.2024.138422.

Sánchez, M., Gómez, C., Avendaño, C., Harmsen, I., Ortiz, D., Ceballos, R., Villamizar-Sarmiento, M. G., Oyarzun-Ampuero, F., Wacyk, J., & Valenzuela, C. (2021). House fly (*Musca domestica*) larvae meal as an ingredient with high nutritional value: Microencapsulation and improvement of organoleptic characteristics. *Food Research International*, 145, 110423. https://doi.org/10.1016/j.foodres.2021.110423.

Schmidt, A., Call, L. M., Macheiner, L., & Mayer, H. K. (2019). Determination of vitamin B12 in four edible insect species by immunoaffinity and ultra-high performance liquid chromatography. *Food Chemistry*, 281, 124–129. https://doi.org/10.1016/j.foodchem.2018.12.039.

Scholliers, J., Steen, L., & Fraeye, I. (2020). Partial replacement of meat by superworm (*Zophobas morio* larvae) in cooked sausages: Effect of heating temperature and insect:Meat ratio on structure and physical stability. *Innovative Food Science & Emerging Technologies*, 66, 102535. https://doi.org/10.1016/j.ifset.2020.102535.

Seo, M., Goo, T. W., Chung, M. Y., Baek, M., Hwang, J. S., Kim, M. A., & Yun, E. Y. (2017). *Tenebrio molitor* larvae inhibit adipogenesis through AMPK and MAPKs signaling in 3T3-L1 adipocytes and obesity in high-fat diet-induced obese mice. *International Journal of Molecular Sciences*, 18(3), 518. https://doi.org/10.3390/ijms18030518.

Shorstkii, I., Comiotto Alles, M., Parniakov, O., Smetana, S., Aganovic, K., Sosnin, M., Toepfl, S., & Heinz, V. (2022). Optimization of pulsed electric field assisted drying process of black soldier fly (*Hermetia illucens*) larvae. *Drying Technology*, 40(3), 595–603. https://doi.org/10.1080/07373937.2020.1819825.

Soetemans, L., Uyttebroek, M., & Bastiaens, L. (2020). Characteristics of chitin extracted from black soldier fly in different life stages. *International Journal of Biological Macromolecules*, 165, 3206–3214. https://doi.org/10.1016/j.ijbiomac.2020.11.041.

Son, Y. J., Choi, S. Y., Hwang, I. K., Nho, C. W., & Kim, S. H. (2020). Could defatted mealworm (*Tenebrio molitor*) and mealworm oil be used as food ingredients? *Foods*, 9(1), 40. https://doi.org/10.3390/foods9010040.

Stull, V. J., Finer, E., Bergmans, R. S., Febvre, H. P., Longhurst, C., Manter, D. K., Patz, J. A., & Weir, T. L. (2018). Impact of edible cricket consumption on gut microbiota in healthy adults, a double-blind, randomized crossover trial. *Scientific Reports*, 8(1), 1–13. https://doi.org/10.1038/s41598-018-29032-2.

Sweers, L. J. H., Keppler, J. K., Feng, S., Aponte Zea, J., van Bokhorst-van de Veen, H., Timmermans, R. A. H., Boom, R. M., Fogliano, V., Lakemond, C. M. M., & Mishyna, M. (2024). High hydrostatic pressure for decontamination of soluble insect proteins prevents protein denaturation better than blanching. *Innovative Food Science & Emerging Technologies*, 96, 103743. https://doi.org/10.1016/j.ifset.2024.103743.

Udomsil, N., Imsoonthornruksa, S., Gosalawit, C., & Ketudat-Cairns, M. (2019). Nutritional values and functional properties of house cricket (*Acheta domesticus*) and field cricket (*Gryllus bimaculatus*). *Food Science and Technology Research*, 25(4), 597–605. https://doi.org/10.3136/fstr.25.597.

Van Huis, A., Van Itterbeeck, J., Klunder, H., Mertens, E., Halloran, A., Muir, G., & Vantomme, P. (2013). Edible insects: future prospects for food and feed security. *FAO Forestry Paper*, 171, 1–201. http://www.fao.org/docrep/018/i3253e/i3253e.pdf.

Wu, R. A., Ding, Q., Lu, H., Tan, H., Sun, N., Wang, K., He, R., Luo, L., Ma, H., & Li, Z. (2020). Caspase 3-mediated cytotoxicity of mealworm larvae (*Tenebrio molitor*) oil extract against human hepatocellular carcinoma and colorectal adenocarcinoma. *Journal of Ethnopharmacology*, 250, 112438. https://doi.org/10.1016/j.jep.2019.112438.

Xu, J., Xiao, S., Wang, J., Wang, B., Cai, Y., & Hu, W. (2023). Comparative study of the effects of ultrasound-assisted alkaline extraction on black soldier fly (*Hermetia illucens*) larvae protein: Nutritional, structural, and functional properties. *Ultrasonics Sonochemistry*, 101, 106662. https://doi.org/10.1016/j.ultsonch.2023.106662.

Xu, X., Ji, H., Belghit, I., Liland, N. S., Wu, W., & Li, X. (2021). Effects of black soldier fly oil rich in n-3 HUFA on growth performance, metabolism and health response of juvenile mirror carp (*Cyprinus carpio* var. specularis). *Aquaculture*, 533, 736144. https://doi.org/10.1016/j.aquaculture.2020.736144.

Yeruva, T., Jayaram, H., Aurade, R., Shunmugam, M. M., Shinde, V. S., Venkatesharao, S. R. B., & Azhiyakathu, M. J. (2023). Profiling of nutrients and bioactive compounds in the pupae of silkworm, *Bombyx mori*. *Food Chemistry Advances*, 3, 100382. https://doi.org/10.1016/j.focha.2023.100382.

Yong, H. I., Kim, T. K., Cha, J. Y., Lee, J. H., Kang, M. C., Jung, S., & Choi, Y. S. (2023). Effects of edible insect extracts on the antioxidant, physiochemical, and microbial properties of Tteokgalbi during refrigerated storage. *Food Bioscience*, 52, 102377. https://doi.org/10.1016/j.fbio.2023.102377.

Yu, X., He, Q., & Wang, D. (2021). Dynamic analysis of major components in the different developmental stages of *Tenebrio molitor*. *Frontiers in Nutrition*, 8, 689746. https://doi.org/10.3389/fnut.2021.689746.

Zhang, F., Cao, C., Kong, B., Sun, F., Shen, X., Yao, X., & Liu, Q. (2022). Pre-dried mealworm larvae flour could partially replace lean meat in frankfurters: Effect of pre-drying methods and replacement ratios. *Meat Science*, 188, 108802. https://doi.org/10.1016/j.meatsci.2022.108802.

Zhang, F., Sun, Z., Li, X., Kong, B., Sun, F., Cao, C., Chen, Q., Zhang, H., Liu, Q. (2023). Ultrasound-assisted alkaline extraction of protein from *Tenebrio molitor* larvae: Extraction kinetics, physiochemical, and functional traits. *Ultrasonics Sonochemistry*, 95, 106379. https://doi.org/10.1016/j.ultsonch.2023.106379.

Zhu, D., Huang, X., Tu, F., Wang, C., & Yang, F. (2020). Preparation, antioxidant activity evaluation, and identification of antioxidant peptide from black soldier fly (*Hermetia illucens* L.) larvae. *Journal of Food Biochemistry*, 44(5), e13186. https://doi.org/10.1111/jfbc.13186.

Zielińska, E., Baraniak, B., & Karaś, M. (2017). Antioxidant and anti-inflammatory activities of hydrolysates and peptide fractions obtained by enzymatic hydrolysis of selected heat-treated edible insects. *Nutrients*, 9(9), 970. https://doi.org/10.3390/nu9090970.

Zulkifli, N. F. N. M., Seok-Kian, A. Y., Seng, L. L., Mustafa, S., Kim, Y. S., & Shapawi, R. (2022). Nutritional value of black soldier fly (Hermetia illucens) larvae processed by different methods. *PLoS One*, 17(2), e0263924. https://doi.org/10.1371/journal.pone.0263924.

7 Regulatory and Legislative Implications of the Use of Insects for Human Consumption

Simone Belluco, Juliane Pinarelli Fazion, and Beatrice La Porta

7.1 INTRODUCTION

The term Policy, according to the Cambridge Dictionary, is defined as "a set of ideas or a plan of what to do in particular situations that has been agreed to officially by a group of people, a business organisation, a government, or a political party". Requiring an agreement by a group of people, a population in the case of a country, they are strongly linked to culture and values. In this framework, food safety policies are not an exception.

The evolution of food safety policies has been driven by the need to address the increasing demands of consumers and the growing public health concerns associated with food safety. This has resulted in a shift towards comprehensive, risk-based approaches applied throughout the food supply chain. The advent of risk-based, farm-to-fork food safety systems in the Organization for Economic Co-Operation and Development (OECD) countries has been informed by international deliberations through the Codex Alimentarius (Hoffmann, 2010). While food security remains the predominant policy

DOI: 10.1201/9781003488156-7

approach, food sovereignty has emerged as a critical response to be considered in policy design and research. Food sovereignty refer to the right for people to control how and what kind of food is produced (Sampson et al., 2021).

However, the interplay between global socio-demographic shifts and evolving demands in the agri-food market (Esposti et al., 2008; Ramírez Nárdiz, 2013; Rollin et al., 2011), coupled with technological advancements in production processes (Sirsi, 2013), the globalisation of trade, and emerging dietary and environmental concerns, has heightened consumer and business operator interest in the introduction of products previously not widely consumed within the European Union (La Porta, 2021a).

Edible insects are a perfect case to discuss such differences. Insects recognize a long history of human consumption in several countries worldwide, with more than 2000 species used as food (Omuse et al., 2024). According to FAO, more than 2 billion people consume insects that, in some countries, represent a staple food and provide an important amount of essential nutrients (FAO, 2013). In other countries, they are considered novel and/or non-traditional food with a need for safety assessment and pre-market approval.

From a legislative point of view, the situation is very heterogeneous and quite anomalous because edible insects range from being a staple food with a long history of consumption in some countries to a novel food with the need for pre-market approval in others.

Differences in the safety "threshold" among countries are based on the modern mechanisms of policymaking, where institutional risk manager set the Appropriate Level of Protection (ALOP) for the population of their territories of competence. In a rationale system, ALOP should guide the choice of Food Safety Objective (FSO) and Performance Objective (PO) in a system able to implement a set of microbiological and chemical criteria driving producers towards the production of safe food and achieving the expected level of protection of the population (Gkogka et al., 2013). It is acknowledged that food safety standards applied in developed countries could heavily affect developing ones, acting as barriers to exports. However, they can also act as catalysts for implementing food safety systems (Henson & Jaffee, 2006).

The need for pre-market approval is justified by the application, in some countries like the European Union (EU), of the precautionary principle.[1] In the EU, for example, its application to novel food drives the current situation where if potential risks from the consumption of new foods are identified, but scientific uncertainty persists, provisional risk management measures necessary to ensure the high level of health protection chosen in the community may be adopted, pending further scientific information for a more comprehensive risk assessment.

Food safety standards need to move forward with recognizing elements other than food safety alone, such as food security, sustainability, and animal welfare, to find the appropriate trade-off (Vågsholm et al., 2020). The growing importance of all health-related aspects, in the One Health perspective, is evident, for example, in the Regulation (EU) 625/2017, where the definition of hazard has been extended to go beyond full safety and towards a comprehensive understanding of process impacts.[2] Thus, for example, "hazard" means any agent or condition with the potential to have an adverse effect on human, animal, or plant health, animal welfare, or the environment, whereas in the previous regulation, only adverse health effects were considered (Hermann, 2009).

7.2 THE LEGAL FRAMEWORK OF NOVEL FOOD IN THE EUROPEAN UNION

"New" products intended for human consumption are classified as "novel foods" under Article 3 of Regulation (EU) 2283/2015 (the Novel Food Regulation or NFR),[3] provided they meet two criteria: (i) they fall under one of the categories listed in the Regulation (the material criterion) and (ii) they have not been significantly consumed within the Union market before 15 May 1997[4] (the temporal criterion).[5]

In force since 1 January 2018, the NFR has redefined the legal framework for novel foods, previously governed by Regulation (EC) 258/1997, to better ensure the free movement of safe and healthy new food products within the EU and provide broader protection for emerging social and economic interests (Russo, 2010). To ensure comprehensive consumer protection and compliance with the EU's stringent food safety standards, the European Commission is the sole institution responsible for deciding on the introduction of a novel food into the EU market[6] and for updating the Union's list of authorized food products. This list serves as an inventory of all food products permitted for trade within the Union, specifying, for each novel food included, the established requirements regarding conditions of use, restrictions, and labelling (Paganizza, 2019).

The establishment of the European Food Safety Authority (EFSA) by Regulation CE 178/2002 marked a critical evolution in the EU's food safety system, centralizing scientific evaluations to promote uniformity across Member States (Hood et al., 2003). As the Union's food safety risk assessor, EFSA's primary function is to provide independent, science-based evaluations, which have played a pivotal role in enhancing food safety across the EU, especially in the novel food sector (Gabbi, 2025), so contributing to the harmonisation of risk assessment methodologies, including those related to emerging risks.[7]

While EFSA's mandate focuses on risk assessment, the responsibility for risk management – namely, the final decision regarding including on not new products in the Union's food list – remains with the European Commission, which is not bound by EFSA's opinions (Majone, 2010). This division of roles ensures a clear distinction between scientific evaluation and policy decisions, creating a system where independent risk assessments guide regulatory actions. The resulting framework is thus characterized by a complex interplay between various actors, wherein those responsible for identifying risks and those tasked with managing them are closely interconnected, particularly given the scientific uncertainty and urgency that frequently characterize the food sector.[8]

To achieve the specified safety, social, and economic objectives, the NFR established a centralized risk assessment and authorisation system for novel foods, moving away from a previous model that granted authorisation solely to the individual applicant. Under the new system, authorisation is granted for the product's market entry, thus enabling any food business operator to introduce the product into the market.

Furthermore, the NFR also established two distinct administrative procedures differing in length and complexity by their articulation based on the characteristics of the under-analysis-product and its demonstrated – or not – previous consumption as traditional food in a third country (La Porta, 2020).

7.2.1 Edible Insects as Novel Food Products

The establishment of a centralized system, governed by decisions of the European Commission and supported by the involvement of EFSA, represents a significant departure of the NFR from the previous framework under Regulation (EC) 258/1997, as implemented by Recommendation 97/618/EC,[9] which delegated responsibility for the food safety evaluation to individual Member States.

The importance of a centralized system becomes evident when considering the divergences in interpreting and applying the Regulation (EC) 258/1997 that previously emerged among EU Member States based on their degree of regulatory progressiveness or conservatism. The NFR has reduced these inconsistencies, and it is extremely clear with reference to the case of edible insects (Belluco et al., 2017).

According to Article 1(e) of Regulation (EC) No 258/1997, within the categories under which a product not commonly consumed before 15 May 1997 would be considered a novel food must also be included "*foods and food ingredients consisting of or isolated from plants and food ingredients isolated from animals*". For years, the absence of the word "*food*" and the omission of the phrase "*consisting of*" in relation to food ingredients of animal origin gave rise to interpretative challenges. These challenges enabled certain Member States to permit the marketing of edible insect species, both in whole form and in products derived from such animals, while excluding ingredients isolated from insects, which were explicitly covered under Regulation (EC) No 258/1997.[10] An interpretation that was confirmed by the 2019 Court of Justice of the European Union's ruling with reference to a case under the application of the 1997 regulation.[11]

In Italy, as in other Union countries, a more restrictive stance was adopted regarding novel foods, rejecting broader interpretations of the Regulation's letters[12] and adopting a zero-tolerance policy about the presence of insects in the Italian market. However, edible insects had been available on the market for over a decade[13] in some EU countries, particularly Northern Europe (Belluco et al., 2017). This position of tolerance, coupled with a fragmented system that allowed national authorities to decide on individual products, created years of significant uncertainty for market operators and internal imbalances within the Union, effectively creating a "two-speed" EU and violating the principles outlined in the Treaty on the Functioning of the European Union.

The Union legislator enacted the NFR to harmonize the application of the legal framework and to address the ongoing violation of the core principle of free trade within the Union market.[14] Any commercial regulation by Member States that directly or indirectly hinders intra-Union trade, either in fact or potentially, must be viewed as a barrier, a measure having an effect equivalent to quantitative restrictions.

Despite the unique characteristics of foods and food ingredients consisting of or derived from insects and their consumption in the EU before the implementation of the new novel food regulation – due to prior interpretative gaps – in 2015, the Union legislator opted to classify insects as food products for which no safe use had been established within Union territory. As a result, Recital 8 of the NFR recommends the inclusion of "*whole insects and their parts*" within the categories of novel foods, as further clarified in Article 3(2)(a)(v) of the NFR.

Nonetheless, the new EU regulation represents a significant advancement by introducing a distinction within the novel food framework between traditional products from third countries – a category into which insects can fall – and entirely new products, a differentiation that was absent under the previously applicable Regulation (EC) 258/1997. This distinction is intended to acknowledge the specific cultural and historical contexts in which certain foods have been safely consumed for extended periods outside the EU, thereby reducing the regulatory burden for products with a well-documented history of safe use. The regulation fosters innovation while ensuring food safety by establishing a distinct pathway for such products. As a result, this differentiation involves the application of a separate procedure for placing a novel food on the market (Bonora, 2016).

This differentiation in the regulatory process finds no precedent either in the sector-specific legislation applied until December 2017 or in the general framework for food regulation.[15] It was only with Article 3(2)(c) of the NFR that a "*traditional food from a third country*" was defined for the first time as any novel product, thereby having to meet both temporal and material criteria.[16]

However, what distinguishes a traditional product from an entirely new one is "*a history of safe food use in a third country*" and its necessary derivation from "primary production",[17] regardless of how it is processed.[18]

The rationale for this differentiation lies in acknowledging the cultural, nutritional, and environmental diversity of food production across different regions, thus allowing third countries to more easily access the EU market while ensuring that food safety standards are met, as expressly emphasized in Recital 15 of the NFR.

While the definition of the first requirement appears clear and immediately applicable, the reference to "*a history of safe food use in a third country*", meaning a country outside the European Union, necessitates a more in-depth interpretative analysis, considering the entirety of the 2015 Regulation (La Porta, 2020).

Regarding the history of safe food use, Recital 15 of Regulation (EU) 2283/2015 specifies that safety must be demonstrated not only by data defining the product's composition but also by "*Those foods should have been consumed in at least one-third country for at least 25 years as a part of the customary diet of a significant number of people. The history of safe food use should not include non-food uses or uses not related to normal diets*". While the Union legislator's decision to establish a period of a minimum of 25 years avoids the more generic reference to a "generation" as seen in the unadopted 2008 draft proposal,[19] it still appears somewhat arbitrary. Furthermore, broad and undefined terms remain, such as "*a significant number of people*" and the reference to a "*customary diet*".[20]

The distinction between traditional products and entirely new foods outlined in the NFR is linked to the definition of a different procedure depending on the type of product the applicant wishes to introduce to the Union market, and it potentially affects edible insects due to their large consumption in many countries worldwide. Although both completely new products and traditional products from third countries must undergo procedures that are "efficient, of limited duration, and transparent," only for products with demonstrated prolonged and safe use in third countries, the applicant can choose between applying the fast-tracked and streamlined procedure, initiated by notifying the Commission,[21] or the ordinary procedure under Article 10 of the NFR, making clear that the simplified process involves reduced documentation and shorter timelines, beginning with a mere notification, as defined under Article 15 and follows of the NFR.[22]

In both procedures, the role of EFSA is significant in relation to novel food risk assessment and the potential consequences of introducing new products into the market, with specific attention to vulnerable population groups.[23] However, the operator's ability to use the streamlined procedure under Article 18 of the NFR for products already consumed safely in third countries is subject to significant limitations, primarily due to the non-applicability of the scientific data protection rules provided for in Article 26 of the NFR.[24]

This protection is justified by the transition from the decentralized system established under the 1997 regulation – where authorisations were granted to the producer – to the centralized system introduced in 2015, which authorizes the product and adds it to a Union-wide food list. Consequently, measures are needed to safeguard applicants' research and development (R&D) investments, which support new submissions. Article 26 of the NFR provides applicants who opt for the standard procedure with protection over the data submitted in their application, preventing its use by third parties for 5 years. This system of protection mirrors the framework governing nutritional and health claims.[25]

To prevent the risk of discouraging private investment in the R&D of new foods, ingredients, and production technologies, the Union legislator in 2015 provided for the possibility that the applicant seeking authorisation for a novel food[26] could, upon justified request, obtain a standard, non-renewable 5-year period during which third parties could not use all new evidence and scientific data submitted by the first applicant under the Article 10 NFR application without previously agreeing with him.

The 5-year period, which begins on the date of authorisation,[27] must also be understood in light of the provisions of Article 28 of the NFR, which governs the authorisation procedure in the event of a parallel application for a health claim authorisation. Where there are simultaneous applications for novel food authorisation and health claim authorisation, the timelines are aligned so that the protection period granted under Article 21 of Regulation (EC) No 1924/2006 cannot exceed the time limit set by the novel food regulation.[28]

The data protection granted under NFR and the resulting exclusive placing of the product on the market for 5 years, benefitting the first applicant, was recognized by the Commission in the case of the first authorisation of an insect food product: dried *Tenebrio molitor* larvae under Commission Implementing Regulation (EU) 882/2021.[29] More broadly, it should be noted that, as of November 2024, no application for recognizing edible insects – animals widely consumed in the food cultures of many third-country populations – has been submitted through the simplified process reserved for traditional foods.[30]

With regard to the application of Regulation (EU) 882/2021 in Italy, challenges have arisen due to the stringent interpretation and application of the "data protection principle" established by Article 26 of the NFR. Restrictions have been placed on the production and marketing of *Tenebrio molitor* by business operators not directly mentioned in the authorizing regulation by the European Commission but who have entered into commercial agreements with the initial applicant, as identified under Article 2 of Regulation (EU) 882/2021, for the commercialisation of the authorized novel food. During the 5-year protection period, the Italian Ministry of Health granted the initial applicant exclusive rights to restrict subsequent applicants from using their scientific data to support independent authorisation requests without consent, thereby ensuring the initial applicant's exclusive control over the use of such data in the authorisation process. According to the Italian Ministry's interpretation, business operators who have received the first applicant's data – whether freely or for a fee – are prohibited from marketing food products derived under Regulation (EU) 882/2021 and may only use the data to submit a new authorisation application before the European Commission. This authorisation process, however, entails substantial time and financial costs, potentially resulting in approval only after the expiration of the 5-year protection period, thus reducing the practical benefit of agreements with the initial applicant and limiting his potential economic advantage in monetizing his legal prerogatives. By invoking food safety considerations to justify its position, the interpretation adopted by the Italian Ministry of Health appears to overlook the true economic rationale behind Article 26 of the NFR, as clarified in Recital 30, thereby constraining the advantages of choosing the more complex authorisation procedure.

Just as the Italian application of data protection rights under Regulation (EU) 882/2021 seems to frustrate the initial intent of the EU legislator and lacks alignment with the practices of other Member States, another aspect of the NFR that appears similarly undermined is the ease of introducing new products through the simplified procedure outlined in Article 14 et seq. of the NFR. Indeed, examining the evaluated applications reveals that the limited use of this shorter procedure has hindered the anticipated support for business operators, particularly those in developing or underdeveloped countries, thereby contravening the Union legislator's original objectives.

The forecast of a different procedure for foods with a history of safe use in non-EU countries, as opposed to entirely new products, can be traced back to the challenges that arose during the application of Regulation (EC) 258/1997. The lack of distinction between products and the requirement for a lengthy, costly, and uncertain authorisation process[31] led to significant application difficulties. The complex compatibility with World Trade Organization (WTO) rules[32] prompted the Court of Justice of the European Union (CJEU)[33] to consider that such a regime on novel foods could hinder the free movement of goods internationally, where the barriers to marketing products from third countries not recognized by the EU as food were disproportionate to the actual risks to consumer health (Hermann, 2009).

Due to the significant difficulties in introducing new foods into the Union under Regulation (EC) 258/1997 (Lahteenmaki-Uutela, 2007), several Latin American countries over the years expressed concerns[34] about the non-compliance of the regulation with WTO rules on the free movement of goods (Volpato, 2015), considering Union regulations as technical barriers that were difficult to overcome (Bronckers & Soopramanien, 2008; Downes, 2013; Streinz, 1998). However, no formal complaint was ever filed with the WTO, and the European Union consistently rejected these concerns, asserting that its regulations complied with the WTO Agreement on Technical Barriers to Trade (the so-called OTC Agreement).[35]

While numerous practical challenges persist, the changes implemented in 2015 have potentially attracted interest from countries in Africa and South America. In these regions, the opportunity to access new markets with traditional products not yet present in the EU, such as insects, could potentially represent new economic opportunities (Volpato, 2015). Nevertheless, the initial optimism observed when the NFR entered into force had to be reconsidered following the subsequent issuance of guidelines by EFSA.[36] These guidelines required submitting information on the continued use of the product supported by scientific publications, expert assessments, monographs, information from international or national organisations, government documents, or commercial sales data. This requirement also provided information about the product's contribution to the local population's overall intake of macro- and micronutrients (La Porta, 2020). Moreover, in the years following the regulation, for products widely consumed abroad, such as insects, applicants generally preferred to follow the standard authorisation procedure even without invoking the data protection provided under Article 26 of the NFR, frustrating the initial optimism regarding an easier and more immediate introduction of insects into the EU market.

By applying EFSA's guidelines, challenges have arisen regarding the introduction of edible insects into the market as part of efforts to promote a more sustainable system and support circular economy processes, with the aim of achieving the objectives of the EU Green Deal, combating climate change, and addressing environmental degradation.[37] Under these guidelines and the EU Implementing Regulations that authorized specific insect species for the market, insects intended for human consumption must not be fed on waste or scraps; rather, they are required to receive the appropriate feed specified in the authorisation procedure to minimize health risks.

7.3 THE REGULATORY FRAMEWORK OF NOVEL FOOD IN NON-EU COUNTRIES

7.3.1 United States

Concerning the farming and commercialisation of edible insects in the United States, the recognition of the ecological benefits of insect farming – such as the reduced use of resources like water and land, along with the high nutritional content they provide (FAO, 2013) – has led to the frequent appearance of insect-based products on the market, with numerous companies engaged in this sector.

Unlike the regulatory framework in the EU, the commercialisation of edible insects in the US has been facilitated by the absence of explicit legal references to insects as food and the lack of specific regulations governing the marketing of "novel" products. Additionally, the divergent application of the precautionary principle between EU food legislation and the U.S. food law has practical implications for the introduction of new products into the market (Cohen & Broad Leib, 2024). The Federal Food, Drug, and Cosmetic Act (FFDCA) provides a broad definition of the term "food".[38] Accordingly, insects intended for human consumption may fall into this definition and, as such, must comply with the food safety and hygiene regulations established by the FFDCA (Boyd, 2017). The U.S. regulatory approach focuses on post-market enforcement rather than pre-market precaution. This difference reflects the broader U.S. preference for industry self-regulation, where regulatory action is typically prompted only by significant risks identified through scientific evidence, relying more on reactive than preventive measures. Concerning novel foods, U.S. regulation does not employ a centralized system akin to the EU. Instead, the U.S. Food and Drug Administration (FDA) primarily oversees food safety through the FFDCA, which categorizes any new food or food ingredient intended for the market as either generally recognized as safe (GRAS) or requiring pre-market approval. Unlike the EU's detailed procedures for novel foods, the U.S. system allows manufacturers to self-certify GRAS status, avoiding direct FDA oversight unless a specific safety concern arises (Broad Leib & Pollans, 2019). Alternatively, GRAS status can be obtained through a scientific evaluation process or based on a substantial history of consumption by a significant number of people prior to 1958.

However, under sections 201(s) and 409 of the FFDCA, any substance intentionally added to food that becomes an integral part of the food or changes its characteristics is classified as a food additive and subject to mandatory FDA approval. According to this law, a substance that is not generally recognized and whose safe use has not been sufficiently demonstrated must undergo a pre-market evaluation by the relevant authority unless it falls within one of the exclusions from the definition of a food additive (Laurenza & Carreño, 2015).

Insect-based products must also meet the general requirements for all food products marketed in the United States, including undergoing specific microbiological testing. When classified as food additives, insect-based products must comply with the requirements set out in § 348 of the FFDCA[39] concerning "unsafe food additives", including provisions related to the "*conditions of safe use; contents; description of production methods and controls; samples; notice of regulation*".

Insect food products must comply with the standards established by the FFDCA to ensure they are produced, stored, and transported under hygienic conditions. It is crucial that insects intended for human consumption are specifically raised for this purpose. Insects raised for alternative uses, such as animal feed or those harvested from natural environments, are not deemed suitable for human consumption.

In addition to the FFDCA provisions, insect-based food products must be produced in regulated facilities in accordance with the manufacturing rules established under Good Manufacturing Practice (GMP), as defined by the FDA. Pre-packaged finished products must also bear labels specifying both the common and scientific names of the insects included, and they must comply with allergen-labelling requirements applicable to shellfish due to the significant similarities in allergenic risk between the two product types. The information provided on the label must be accurate, truthful, and not misleading or incomplete in any way, as it could potentially mislead consumers. In contrast to the European Union, which adopted a more lenient approach, the United States has approached the commercialisation of edible insects with cautious openness. This has allowed many entrepreneurs to explore the potential of this market, offering consumers access to products that are not traditionally part of Western diets.

7.3.2 Canada

Canadian legislation is, for the general approach towards novel food, similar to the European one as, also in this case, the definition comprises a material criterion, defined by the category listed in the regulations[40] and a temporal criterion.[41] Food that meet both criteria are novel and need a pre-market assessment. However, two crucial differences with the European Union exist in the temporal criterion. The first relies on the fact that the criterion is not purely temporal, as "safe use" is mentioned (Note 58) and not only use. "Safe use" is otherwise mentioned in the "traditional food from third countries" authorisation pathway that, however, as discussed above is of uncertain application in the case of insects. The second one regards the way the temporal criterion is evaluated because this requirement can also be demonstrated by relying on experiences from third countries. The reliability of the source of information is appraised, giving higher confidence to countries with a jurisdiction similar to the Canadian one (Belluco et al., 2017) and not, as in the EU, considering exclusively data referring to the EU itself.

Thus, insects could be defined as novel food within the category a) if no history of safe use as food is demonstrated or falls outside the definition and consequently able to access the market if such history is provided. Despite the similarities with the EU NFR, a basic difference exists in the ratio of legislation, as in Canada, there is a single binary path (novel food or not), whereas in the EU NFR, there is a first binary path, with a second branching in case of novel food where an additional choice should be made to follow the full authorisation procedure or the streamlined one for traditional food.

Details on the demonstration of a history of safe use for food of animal origins are not available among accessible online information. However, looking into different food categories, it is possible to suppose that a number of information would be needed to support a claim that a product has a history of safe use. The first one regards the respect of the pure temporal criterion and asks for the provision of historical evidence indicating ongoing, frequent consumption by a cross-section of the population where it has been used over several generations. Compared to the EU definition of "history of safe consumption", it is stricter in terms of the required number of generations (several vs. 25 years) and lighter in terms of the portion of the population consuming it (cross-section vs. significant part). Despite the different wording, the frequency of consumption should be regular in both cases (ongoing, frequent vs. customary), and limited, short-term uses should not be admitted. Evidence may be obtained, for example, from scientific publications, patents, non-scientific publications, books, cookbooks, and/or affidavits from two or more independent, reputable authorities that include well-documented accounts of food use and knowledge sources.

To complete the criterion regarding its non-temporal component, the safety of use should be demonstrated. Information should be provided on the adverse effects following consumption in relevant countries.

As shown in Table 7.1, several insect species and derived products are already considered as non-national in Canada. Species outside this list, such as grasshoppers or black soldier flies, require a novel food determination during which companies must demonstrate a history of safe consumption.

In the case of insect species resulting as novel food, after the novelty determination process, a complete demonstration of the product's nutritional, chemical, microbiological, and toxicological safety before going to market.[42] This should include reliable, high-quality information that has to be provided by the applicant, including a description of how the food is produced, the expected consumption level within the Canadian population, and the composition of the food based on randomly selected, statistically valid samples with particular attention to the presence of compounds in the food which may have implications for the health of any groups of the Canadian population.

After the Insect's approval, Section 7 of the Canadian Food and Drugs Act states that establishments that wish to process insects should operate under good manufacturing practices and have a sampling program in place to proceed with microbiological analysis.

Beyond this federal requirement, each province is responsible for agricultural production standards and food safety guidelines (Lähteenmäki-Uutela et al., 2017).

TABLE 7.1 List of Non-Novel Determinations for Food and Food Ingredients.

DATE	*NAME OF THE FOOD OR FOOD INGREDIENT*	*DESCRIPTION*	*RATIONALE FOR NON-NOVEL DETERMINATION*
2016-07-18	Cricket (Acheta domesticus) powder	A powder made from whole crickets	Ingredient has a history of safe use as a food
2016-07-18	C-fu™ – made from a process that extracts and restructures the proteins in yellow mealworm larvae (*Tenebrio molitor*), adult house crickets (*Acheta domesticus*), and superworm larvae (*Zophobas morio*)	Food with high nutrient content, exhibiting similar texture and appearance to tofu, made from extracts of protein from yellow mealworm larvae, adult house crickets, and superworm larvae.	Product has a history of safe use as food
023-12-11	ŸnMeal™	A protein-rich flour made from larvae of Tenebrio molitor (yellow mealworm)	Product has a history of safe use as a food
2017-04-18	Banded cricket (Gryllodes sigillatus) flour	A type of flour made from banded crickets	Ingredient has a history of safe use as a food
2017-04-18	Cricket (Gryllodes sigillatus) powder	A powder made from whole crickets	Ingredient has a history of safe use as a food
2017-04-18	Ground banded crickets (Gryllodes sigillatus)	A powder made from whole ground banded crickets	Ingredient has a history of safe use as a food
2017-04-18	Roasted banded crickets (Gryllodes sigillatus)	Whole food made from roasted banded crickets	Ingredient has a history of safe use as a food
2020-03-03	Silkworm (Bombyx mori)	Whole or minimally processed silkworm insect	Ingredient has a history of safe use as a food
2020-03-03	Silkworm (Bombyx mori) powder	A powder made from whole silkworm insect	Ingredient has a history of safe use as a food
2021-06-21	Silk protein isolated from Bombyx mori cocoons.	Silk protein isolated from Bombyx mori cocoons.	Ingredient has a history of safe use as a food
2020-02-21	Lesser mealworm or buffalo worm (Alphitobius diaperinus)	Whole or minimally processed lesser mealworm or buffalo worm	Ingredient has a history of safe use as a food

Source: Adapted from https://www.canada.ca/en/health-canada/services/food-nutrition/genetically-modified-foods-other-novel-foods/requesting-novelty-determination/list-non-novel-determinations.html

7.3.3 Australia and New Zealand

The food safety system in Australia and New Zealand presents similarities to that of Europe, being governed by a highly integrated system coordinated by Food Standards Australia New Zealand (FSANZ). This agency is responsible for developing scientific and regulatory standards concerning food products' composition, labelling, and safety. These standards are founded on risk assessments and scientific evidence, providing a coherent regulatory framework for both countries.

Under the current regulations in Australia and New Zealand, edible insects can be classified into three categories: 1) traditional foods, 2) non-traditional foods, and 3) novel foods (Lähteenmäki-Uutela et al., 2017). According to Standard 1.1.2–8 of the Australia New Zealand Food Standards Code,[43] a "non-traditional food" is defined as a product that lacks an established history of human consumption in these countries. Novel food is a specific category under non-traditional foods that requires a risk assessment for public health and safety prior to their market introduction. Therefore, if a non-traditional food does not raise safety concerns, it may not require assessment and, consequently, may not be classified as a novel food.

Schedule 25–2 of the Australia New Zealand Food Standards Code[44] contains a list of novel foods approved for human consumption and marketing. This list details the conditions of use, restriction, and labelling requirements for each novel food product. To market a novel food or a product containing a derived ingredient, companies must submit an application to FSANZ, where it will undergo a comprehensive pre-market safety assessment. Once authorized, the novel food will be listed in Schedule 25, allowing its sale under specific conditions.

It is essential to determine whether a food should be classified as novel. It is possible to consult the Advisory Committee on Novel Foods (ACNF)[45] for a recommendation. The ACNF assesses whether a food meets the criteria for being defined as "non-traditional food" under the standard and subsequently evaluates whether an assessment is necessary to ensure the safe use of the food. If such an assessment is deemed necessary, the food will be classified as "novel food".

Under Section 1.5.1–3 of the Australia New Zealand Food Standards Code, applicants for novel food authorization may also request an exclusive sales permit for up to 15 months, giving them limited market exclusivity. After this period, the permit becomes generic and is no longer specific to a brand.

In the case of edible insects, in accordance with the recommendations of the ACNF, three insect species have been classified as non-traditional foods in Australia and New Zealand: *Zophobas morio* (super meal moth), *Tenebrio molitor* (meal moth), and *Acheta domesticus* (house cricket). As no safety issues have been identified, these species have not been considered novel foods. However, they must be marketed with labelling that clearly indicates their true nature. It is also recommended that the labelling indicate the potential risk of allergenicity for individuals allergic to shellfish and other sensitive persons, particularly regarding the consumption of crickets or their derivatives.[46]

7.3.4 Singapore

Singapore Food Agency is the young (founded in 2019) governmental organisation of Singapore whose mission is to ensure and secure a supply of safe food. Singapore is a small but populous nation. Its government is strongly committed to reducing food dependency from neighboring countries and supporting the development of businesses aimed at supplying animal-based food. Edible insects are considered to be one of the possible new protein sources. The SFA approach towards novel food, in general, made global headlines after being the first agency to approve lab-grown meat.

This approach aligns with the "smart nation" concept that drives Singapore's vision. A parallelism with the approach taken towards digital technology has been noticed in the domains of standardisation, regulation, intellectual property, data-gathering, and training. It has also been anticipated that these shifts in procurement, regulation, and consumption of foods may ultimately challenge traditional food pathways and Singapore's food identity (Stevens & Ruperti, 2024).

In 2019, SFA introduced the novel food regulatory framework. Similarly to other countries, the framework requires companies to seek pre-market allowance for novel food. Companies producing novel food products are required to conduct and submit safety assessments of their products for SFA's review before they are allowed for sale. SFA will review these safety assessments to ascertain that potential food safety issues have been addressed. To support the framework, SFA established the SFA Novel Food Safety Expert Working Group, a multidisciplinary team of experts, to provide scientific advice and ensure that novel food safety assessments are rigorously reviewed.[47]

The case of edible insects attracted particular attention. To address the request for insect farming and consumption, SFA engaged with scientists all around the world, exploring the work done by other food safety agencies and consulting available evidence and experiences. The result of this process was a list of species that are considered safe (Table 7.2). Species not present in the list need specific authorisation. Applicant should provide SFA several information to allow the assessment: scientific name of insect and insect-like species and intended life stage for human consumption; details of the farming and processing method, including the type of substrate (i.e., material used for feeding insects); evidence of history of use as a food in a country other than Singapore; information relied on to establish that the species and life stage are safe for human consumption (e.g., scientific literature, laboratory reports); and Singapore address or entity of applicant.

Interestingly, during the evaluation process, SFA also inquired about the Fatwa committee of Singapore to understand the Muslim position towards insects. This sounds like a very inclusive approach, with Muslims around 15% of the population. It is known that Muslim have different positions regarding the Halal state of insects according to the relevant school of taught. Despite some of them recognizing insects as unclean, the Fatwa committee took into account different interpretations, leading to the inclusion of edible insects in Halal food (Islamic religious council of Singapore, 2024).

The production of edible insects should follow a series of requirements. Wild harvested insects should only be used as a supplemental breeding stock. Applicants should establish and submit a Hazard Analysis Critical Control Point (HACCP)/Food Safety Management System (FSMS) plan in their production and post-harvest processes to identify the hazard and the control measures. The applicant is required to implement the HACCP/FSMS as

TABLE 7.2 List of Edible Insects Considered as Safe and Allowed by SFA.

ORDER	*COMMON NAME*	*LIFE STAGE PERMITTED*	*LATIN NAME*
Orthoptera	• House cricket • Banded cricket – • Common/field cricket • Black/field cricket/ two-spotted cricket • African migratory locust • American desert locust/Desert locust • Grasshopper	Adult stage only	• *Acheta domesticus* • *Gryllodes sigillatus* *Teleogryllus testaceus/ Teleogryllus mitratus* *(Gryllus testaceus)* *Gryllus bimaculatus* *Locusta migratoria* *Migratorioides* *Schistocerca americana/ Schistocerca gregaria* *Oxya japonica*
Coleoptera	• Superworm beetles/ Giant mealworm beetle/King mealworm • Mealworm • Lesser mealworm • Whitegrub • Giant Rhino beetle grub	Larva stage only	*Zophobas atratus/ Zophobas morio* *Tenebrio molitor* *Alphitobius diaperinus* *Protaetia brevitarsis* *Allomyrina dichotoma/ Trypoxylus dichotomus*
Lepidoptera	• Greater wax moth/ Honeycomb moth • Lesser wax moth • Silk moth/silkworm	Larva stage only (also pupa for *Bombyx mori*)	*Galleria mellonella* *Achroia grisella* *Bombyx mori*
Hymenoptera	Western honey bee/ European honey bee	Larva and adult stage	Apis mellifera

submitted throughout the license term. The substrate used for feeding insects should not be contaminated with any substances that may cause the insects to be unsafe for human consumption. Manure, decomposing organic material, and materials originating from diseased animals or fish should not be used as feeding substrates. For materials of ruminant origin, only deboned meat cuts and dairy (e.g., milk and cheese) products are allowed. For food waste to be used as an insect substrate, the food waste should not be in a decaying or decomposing state. Applicants should ensure that there is a system to screen inputs into the waste stream to ensure that the substrate will not contain disallowed materials. Good hygiene practices should be implemented during post-harvest handling to prevent cross-contamination. Insects for human consumption should have been subjected to sufficient heat treatment, or an equivalent bactericidal process, to kill pathogens prior to consumption and are safe for consumption.[48]

7.4 DISCUSSION

7.4.1 Balancing Safety and Sustainability as a Challenge for Policymakers+

The concept of sustainability and sustainable development, intrinsically linked to the broader notion of "long-term durability", primarily seeks to ensure the prudent and conscientious use of resources to guarantee that future generations can enjoy appropriate living conditions.[49]

Sustainability and its communication present an engaging discussion area, particularly concerning novel foods, including insects. These products are often developed to respond to market demands for increasingly environmentally friendly food and are aligned with consumers' ethical, social, health, and economic needs. Additionally, the EU NFR indirectly references the significance of sustainability and the role of novel foods as key instruments in achieving it.[50]

The reference to reducing the environmental impact of food production in Recital 29 of the NFR establishes a clear link to the concept of sustainability and, consequently, corporate social responsibility.[51] Although, as noted, there is no universally accepted definition of "sustainable" (La Porta, 2021b), the extensive literature and national intervention plans for specific agricultural and food sectors help identify key parameters interrelated to one another: environmental protection, economic growth, and social development.[52] Referring to insects, when farmed for human consumption, due to their significantly lower consumption of valuable resources such as water and land, could potentially be considered a more sustainable source of protein compared to other animals used for similar nutritional purposes, such as cows or pigs (Belluco et al., 2022; Guiné et al., 2021).

Despite sustainability being a relevant parameter in evaluating a novel product, it is not considered part of the risk assessment for including a new product on the Union food list. Data regarding a product's potential environmental and economic sustainability are not examined by EFSA or the European Commission, and despite the references in Recital 29, the Union legislator remains indifferent to agro-food resource protection issues. Consequently, during the evaluation for market approval, the degree of sustainability of the new food does not appear to carry substantial relevance.

Although regulations require evaluating significant changes in nutritional value or potential nutritional disadvantages compared to substitute or analogous products already on the market, the efforts of the 2015 legislator appear limited and somewhat out of step with the current regulatory framework, which places sustainability issues at its center.

Therefore, animal welfare regulations, which may be relevant to insect-based food evaluations, as well as the European New Green Deal policies or the One Health Strategy, do not seem to be concretely integrated into the product evaluation process. To date, only food safety aspects are considered in the authorisation process, excluding broader issues.

The primary focus of the EFSA risk assessment remains on the hygienic and sanitary safety of food and informational safety, defined as "*adequate and complete communication to the consumer regarding the characteristics, methods, and quantities of food consumption*" (Costato & Rizzioli, 2010; Russo, 2010). Thus, the assessment focuses on the composition of the food and its toxicological and hygienic safety, understood as the absence of alterations in the production process and/or the use of the product and the lack of exogenous factors that may pose physical and chemical or biological risks to the consumer, as well as informational safety[53] and nutritional safety, understood as the absence of nutritional disadvantages for the consumer.

Despite the centrality of food safety in a broad sense, the challenge of balancing this with "sustainability" presents a complex and unavoidable issue not only for the legislator – who seems to have paid little attention to this matter in the regulation of novel foods – but also for the interpreters of the regulation, particularly the European Commission and EFSA.

On the one hand, public health protection requires strict food safety standards to ensure that new products do not pose risks to consumers. On the other hand, increasing pressure for more sustainable food production models requires the application of regulations that positively consider aspects such as innovation and the adoption of environmentally friendly technologies without compromising the high safety standards required.

Given EFSA's mandate, with the express focus on "*scientific evidence demonstrating that the novel food does not pose a safety risk to human health*" during the authorisation procedure established under Article 10 of the NFR, and considering the content of the EFSA guidelines on novel foods, it appears highly challenging to integrate an evaluation of sustainability (that takes care of social, economic, and environmental aspects) into the established rigid framework. Should EFSA extend its assessment beyond its competence, it risks exceeding its mandate. Moreover, any requests for additional information related to sustainability from the applicant under Articles 11(4) and 17(4) of the NFR or delays in issuing its opinion due to such considerations could lead to EFSA violating principles governing public administration efficiency and transparency. This could result in liability in cases of breaches of the law by public administration, a fundamental guarantee of rights, and an essential element of the rule of law (European Parliamentary Research Service, 2018).

However, it is also true that Article 12 of the NFR allows the Permanent Committee on Plants, Animals, Food, and Feed (PAFF) to intervene during the novel food authorisation process, with the task of ensuring compliance with safety standards while contributing to the assessment of ethical and socio-economic aspects. Nevertheless, the consultation of this body remains limited, non-binding, and less influential than in the case of GMOs, where, under Regulation (EC) 1829/2003, the intervention of Member State representatives within the Committee, issuing a qualified majority opinion, is fundamental for the Commission to authorize a product without referring the matter to the Council.

Moreover, Regulation (EU) 625/2017 on official controls introduced updated definitions of "risk" and "hazard" in Article 3. It defines hazard (23) as "*any agent or condition with the potential to have an adverse effect on human, animal, or plant health, animal welfare, or the environment*" and risk (24) as "*a function of the probability of an adverse effect on human, animal, or plant health, animal welfare, or the environment, and of the severity of that effect, consequential to a hazard*". The inclusion of an explicit reference to potential adverse effects not only on humans but also on animals, plants, and the environment aligns with the unified approach of the new regulation and its intrinsic reference to sustainability. It aims to move beyond the "farm-to-fork" approach of the Regulation (CE) 178/2002, adopting a "life-cycle" perspective that considers interventions across all sectors, regardless of their immediate impact on the food industry (Albisinni, 2017).

While not directly applicable to EFSA's risk assessments for novel food products, this 2017 cross-sector regulation could influence the broader application and interpretation of food law, impacting all EFSA activities.

Although Regulation (EU) 625/2017 does not apply immediately to the novel food authorisation process, and sustainability is not explicitly referenced in the novel food risk assessment procedure, it would be appropriate for the legislator to intervene and achieve a balance between scientific evidence on food safety and the environmental and social impacts associated with production methods or the novel food product itself.

By the way, integrating sustainability principles into the European regulatory framework for novel foods is feasible without requiring a new regulation, as targeted adjustments could be made to the existing system based on Article 35(2) of the NFR.

Such an integration would represent a step towards creating a more resilient food system, one that protects human health while safeguarding future generations. The goal would be to reconcile these needs without imposing excessive barriers to innovation, promoting a more comprehensive and structured risk assessment based on transparent and internationally harmonized criteria, beginning with a clear and unequivocal definition of "sustainability".

Many of the findings made so far on the European regulatory framework are also valid for other countries. The focus on NFR is based on the restrictive nature of the measures contained therein, which makes it a valid starting point for observing its risks and benefits. In other countries, small differences allow a different perspective to be appreciated on the novel food concept, which, although mainly based on food safety, allows other aspects to be valued. In Canada and New Zealand, for example, the tradition and third country epidemiological evidence are being valued more, allowing a food with a history of consumption that has been around for thousands of years in various regions of the world to be not considered as totally new. This flexibility has meant, for example, starting from a list of allowed insects, limiting the need for pre-market evaluation to species with a less extensive consumption tradition. The US and Singapore have different approaches. The US is in fact allowing safety assessment directly to companies that want to market the product. This is in line with the widespread regulatory approach that places primary responsibility on Food Business Operators, but it may sometimes make the assessment less thorough with corrections necessary at later stages after placing on the market. Singapore, on the other hand, adopted a proactive approach, in line with the main mission of the food policy of this city state that aim at reducing its dependence from import. This approach based in direct involvement of governmental agency on the evaluation of edible insects has allowed companies to have a list of species already evaluated and authorized. The direct involvement of the institutional risk manager at least in a starting phase of authorisation allows to broaden the level of considerations beyond food safety alone with a more integrated vision of the desired food system.

NOTES

1 Article 7 of Regulation 178/2002. In specific circumstances where, following an assessment of available information, the possibility of harmful effects on health is identified but scientific uncertainty persists, provisional risk management measures necessary to ensure the high level of health protection chosen in the community may be adopted, pending further scientific information for a more comprehensive risk assessment. Measures adopted on the basis of paragraph 1 shall be proportionate and no more restrictive of trade than is required to achieve the high level of health protection chosen in the community, regard being had to technical and economic feasibility and other factors regarded as legitimate in the matter under consideration. The measures shall be reviewed within a reasonable period of time, depending on the nature of the risk to life or health identified and the type of scientific information needed to clarify the scientific uncertainty and to conduct a more comprehensive risk assessment.

2 Regulation (EU) 2017/625 of the European Parliament and of the Council of 15 March 2017 on official controls and other official activities performed to ensure the application of food and feed law and rules on animal health and welfare, plant health and plant protection products. The legislator explicitly states (*Considerandum 28*) that "Certain definitions currently set out in Regulation (EC) No 882/2004 should be adapted to take account of the broader scope of this Regulation."

3 According to Article 3 par. 2 of the NFR,

> (a) 'novel food' means any food that was not used for human consumption to a significant degree within the Union before 15 May 1997, irrespective of the dates of accession of Member States to the Union, and that falls under at least one of the following categories: (i) food with a new or intentionally modified molecular structure, where that structure was not used as, or in, a food within the Union before 15 May 1997; (ii) food consisting of, isolated from or produced from microorganisms, fungi or algae; (iii) food consisting of, isolated from or produced from material of mineral origin; (iv) food consisting of, isolated from or produced from plants or their parts, except when the food has a history of safe food use within the Union and is consisting of, isolated from or produced from a plant or a variety of the same species obtained by: traditional propagating practices which have been used for food production within the Union before 15 May 1997; or non-traditional propagating practices which have not been used for food production within the Union before 15 May 1997, where those practices do not give rise to significant changes in the composition or structure of the food affecting its nutritional value, metabolism or level of undesirable substances; (v) food consisting of, isolated from or produced from animals or their parts, except for animals obtained by traditional breeding practices which have been used for food production within the Union before 15 May 1997 and the food from those animals has a history of safe food use within the Union; (vi) food consisting of, isolated from or produced from cell culture or tissue culture derived from animals, plants, microorganisms, fungi or algae; (vii) food resulting from a production process not used for food production within the Union before 15 May 1997, which gives rise to significant changes in the composition or structure of a food, affecting its nutritional value, metabolism or level of undesirable substances; (viii) food consisting of engineered nanomaterials as defined in point (f) of this paragraph; (ix) vitamins, minerals and other substances used in accordance with Directive 2002/46/EC, Regulation (EC) No 1925/2006 or Regulation (EU) No 609/2013, where: a production process not used for food production within the Union before 15 May 1997 has been applied as referred to in point (a) (vii) of this paragraph; or they contain or consist of engineered

nanomaterials as defined in point (f) of this paragraph; (x) food used exclusively in food supplements within the Union before 15 May 1997, where it is intended to be used in foods other than food supplements as defined in point (a) of Article 2 of Directive 2002/46/EC.

4 Under the previous regulation, which is still applicable, the Court of Justice of the European Union, First Chamber, in its ruling of 09/06/2005, Case No. 211, HLH W. GmbH and another v. Germany and another, clarified,

> [T]hat food or that food ingredient has not been consumed in a significant quantity by humans in any of the Member States before the reference date. 15 May 1997 is the reference date for the purpose of determining the extent of human consumption of that food or food ingredient.
>
> In Rassegna diritto Farmaceutico (2005), 6, p. 1338.

5 It should be noted that, for a food product not to be considered a novel food, its use must have occurred before 15 May 1997, in the individual Member States, regardless of the date of their accession to the European Community.

6 Bear in mind that, according to Article 3 of the NFR, even food products commonly consumed in the EU could be considered novel if techniques or methodologies are adopted to alter their composition or structure in a manner that impacts their nutritional value, metabolism, or the presence of undesirable substances, so obtaining a final product that is different from the one already on the market.

7 A definition of "emerging risk" is provided by EFSA, 10.7.2007, EFSA/SC/415 Final.
On the topic, A. Kocharov, *EFSA and Identification of Emerging Risks*, European Food & Feed Law Review, 3, 2010, 144.

8 This intrinsic uncertainty can be overcome within a coherent decision-making process only by acknowledging that risk analysis comprises three interconnected aspects: scientific assessment, risk management, and risk communication.

9 European Commission Recommendation, *Concerning the scientific aspects and the presentation of information necessary to support applications for the placing on the market of novel foods and novel food ingredients and the preparation of initial assessment reports under Regulation (EC) No 258/97 of the European Parliament and of the Council*, July 29, 1997, OJ L 253, 16.9.1997.

10 In this regard, in 2014, the Belgian Federal Agency for the Safety of the Food issued Circular PCCB/S3/ENE/KBE/1158552 concerning the farming and commercialization of insects and insect-based food products intended for human consumption.

11 The European Court of Justice in Case C-526/19 Entoma SAS v Ministre de l'Économie et des Finances, Ministre de l'Agriculture et de l'Alimentation, ruled that

> Article 1(2)(e) of Regulation (EC) No 258/97 of the European Parliament and of the Council of 27 January 1997 concerning novel foods and novel food ingredients, as amended by Regulation (EC) No 596/2009 of the European Parliament and of the Council of 18 June 2009, must be interpreted as meaning that foods consisting of whole animals intended to be consumed as such, including whole insects, do not fall within the scope of that regulation.

12 The Italian Ministry of Health, through its Ministerial Circular dated October 29, 2013, titled "Controlli ufficiali in merito all'uso di insetti in campo alimentare con specifico riferimento all'applicabilità del Reg. CE 258/2007 sui "novel food", clarified that insects, as products originated by animals, fell within the scope of Regulation (EC) No 258/1997, and, as such, could not be placed on the Italian market without they had completed the necessary authorization procedure, according to the provisions of the aforementioned Regulation and Recommendation 97/618/EC. Similarly, the Ministère de la Santé of Luxembourg that, in its

press release CS.2014–584 dated December 22, 2014, stated: "*les insectes comestibles non autorisés au Luxembourg*".

13 Belgium's Federal Agency for the Safety of the Food Chain, in its opinion "Food safety aspects of insects intended for human consumption", published on 29 January 2014 and approved by the Belgian Scientific Committee of the FASFC (Belgium's Federal Agency for the Safety of the Food Chain) on 12 September 2014, and validated by the Board of the Superior Health Council on 3 September 2014, while adopting a highly tolerant stance, highlighting the potential public health risks associated with the consumption of insects. These include microbiological, chemical, physical, and parasitological risks, and risks to humans associated with the consumption of insects. In this opinion, the FASFC, in line with earlier statements by the FAO, acknowledged the significant potential of insects as an alternative protein source for humans and expressly tolerated the domestic trade of certain species. The opinion also addressed microbial, chemical (including allergens), and physical risks specifically related to insect consumption, emphasizing the need for further research in the field. This is because the limited scientific data on food safety concerning insects does not allow for assurance of the absence of risks related to the widespread adoption of entomophagy. It recommended cooking the product and adopting proper storage conditions, as well as providing information to consumers through labels containing warnings about potential allergic reactions, especially for individuals already sensitive to shellfish or dust mites.

14 This principle was first established by the Court of Justice in the famous Cassis de Dijon ruling of 20 February 1979, Case C-120/78, published in the Official Journal of the European Communities C-256 of 3 October 1980, p. 23.

15 Here referring to Regulation CE 178/2002. On the topic, F. ALBISINNI (2017), *Strumentario di diritto alimentare europeo*, UTET.

16 The 2015 Regulation dedicates the entire Section II of Chapter III to the specific rules for traditional foods from third countries, where it defines the procedures for the authorization of a novel food. With the Commission Implementing Regulation (EU) 2017/2468, the Union legislator established the administrative and scientific requirements for traditional foods from third countries under Regulation (EU) 2015/2283.

17 Under Article 3, point 17, of Regulation (EC) No. 178/2002, to which Regulation (EU) 2015/2283 refers, "Primary production" means "*all stages of the production, rearing, or growing of primary products, including harvesting, milking, and livestock production prior to slaughter, and includes hunting, fishing, and the harvesting of wild products*".

18 According to Recital 16 of Regulation (EU) 2015/2283 on novel foods:

> Foods from third countries which are regarded as novel foods in the Union should only be considered as traditional foods from third countries when they are derived from primary production as defined in Regulation (EC) No 178/2002, regardless of whether or not they are processed or unprocessed foods.

19 Brussels, 14.1.2008 COM(2007) 872 final 2008/0002 (COD) Proposal for a Regulation of the European Parliament and of the Council on novel foods and amending Regulation (EC) No 1331/2008 [common procedure] at Article 3(2)(b) referred to "*traditional food from a third country" meaning novel food with a history of food use in a third country, meaning that the food in question has been and continues to be part of the normal diet for at least one generation in a large part of the population of the country.* However, in Regulation (EU) 2015/2283, Article 3(2)(c), such products are instead defined as "*a novel food as defined in paragraph (a) of this article . . ., derived from primary production . . . with a history of safe use as food in a third country.*" According to letter (b), the term "*history of safe use as food in a third country*" *means that the safety of the food in question has been confirmed with compositional data and from experience of continued use for at least 25 years in the customary diet of a significant number of people in at least one third country, prior to a notification referred to in Article 14.*

20 The 25 years was decided differently from the original proposal to amend Regulation (EC) No. 258/1997 on novel foods, which was put forward in 2008 and stalled in 2013 due to the inability to reach a shared position among all Member States. For further details on the topic, see: I. GERSTBERGER (2008), *The Proposal for a Revised Novel Food Regulation – An Improvement for the Worse?*, in *European Food & Feed Law Review*, vol. 3, n. 4, 213.

21 According to Recital 22 of Regulation (EU) 2015/2283 on novel foods:

> As regards traditional foods from third countries having a history of safe food use, the applicants should be able to opt for a faster and simplified procedure to update the Union list if no duly reasoned safety objections are expressed.

The notification required for traditional products from third countries replaces the authorization request required for entirely new products.

22 In this latter case, the operator submits a simple notification containing the information specified in Article 14 of the NFR, namely:

> The notification shall include the following information: (a) the name and address of the applicant; (b) the name and description of the traditional food; (c) the detailed composition of the traditional food; (d) the country or countries of origin of the traditional food; (e) documented data demonstrating the history of safe food use in a third country; (f) a proposal for the conditions of intended use and for specific labelling requirements, which do not mislead the consumer, or a verifiable justification why those elements are not necessary.

23 Recital 23 of the NFR.

24 Reference is made to the provisions of Article 5 of Commission Implementing Regulation (EU) 2017/2469 of 20 December 2017, which establishes the administrative and scientific requirements for applications under Article 10 of the NFR. It is also noted that applications for the authorization of a novel food under Article 10 of the NFR, submitted after 26 March 2021, must comply with the guidelines produced by EFSA (EFSA, Guidance on the preparation and submission of an application for authorization of a novel food in the context of Regulation (EU) 2015/22831 (Revision 1) in EFSA Journal 2021;19(19):6555). Particular attention should be given to the information in Section 1 on the characterization of the novel food, technical, and scientific data.

25 According to Article 3(1) of Commission Implementing Regulation (EU) 2017/2469 on novel foods, the application for authorization consists of (a) a cover letter, (b) a technical dossier, and (c) a summary of the dossier. In the case of an authorization application for traditional products from third countries, under Article 4(1) of Commission Implementing Regulation (EU) 2017/2468,

> The application is submitted electronically to the Commission and consists of the following elements: (a) cover letter; (b) technical dossier (which includes administrative and scientific data); (c) summary of the dossier; (d) duly reasoned safety objections, as referred to in Article 15(2) of Regulation (EU) 2015/2283; (e) the applicant's response to the duly reasoned safety objections.

Under Article 1 of Regulation (EC) No 258/1997, it was foreseen that the authorization would be granted only to a single entity based on the documents and scientific evidence submitted by that entity, and once authorized, the product could not be marketed by another company other than the one that had initiated the procedure. However, under Article 3(4) of Regulation (EC) No 258/1997, other interested entities were also permitted to market new products or food ingredients that,

> By way of derogation from paragraph 2, the procedure laid down in Article 5 shall apply to foods or food ingredients referred to in Article 1 (2) (b), (d) and (e) which, on the basis of the scientific evidence available and generally recognized or on the

basis of an opinion delivered by one of the competent bodies referred to in Article 4 (3), are substantially equivalent to existing foods or food ingredients as regards their composition, nutritional value, metabolism, intended use and the level of undesirable substances contained therein.

For these so-called equivalent products, a more streamlined and faster procedure could be initiated. However, in the case of scientific evaluation based on an appropriate analysis of existing data, additional evidence could be required, leading to the initiation of the authorization procedure, as outlined in Article 4 of the 1997 Regulation, if differences in the assessed characteristics emerged compared to the product for which equivalence was claimed.

For a more comprehensive analysis of the topic, see, among others: M. Hermann, *The impact of the European Novel Food Regulation on trade and food innovation based on traditional plant foods from developing countries*, in *Food Policy*, 34, 2009, 499 ss.; R. Hyde et al., *European Novel Foods policy at a critical juncture: drawing lessons for future Novel Food governance through a retrospective examination of Regulation 258/97*, in *Food and Drug Law Journal*, 72, 2017, 480 ss.; A.C. Huggett, C. Conzelmann, *EU regulation on novel foods: Consequences for the food industry*, in Trends in Food Science & Technology, 8, 1997; D.A Jonas. ed al., *The Safety Assessment of Novel Foods,* guidelines prepared by *ILSI Europe Novel Food Task Force in Food and Chemical Toxicology Journal*, 34, 1996.

26 Here referring to submission made under Article 10 of the NFR.

27 For a clearer definition of the time period during which data protection is recognized and to ensure full awareness of it by third parties, Article 27 of the NFR stipulates that, at the time a new product is added to the Union list, the following information must also be included alongside various relevant product details: (a) "the date of inclusion of the novel food in the Union list"; and (e) "the expiry date of the data protection referred to in Article 26".

28 Where there are simultaneously pending applications for novel food authorization and health claim authorization, it provides for an alignment of timelines so that the protection period granted under Article 21 of Regulation (EC) 1924/2006 cannot exceed the time limit set by the novel food regulation. Regulation (EU) 2015/2283, by referring to Regulation (EC) No 1924/2006 on nutrition and health claims made on food, provides that, where the authorization procedure for a health or nutritional claim runs parallel with the procedure initiated following the submission of an application for the authorization of the placing of a novel food on the market, the applicant may suspend the procedure for the novel food pending the conclusion of the procedure for the health claim recognition.

The provisions of Article 28 of the NFR are considered to reflect the Union legislator's attention to the criticisms raised by scholars and operators in the field regarding the provisions of the regulation on nutritional and health claims, which form the basis for the 2015 data protection regulations. To align the protection periods provided by the various regulations, Article 28 of the NFR governs the procedure in the case of a parallel application for the authorization of a health claim. As already indicated in Recital 31 of the NFR, the legislator recognized the need to overcome the difficulties arising from the misalignment of timelines. According to Article 28 NFR, upon the applicant's request, the Commission may suspend the ongoing novel food authorization procedure. However, as this suspension is activated only at the applicant's initiative, it can be interrupted at any time, and the applicant may withdraw the request, thereby allowing the resumption of the authorization procedure according to the normal course. Once the suspension is approved, the starting date of the suspension must be communicated to the applicant, and the suspension may last no longer than 7 months from the publication date of the Authority's opinion (Article 12(1) NFR), during which the Commission shall submit to the Standing Committee on Plants, Animals, Food, and Feed – operating under Article 30(1) of the NFR – a proposal for an implementing act to authorize the placing of the product on the market. The suspension period ends when the Commission receives the Authority's opinion on the health claim in accordance with Regulation (EC) No

1924/2006 (Article 28(4) NFR); from this moment, the deadline under Article 12(1) of the NFR starts anew.

29 The reference to exclusivity under Regulation (EU) 2021/882 should be interpreted broadly, as it is doubtful that a sales block will be imposed on those entities that have so far benefitted from the moratorium provided for in Article 35(2) of the NFR. These entities, under Regulation (EC) No 258/1997, were already producing and marketing their products in certain Member States that tolerated such activities and, under the cited article, have been allowed to continue doing so, but only within those specific geographical areas. Following the exclusive market authorization granted to the applicant under NFR, the beneficiary will be able to market the product throughout the entire Union territory.

30 The Commission reviewed several authorization applications for the marketing of various insect species, both used in their entirety and as ingredients. Some of the applications submitted include requests for the application of Article 26 of the NFR, thereby seeking data protection for the information presented (these applications pertain to "*whole and ground Alphitobius diaperinus (lesser mealworm) larvae products; dried Tenebrio molitor (mealworms) and dried Gryllodes sigillatus (crickets)*"). For these products, therefore, it will not be possible to envisage free use by parties other than the applicant for a period of 5 years from the date of authorization, under the aforementioned regulation, unless an agreement is reached with the original applicant, or unless a new request based on demonstrated proprietary data is submitted by another party.

For three insect species, specifically *Acheta domesticus*, *Locusta migratoria*, and *Tenebrio molitor*, the application does not include a request for data protection. This implies that these animals, if recognized as food, may also be used by entities other than the original applicants without any restrictions, except for compliance with the indicated safety requirements, including the substrate verified through analysis, and any applicable limits and precautions. Furthermore, adherence to the applicable HACCP regulations will be required.

31 Under the previous regime, the average duration of an authorization request was approximately 3 years before a decision could be reached, and the average cost of the procedure was very high, thus requiring applicants to invest significant sums of money. These figures were presented during the beauty contest "*Etichettatura, product recall e novel food*" organized by 4C Legal in Milan on November 23, 2016.

32 European Commission, *Evaluation Report on the Novel Food Regulation 258/97 Concerning Novel Foods and Food Ingredients*, 22 January 2004.

33 Judgment of the Court (Second Chamber) of 15 January 2009. Case C-383/07. M-K Europa GmbH & Co. KG v. Stadt Regensburg. Reference for a preliminary ruling: Bayerischer Verwaltungsgerichtshof – Germany. Preliminary ruling – Regulation (EC) No 258/97 – Article 1, paragraphs 1–3 "Novel foods and novel food ingredients." In European Court Reports 2009 I–00115.

34 Peru's response to the Communication of the European Community regarding Regulation (EC)258/97 on Novel Foods, G/SPS/GEN/699, 8 June 2006.

35 Agreement concluded within the WTO on Technical Barriers to Trade, Tokyo, 12 April 1979, in force since 1 January 1980.

36 EFSA, Guidance on the preparation and presentation of the notification and application for authorisation of traditional foods from third countries in the context of Regulation (EU) 2015/2283, 2016. According to Article 29(6) letter (b) of Regulation CE 178/2002, which establishes the general principles and requirements of food law, the European Food Safety Authority is tasked with setting

> the guidelines governing the scientific evaluation of substances, products or processes which are subject under Community legislation to a system of prior authorisation or entry on a positive list, in particular where Community legislation makes provision for, or authorises, a dossier to be presented for this purpose by the applicant.

37 Communication from the Commission "The European Green Deal", COM(2019) 640 final, Brussels, 11.12.2019.

38 According to the Food, Drug, and Cosmetic Act Title 21 §321(f), *the term "food" means (1) articles used for food or drink for man or other animals, (2) chewing gum, and (3) articles used for components of any such article.*

39 Food, Drug, and Cosmetic l Title 21 § 348, https://www.gpo.gov/fdsys/pkg/USCODE-2010-title21/pdf/USCODE-2010-title21-chap9-subchapIV-sec348.pdf

40 The Canadian Food and Drug Regulations, Section B, Division 28

(b) a food that has been manufactured, prepared, preserved, or packaged by a process that
  (i) has not been previously applied to that food and
  (ii) causes the food to undergo a major change; and

(c) a food that is derived from a plant, animal, or microorganism that has been genetically modified such that
  (i) the plant, animal, or microorganism exhibits characteristics that were not previously observed in that plant, animal, or microorganism,
  (ii) the plant, animal, or microorganism no longer exhibits characteristics that were previously observed in that plant, animal, or microorganism, or
  (iii) one or more characteristics of the plant, animal, or microorganism no longer fall within the anticipated range for that plant, animal, or microorganism (aliment nouveau).

41 (a) a substance, including a microorganism, that does not have a history of safe use as a food;

42 Guidelines for the Safety Assessment of Novel Foods. Food Directorate. Health Products and Food Branch. Health Canada. June, 2006. Updated: July 2022 https://www.canada.ca/en/health-canada/services/food-nutrition/legislation-guidelines/guidance-documents/guidelines-safety-assessment-novel-foods–2006.html

43 According to Standard 1.1.2–8 of the Food Standards Code of Australia and New Zealand, *novel food* means a non-traditional food that requires an assessment of the public health and safety considerations having regard to: (a) the potential for adverse effects in humans, or (b) the composition or structure of the food, (c) the process by which the food has been prepared, or (d) the source from which it is derived, (e) patterns and levels of consumption of the food, or (f) any other relevant matters. *Non-traditional food* means: (a) a food that does not have a history of human consumption in Australia or New Zealand, (b) a substance derived from a food, where that substance does not have a history of human consumption in Australia or New Zealand other than as a component of that food, or (c) any other substance, where that substance, or the source from which it is derived, does not have a history of human consumption as a food in Australia or New Zealand. Either of the following: (a) the presence of a food in a food for special medical purposes and (b) the use of a food as a food for special medical purposes – does not constitute a history of human consumption in Australia or New Zealand in relation to that food for the purposes of this section.

44 Australia New Zealand Food Standards Code, Schedule 25, Permitted novel foods. https://www.legislation.gov.au/F2015L00440/latest/text

45 https://www.foodstandards.gov.au/business/novel/novelcommittee

46 Recent information suggests that eating crickets or foods made from them may increase the risk of allergenicity in people who are crustacean-allergic or other sensitive people, as highlighted in the update to the Record of Views from the March 2019 ACNF meeting. https://www.foodstandards.gov.au/sites/default/files/2024-08/Record%20of%20Views%20updated%20August%202024.pdf

47 https://www.sfa.gov.sg/regulatory-standards-frameworks-guidelines/novel-food-framework/overview-of-novel-food-regulatory-framework

48 https://www.sfa.gov.sg/food-safety-tips/food-risk-concerns/risk-at-a-glance/insects.

49 As early as 1987, the World Commission on Environment and Development (WCED) defined "sustainable" as a type of development capable of "meeting the needs of the present without compromising the ability of future generations to meet their own needs".

On sustainability: I. CANFORA, L. COSTANTINO, A. JANNARELLI (edited by) (2017), Cacucci Ed. Quaderni di diritto privato europeo, 317; S. CARMIGNANI (2013), *Agricoltura e pluridimensionalità dello sviluppo sostenibile*, Il Trattato di Lisbona e la nuova PAC. Atti del convegno di Bari 27–28 marzo 2014, C. CONSALVO CORDUAS, *Sostenibilità ambientale e qualità dello sviluppo*, Ed. Nuova Cultura; P. FOIS (edited by) (2007), *Il principio dello sviluppo sostenibile nel diritto internazionale ed europeo dell'ambiente*, XI Convegno Alghero, 16–17 June 2006, Edit. scientifica, 223; F. FRACCHIA (2010), *Sviluppo sostenibile e diritti delle generazioni future*, Rivista quadrimestrale di diritto ambientale, 0; V. PEPE (2001), *Lo sviluppo sostenibile tra diritto internazionale e diritto interno*, in Rivista giuridica dell'ambiente, 212.

50 According to Recital 29 of the NFR, the legislator highlights, "*New technologies and innovations in food production should be encouraged as they could reduce the environmental impact of food production, enhance food security and bring benefits to consumers as long as the high level of consumer protection is ensured*".

51 For the concept of Corporate Social Responsibility (CSR), reference is made, among others, to P. COLANERI (2010), *Il fenomeno della responsabilità sociale di impresa in agricoltura*, in Economia & Diritto Agroalimentare, XV, 225.

52 The parameters are for the first time considered within the Brundtland Report (also known as *Our Common Future*), a 1987 document drafted by the World Commission on Environment and Development, in which the concept of sustainable development was introduced.

53 Informational safety is to be understood as the adequate and complete communication to the consumer regarding the characteristics of the food, as well as the recommended methods and quantities for its consumption.

REFERENCES

Albisinni, F. (2017). *Strumentario di diritto alimentare europeo*. UTET. https://shop.wki.it/libri/strumentario-di-diritto-alimentare-europeo-s20998/

Belluco, S., Halloran, A., & Ricci, A. (2017). New protein sources and food legislation: The case of edible insects and EU law. *Food Security*, *9*(4), 803–814. https://doi.org/10.1007/s12571-017-0704-0

Belluco, S., Kinyuru, J., & Fisher, B. L. (2022). Editorial: Insects: A source of safe and sustainable food? *Frontiers in Sustainable Food Systems*, *6*, 887289. https://www.frontiersin.org/articles/10.3389/fsufs.2022.887289

Bonora, G. (2016). I novel foods nel reg. (UE) n. 2015/2283 e gli insetti: Una possibile evoluzione dei costumi alimentari? *Rivista di Diritto Alimentare*, *1*, 42–54.

Boyd, M. (2017). Cricket soup: A critical examination of the regulation of insects as food. *Yale Law & Policy Review*, *36*(17). https://ssrn.com/abstract=3198913

Broad Leib, E. M., & Pollans, M. J. (2019). *The new food safety*. https://doi.org/10.15779/Z38QR4NR0D

Bronckers, M., & Soopramanien, R. (2008). The impact of WTO law on European food regulation. *European Food and Feed Law Review*, *3*(6), 361–375.

Cohen, P. A., & Broad Leib, E. M. (2024). Ingesting risk – The FDA and new food ingredients. *The New England Journal of Medicine*, *391*(10), 875–877. https://doi.org/10.1056/NEJMp2403165

Costato, L., & Rizzioli, S. (2010). *Sicurezza alimentare*. Digesto Discipline Privatistiche. Sezione Civile.

Downes, C. (2013). The rise and fall of the new Eu novel food regulation: The complex influence of the Wto Sps agreement. *Asian Journal of WTO & International Health Law and Policy, 307*.

Esposti, R., Lucatelli, S., & Peta, E. A. (2008). *On the role of consumers in shaping union food law*. https://www.mise.gov.it/images/stories/recuperi/Sviluppo_Coesione/3_.pdf

European Parliamentary Research Service. (2018). *Action for damages against the EU*. https://www.europarl.europa.eu/RegData/etudes/BRIE/2018/630333/EPRS_BRI(2018)630333_EN.pdf

FAO. (2013). *Edible insects: Future prospects for food and feed security*. FAO Forestry paper. https://www.fao.org/4/i3253e/i3253e.pdf

Gabbi, S. (2025). The European food safety authority. In van der Meulen, B., Wernaart, B. & Kramer, L. (eds.), *Elgar concise encyclopedia of food la. Cheltenham*, UK: Edward Elgar Publishing. https://doi.org/10.4337/9781802209709.00026.

Gkogka, E., Reij, M. W., Gorris, L. G. M., & Zwietering, M. H. (2013). Risk assessment strategies as a tool in the application of the Appropriate Level of Protection (ALOP) and Food Safety Objective (FSO) by risk managers. *International Journal of Food Microbiology*, *167*(1), 8–28. https://doi.org/10.1016/j.ijfoodmicro.2013.04.013

Guiné, R. P. F., Correia, P., Coelho, C., & Costa, C. A. (2021). The role of edible insects to mitigate challenges for sustainability. *Open Agriculture*, *6*(1), 24–36. https://doi.org/10.1515/opag-2020-0206

Henson, S., & Jaffee, S. (2006). Food safety standards and trade: Enhancing competitiveness and avoiding exclusion of developing countries. *The European Journal of Development Research*, *18*(4), 593–621. https://doi.org/10.1080/09578810601070753

Hermann, M. (2009). The impact of the European novel food regulation on trade and food innovation based on traditional plant foods from developing countries. *Food Policy*, *34*(6), 499–507.

Hoffmann, S. (2010). *Food safety policy and economics: A review of the literature*. https://www.semanticscholar.org/paper/Food-safety-policy-and-economics%3A-a-review-of-the-Hoffmann/8ff04cad228613f7d743799ea3d4faa36c8aa1e9

Hood, C., Rothstein, H., & Baldwin, R. (2003). *The government of risk: Understanding risk regulation regimes*. https://academic.oup.com/book/40484

Islamic Religious Council of Singapore. (2024). *From lab to table: Novel food from an Islamic perspective*. Islamic Religious Council of Singapore.

La Porta, B. (2020). Il Regolamento europeo in materia di novel food: Riflessioni sugli alimenti tradizionali dei Paesi terzi. *Cultura e Diritti, 1*.

La Porta, B. (2021a). *Gli insetti edibili alla prova del mercato. Il regolamento di esecuzione (UE) 2021/882 relativo all'immissione in commercio della larva di Tenebrio molitor essiccata: Criticità*. https://www.research.unipd.it/handle/11577/3548524

La Porta, B. (2021b). *Riflessioni per una definizione di "prodotto agricolo sostenibile."* https://iris.unipa.it/handle/10447/549799

Lahteenmaki-Uutela, A. (2007). *European novel food legislation as a restriction to trade*. 106th Seminar, October 25–27, 2007, Montpellier, France, Article 7909. https://ideas.repec.org//p/ags/eaa106/7909.html

Lähteenmäki-Uutela, A., Grmelová, N., Hénault-Ethier, L., Deschamps, M.-H., Vandenberg, G. W., Zhao, A., Zhang, Y., Yang, B., & Nemane, V. (2017). Insects as food and feed: Laws of the European Union, United States, Canada, Mexico, Australia, and China. In *Insects as food and feed European food and feed law review* (Vol. 12, pp. 22–36). https://brill.com/view/journals/jiff/1/1/article-p3_3.xml

Laurenza, E. C., & Carreño, I. (2015). Edible insects and insect-based products in the EU: Safety assessments, legal loopholes and business opportunities. *European Journal of Risk Regulation*, *6*(2), 288–292. https://doi.org/10.1017/S1867299X00004591

Majone, G. (2010). Foundations of risk regulation: Science, decision-making, policy learning and institutional reform. *European Journal of Risk Regulation*, *1*(1), 9–32. https://doi.org/10.1017/S1867299X00000027

Omuse, E. R., Tonnang, H. E. Z., Yusuf, A. A., Machekano, H., Egonyu, J. P., Kimathi, E., Mohamed, S. F., Kassie, M., Subramanian, S., Onditi, J., Mwangi, S., Ekesi, S., & Niassy, S. (2024). The global atlas of edible insects: Analysis of diversity and commonality contributing to food systems and sustainability. *Scientific Reports*, *14*(1), 5045. https://doi.org/10.1038/s41598-024-55603-7

Paganizza, V. (2019). *Bugs in law. Insetti e regole: Dai campi alla tavola*. CEDAM.

Ramírez Nárdiz, A. (2013). *La participación de los consumidores y usuarios en la elaboración de la regulación alimentaria en el ámbito comunitario: Una reflexión crítica*. UE, Sociología y Derecho alimentarios: estudios jurídicos en honor de Luis González Vaqué, 147–165. https://dialnet.unirioja.es/servlet/articulo?codigo=4233434

Rollin, F., Kennedy, J., & Wills, J. (2011). Consumers and new food technologies. *Trends in Food Science & Technology*, *22*(2), 99–111. https://doi.org/10.1016/j.tifs.2010.09.001

Russo, L. (2010). La sicurezza delle produzioni "tecnologiche". *Rivista di Diritto Alimentare*, *2*.

Sampson, D., Cely-Santos, M., Gemmill-Herren, B., Babin, N., Bernhart, A., Bezner Kerr, R., Blesh, J., Bowness, E., Feldman, M., Gonçalves, A. L., James, D., Kerssen, T., Klassen, S., Wezel, A., & Wittman, H. (2021). Food sovereignty and rights-based approaches strengthen food security and nutrition across the globe: A systematic review. *Frontiers in Sustainable Food Systems*, *5*. https://doi.org/10.3389/fsufs.2021.686492

Sirsi, E. (2013). Agri-food technologies and the law. In Law and technology: The challenge of regulating technological development (RoboLaw series, 1, pp. 231–252). https://doi.org/10.1400/219023

Stevens, H., & Ruperti, Y. (2024). Smart food: Novel foods, food security, and the smart nation in Singapore. *Food, Culture & Society*, *27*(3), 754–774. https://doi.org/10.1080/15528014.2022.2163455

Streinz, R. (1998). The novel foods regulation – A barrier to trade in the international movement of goods? *European Food Law Review*, *3*, 266.

Vågsholm, I., Arzoomand, N. S., & Boqvist, S. (2020). Food security, safety, and sustainability – Getting the trade-offs right. *Frontiers in Sustainable Food Systems*, *4*, 16. https://doi.org/10.3389/fsufs.2020.00016

Volpato, A. (2015). La riforma del regolamento sui Novel Food: Alla ricerca di un impossibile equilibrio? *Rivista di Diritto Alimentare*, *4*, 26–43.

Technological Advances in Insect-Derived Food Ingredients

8

Ermolaos Ververis, Anthia Matsakidou, Fernando Rivero-Pino, and Adamantini Paraskevopoulou

8.1 INTRODUCTION

Entomophagy, the consumption of insects as food, has been an integral part of the human diet, on a daily or occasional basis, particularly among populations in tropical and subtropical regions of Asia, Africa, Central and South America, and Oceania (Huis et al., 2013; Olivadese & Dindo, 2023; Tagawa et al., 2022). In contrast, the majority of the populations in Europe and North America rarely consume insects. Several national and cross-cultural studies indicate that insects in Western societies remain outside the mainstream dietary habits, with their consumer acceptance being strongly influenced by prior knowledge and familiarity (Boehm et al., 2021; Kröger et al., 2022; Lorini et al., 2021).

8.1.1 Importance of Insect-Derived Ingredients

Entomophagy has recently gained global interest due to its potential to address food security, environmental sustainability, and nutritional challenges (Batat & Peter, 2020; Meyer-Rochow & Jung, 2020; Moruzzo et al., 2021). Certain edible insects hold

DOI: 10.1201/9781003488156-8

promising nutrient profiles regarding protein content, fatty acids, and micronutrients, making them a prospective alternative dietary source (Rumpold & Schlüter, 2013; Zhou et al., 2022). Besides, farming insects presents substantial environmental benefits like lower greenhouse gas emissions, reduced land use, and efficient feed conversion, positioning it as an environmentally friendly food production choice (Aguilar-Toalá et al., 2025; Matandirotya et al., 2022; Van Huis & Oonincx, 2017; Veldkamp et al., 2022; Wade & Hoelle, 2020).

In countries without an entomophagy tradition, such as Germany, the Netherlands, Belgium, Italy, Greece, Poland, and the US, entomophagy is often associated with novelty, curiosity, or environmental benefits, but long-term adoption is hindered by food neophobia and psychological aversion (Boehm et al., 2021; Gkinali, Matsakidou, Michailidis et al., 2024; Hartmann et al., 2015; Hartmann & Siegrist, 2016; Ranga et al., 2024; Sikora & Rzymski, 2025; Tan et al., 2015; Tzompa-Sosa, Moruzzo et al., 2023; Tzompa-Sosa, Sogari et al., 2023). The most prevalent obstacle to insect consumption among Western populations is the feeling of disgust and aversion associated with eating insects (Deroy et al., 2015), with different factors affecting acceptance among countries. For example, a recent comparative survey in Greece and Ireland, two European countries with distinct culinary traditions, showed that Irish participants were more willing to purchase and consume foods containing insects. Greek participants frequently reported psychological barriers and perceived lack of benefits, whereas Irish participants predominantly reported concerns over safety and sensory properties (Ranga et al., 2025).

Research exploring consumers' opinions from different regions of Europe or North America consistently suggests that repulsion toward entomophagy can be mitigated by incorporating insects as ingredients into familiar foods in an invisible form. Dry, granulated insect powders, as well as other insect-derived fractions, could represent a practical approach to enhancing acceptability by integrating them into products such as burger patties, baked goods, and condiments (Castro Delgado et al., 2020; Gkinali, Matsakidou, Michailidis et al., 2024; Megido et al., 2016). Food industry and academia are increasingly investigating the use of insects and products thereof in new or conventional food products, including protein bars, bakery goods, pasta, and meat imitates (Ayieko et al., 2021; Castro Delgado et al., 2020; Kim, Yong, Kim et al., 2019; Wendin et al., 2019).

To obtain insect-based food products of higher marketability, cost-effective production of insect-derived ingredients has to be facilitated, further facilitating their integration into various foods of added value (Collins et al., 2019; Huis et al., 2021; Pippinato et al., 2020). Several studies are currently focused on creating novel ingredients for industrial use derived from insects. Considering the high protein content of various insect species, protein extraction is a primary goal for industrial usage. Extensive efforts have been directed toward the isolation of protein-rich fractions and studying their functional properties across different insect species (Pan et al., 2022). Moreover, other fractions such as lipids and chitin can also be recovered and utilized as food ingredients (Lorrette & Sánchez, 2022; Tan et al., 2021). Micronutrients and non-nutrient components derived from insects are being investigated for their bioactivity and potential as functional food constituents (Chalghaf et al., 2023; Chantakun et al., 2024; da Costa Rocha et al., 2021; Gkinali, Matsakidou, Vasileiou et al., 2022; Huis et al., 2021; Lee et al., 2022; Ma et al., 2024; Mishyna et al., 2019; Psarianos et al., 2023; Rumpold & Schlüter, 2013).

8.1.2 Scope and Objectives of the Chapter

This chapter discusses recent developments in the production and application of insect-derived ingredients in the food sector. Functionality characteristics of these ingredients are explored, alongside prevalent and emerging extraction and processing methods of components such as proteins, lipids, and chitin. The current body of evidence on the effects of different processing methods on the nutritional quality, techno-functional characteristics, and sensory properties of insect-derived food ingredients is summarized. Additionally, the chapter investigates practices for and trends of incorporating these ingredients into different food matrices. The respective regulatory, safety, and technological advances and challenges are presented, and future research directions are proposed.

8.2 INSECTS AS TECHNO-FUNCTIONAL FOOD INGREDIENTS

Traditionally, insects have been consumed whole usually after pretreatment and cooking, with these steps being shaped based on different factors such as the insect species, developmental stage, collection practices, available resources, and cultural context. Ethnic groups employ diverse culinary techniques to enhance food safety, hygiene, flavor, shelf-life, and digestibility of insects (Adámek et al., 2019; Detilleux et al., 2024; Harris, 2016; Mozhui et al., 2020). Common preparation methods include sun-drying, steaming, boiling, roasting, smoking, frying, stewing, pickling, and curing. Herbs, salt, and spices, along with side ingredients such as vegetables, are used to enhance flavor. In some cases, certain insects are consumed raw too (Hlongwane et al., 2022; Liang et al., 2024; Melgar-Lalanne et al., 2019; Olivadese & Dindo, 2023).

Given that insects, when used in a nonreadily recognizable form as food ingredients, are more acceptable to consumers, the use of insect-derived ingredients has gained increasing attention. Consequently, their techno-functional properties, such as emulsification, texturization, and binding, are being explored, along with their foaming ability and water/lipid retention capacity. Additionally, their potential to enhance color, mouthfeel, and flavor is being investigated. Moreover, these ingredients could enhance antioxidant properties within the food matrices, contribute to nutrient fortification, and exhibit prebiotic effects that promote gut health. Thus, beyond the indisputable requirement for food safety, insect-derived ingredients must primarily meet food technological needs in food applications, with their techno-functional properties being crucial for their effective use.

8.2.1 Techno-Functional Properties of Insect-Derived Food Ingredients

A growing body of research explores the functional properties of insect proteins, aiming to optimize their application in food systems. Particular attention has been given to emulsification, including the emulsifying capacity and stability of protein-rich preparations from various insect species under different storage and processing conditions (Gkinali et al., 2023; Gould & Wolf, 2018; Jeong et al., 2021; Kim, Lee et al., 2022; Kim, Yong et al., 2022; Kim, Yong, Jeong et al., 2019; Kim et al., 2023; Mba et al., 2021; Mishyna et al., 2019; Mshayisa et al., 2022; Ndiritu et al., 2017; Purschke et al., 2018; Zhang et al., 2022). The emulsifying properties of insect-derived ingredients can be essential for achieving the desired texture in products such as sauces, dressings, and spreads (Gkinali, Matsakidou, Moschakis et al., 2024).

Additionally, studies have highlighted the promising foaming properties of proteins from various insect species (Jeong et al., 2021; Kim, Lee et al., 2022; Kim, Yong et al., 2022; Kim, Yong, Jeong et al., 2019; Kim et al., 2023; Mintah et al., 2020; Mishyna et al., 2019; Mshayisa et al., 2022; Ndiritu et al., 2017; Purschke et al., 2018; Zhang et al., 2022).

In contrast, fewer studies have examined the gelation properties of insect-derived proteins, leaving a substantial gap in understanding their ability to form stable gels (Dion-Poulin et al., 2020; Jeong et al., 2021; Ndiritu et al., 2017; Purschke et al., 2018). For example, Dion-Poulin et al. (2020) reported that protein denaturation during the drying process prevented hydrolysates made from pressure-treated insect meals from forming gels at concentrations of 3% or 10% (w/v). According to this work, while insect proteins might be useful components, their gelling capabilities might not be as strong as those of other more traditional gelling agents (Dion-Poulin et al., 2020). Temperature, pH, and salinity all have an impact on protein solubility, which subsequently affects the rheological characteristics of protein solutions, including their gelation capacity (Kim, Yong, Jeong et al., 2019). Additionally, the gelation profile of insect-derived proteins might be enhanced when combined proteins from sources (Kim et al., 2021).

The color of insect-derived ingredients can be a critical factor influencing consumer acceptance and overall marketability of these products (Ribeiro et al., 2024). The color of insect powders can vary depending on factors like the species and the processing methods used. For example, powder derived from H. illucens (black soldier fly) and A. domesticus (house cricket) typically exhibits a light brown to beige hue (Lucas-González et al., 2019; Mshayisa et al., 2022).

Moreover, insects and ingredients thereof have the potential to contribute unique textural characteristics to food products. Their incorporation can introduce a desirable crunchiness, thereby enhancing the sensory experience associated with various products and dishes (Kozlu et al., 2024; Wendin et al., 2019). In addition to crunchy textures, finely ground insect powders can contribute to a smooth and creamy consistency in various food formulations. Properties such as water absorption capacity and oil retention of insect powders can be important for textural enhancement in products such as bakery goods and meat analogues. For instance, the use of mealworm powder has been shown to modify the textural properties of dough, improving its elasticity and overall mouthfeel (Adedeji et al., 2022). Khatun et al. (2022) reported that incorporating house cricket flour into chapattis resulted in improved chewiness, demonstrating the positive impact of insect proteins on the textural profile of bakery products. Krawczyk et al. (2024) explored the incorporation of insect protein in meat analogs, noting that overall mouthfeel and creaminess of burgers were improved.

The flavor of insect-derived ingredients can be influenced by various factors, including the species of insect, their diet, the specific components of the ingredient (e.g., amino acids, fatty acids), and the processing methods employed (Dalle Zotte et al., 2019; Meyer-Rochow et al., 2021; Mishyna et al., 2020; Perez-Santaescolastica et al., 2022; Ribeiro et al., 2024; Wendin et al., 2019). For example, it has been reported that crickets and grasshoppers have a mild flavor (Ramos-Elorduy, 1998), termites taste nutty (Igwe et al., 2011; Ramos-Elorduy, 1998), and ants span from sweet and nutty flavors (Ramos-Elorduy, 1998) to lemon and lime (Halloran et al., 2015).

Certain edible insect species are gaining recognition for their potential as sources of bioactive peptides that exhibit antioxidant, anti-inflammatory, antihypertensive, and antimicrobial properties (Hall, 2020; Rivero-Pino, Ververis et al., 2024; Teixeira et al., 2023; Zhang et al., 2025; Zielińska et al., 2018). Enzymatic hydrolysis is a key process to enhance the bioactivity of these peptides, improving their bioavailability and functional characteristics (Ma et al., 2023; Nongonierma et al., 2018; Tejada et al., 2022). The antioxidant activity of insect-derived peptides can be attributed to mechanisms such as scavenging of reactive oxygen species (ROS), chelating metal ions, and inhibiting lipid oxidation (Nino et al., 2021; Wu et al., 2018). Hydrophobic amino acids like valine and lysine found in peptide sequences have been associated with enhanced antioxidant activity of these peptides (Lu et al., 2022). Moreover, numerous insects produce antimicrobial peptides (AMPs) with activity against bacteria, fungi, and viruses (Mastore et al., 2021; Van Moll et al., 2022; Wu et al., 2018). The inclusion of insect-derived AMPs in foods could improve their safety, quality, and shelf life. Research has shown that these peptides can have an important role in reducing microbial counts within food matrices, thereby inhibiting spoilage (Lange & Nakamura, 2023; Veldkamp et al., 2022). In addition to peptides, certain edible insects contain phenolic compounds, which contribute to their overall antioxidant activity. Studies show that fermented insect powders may contain elevated levels of total polyphenols, enhancing their potential as dietary antioxidants (Mendoza-Salazar et al., 2021).

Overall, the techno-functional properties of edible insect proteins remain underexplored. Further research is needed to fully characterize their emulsifying, foaming, and gelling behaviors, as well as antimicrobial and antioxidant activities, which would enable their more effective and targeted integration into food formulations. Expanding this knowledge base is essential for advancing the industrial application of insect-based ingredients and overcoming consumer resistance through incorporation into familiar food products.

8.3 PROCESSING TECHNIQUES FOR OBTAINING INSECT-DERIVED FOOD INGREDIENTS

A wide range of ingredients can be obtained from insects to be utilized in the development of novel, insect-containing food products, enhancing their nutrient content, palatability, and acceptability. However, it is expected that the processes of isolation, handling, stabilizing, storing, distributing, and processing of these obtained ingredients, before their incorporation into new food formulations, must be improved to obtain cost-effective results (Cruz-López et al., 2024; Espinosa-Solís et al., 2024; Laroche et al., 2019; Messina et al., 2019; Psarianos, Aghababaei et al., 2025; Psarianos, Marzban et al., 2025; Psarianos et al., 2023). Multiple studies have been conducted for this purpose, investigating both conventional and novel processing methods and techniques to isolate specific macro- and micro-ingredients from different insect species and developmental stages (Cruz-López et al.,

2024; Espinosa-Solís et al., 2024; Laroche et al., 2019; Messina et al., 2019; Psarianos, Aghababaei et al., 2025; Psarianos, Marzban et al., 2025; Psarianos et al., 2023). Proteins, lipids, and chitin have been the primary targets of these studies as they are abundant among most edible insect species (Rahman et al., 2024). Such attempts typically aim to increase the extraction and recovery yield of these ingredients, preserve or improve their nutritional value and techno-functional properties, and ultimately add value to the final products (Rahman et al., 2024).

8.3.1 Pretreatment Methods for Insect Biomass

As with all livestock, edible insects must undergo several pretreatment stages before entering the main treatment production lines. When insects are harvested, they are often collected alongside feed/substrate residues, frass (feces and exoskeletal remains), and deceased insects. Therefore, depending on the species and their stage of growth, the substrate and frass are usually removed through sieving before any further treatment, while deceased insects are separated either through manual visual inspection or automated detection systems. Moreover, a typical pretreatment process involves insects fasting for 24–48 hours to allow them to empty their digestive tract of feces (Queiroz et al., 2023). In some cases, wings and legs are removed (Purschke et al., 2018).

The enzymatic activity of insects has been reported to be remarkably high and particularly sensitive to killing conditions. Blanching has been suggested as a common treatment to deactivate enzyme activity, reduce microbial load, and improve stability (Liang et al., 2024; Singh et al., 2020). However, a notable drawback of blanching is its tendency to reduce lipid recovery in the form of a cream layer, thus decreasing fat extraction yield (Azzollini et al., 2020). Freezing is considered the most suitable and ethical method for insect killing, although it can result in limited enzyme inhibition and reduced product quality, particularly when slow freezing is applied. The use of liquid nitrogen as a technique for rapid freezing mitigates these effects. Microfiltration, an alternative to thermal treatment, can protect the final protein-rich preparations from microbial growth, leaving them, however, susceptible to enzyme degradation (Sweers et al., 2023).

Drying is an essential pretreatment step, usually applied before insect processing to reduce costs and increase convenience. Various drying methods, including solar-drying, oven-drying, smoke-drying, freeze-drying, microwave-drying, fluidized bed drying, and infrared-drying, have been applied to edible insect pretreatment to lower water activity and microbial load of the insects (Bogusz et al., 2024; Hurtado-Ribeira, Hernández et al., 2023; Hurtado-Ribeira, Villanueva-Bermejo et al., 2023; Megido et al., 2018; Melgar-Lalanne et al., 2019). However, these methods can deteriorate quality attributes such as color, induce lipid oxidation, as well as impact techno-functional properties like protein solubility. Rapid freezing with liquid nitrogen followed by drying has been reported to result in a higher degree of protein denaturation, an improved amino acid profile, and enhanced techno-functional properties of protein extracts, compared to microwave- or freeze-dried *Tenebrio molitor* meal (Zhang et al., 2022). Blanching followed by oven-drying was found to protect the *Hermetia illucens* larvae from lipolysis, which was extensive when non-thermal treatment was used (Hurtado-Ribeira, Villanueva-Bermejo et al., 2023). On the other hand, fat was entrapped in the protein matrix during the blanching of

H. illucens due to protein aggregation, thereby inhibiting fat extraction during mechanical separation (Azzollini et al., 2020). Bogusz et al. (2024) suggested pretreatment with pulsed electric field (PEF) before infrared-convective drying or convective drying of *T. molitor* larvae to better preserve their nutritional value.

8.3.2 Defatting Processes and Techniques

In general, defatting of the insect biomass is a prerequisite for protein extraction. The extracted fat and fat-soluble bioactive compounds can find applications both in foods and cosmetics. It has been confirmed that defatting can enhance the techno-functional properties of insect meals (Zielińska, 2022). Several researchers have reported that the solvent employed during lipid extraction affects the quality attributes and the nutritional value of the resulting protein fraction to a large extend (Kim et al., 2021; del Hierro et al., 2021). The sensory attributes of the insect-containing food products can be improved after defatting as well (Ribeiro et al., 2019; Zielińska, 2022).

The defatting method can influence the properties of the extracted ingredients. Mechanical pressure applied to *H. illucens* larvae yielded a high lipid extraction rate, particularly in the absence of blanching. A key advantage of this extraction method is that minimal water usage eliminates the need for an evaporation step (Azzollini et al., 2020). Cold pressure extraction of *H. illucens* resulted in a higher essential amino acid content. Furthermore, the proteins obtained using cold pressure extraction demonstrated superior emulsifying properties to those extracted after defatting with organic solvents (Kim, Lee et al., 2022).

Aqueous extraction is considered a green method for the simultaneous extraction of lipids and proteins from edible insects. Generally, aqueous extraction enhances the nutrient profile of the recovered lipid fraction. Mba et al. (2021) studied aqueous extraction of lipids and proteins from *Rhynchophorus phoenicis* larvae and found that increasing pH values improved lipid extraction yields. Pranil et al. (2024) examined the aqueous extraction from *Gryllus bimaculatus* and *Acheta domesticus* using boiling or ultrasound-assisted treatment, with the latter being more effective. Furthermore, Wang et al. (2023) synthesized a complex of chitosan cross-linked Fe_3O_4 nanoparticles (CNPs), metal ions, and protease that was used for the recovery of oil and protein from silkworm pupae to decrease the aqueous extraction processing time and temperature. The method also improved the oxidation stability of the extracted lipids.

Kim et al. (2021) compared the use of methanol, ethanol, and *n*-hexane as potential organic solvents for fat extraction from *Protaetia brevitarsis* larvae with aqueous fat extraction. They demonstrated that hexane was superior in extraction efficiency and had a less detrimental effect on the protein fraction. However, their study highlighted that ethanol, despite being less effective in lipid extraction, could be a valuable "greener" alternative to hexane, as it demonstrated similar protein quality. Mahmoudi-Kordi et al. (2022) also investigated the optimal conditions of an ethanol-assisted aqueous fat extraction from *Cicadatra querula*, as a green method of defatting. Their study showed that a relatively efficient extraction (yield of 69.42%) was possible by optimizing the pH value of the ethanol–water ratio (33.59% v/v) to 5.95 and the liquid/solid ratio to 12.30 mL/g. On the other hand, Amarender et al. (2020) found that ethanol extraction compared to hexane resulted in higher protein content (73.72 and 72.08%w/w, respectively) and lower lipid content (9.27 and 11.98%w/w, respectively) in the defatted cricket (Gryllidae) meal. Similarly, ethanol was the most efficient organic solvent among acetone, diethyl ether, hexane, and petroleum ether in freeze-dried edible crickets (*A. domesticus* and *Gryllodes sigillatus*) defatting (Ribeiro et al., 2019). Mixtures of ethanol with hexane or isopropanol improved both the fat extraction rate and the quality of fat from lesser mealworm (*Alphitobius diaperinus* larvae). The obtained fat exhibited enhanced oxidation stability and a high content of bioactive compounds such as carotenoids, phenolics, and vitamin E (Gharibzahedi & Altintas, 2023). Furthermore, fat extraction from *G. bimaculatus* with ethanol, hexene, and acetone interfered particularly with the recovered proteins' foaming ability (Jeong et al., 2021).

Beyond investigating the effects of different solvents on fat extraction from edible insects, several novel techniques have been employed to increase process yield or to improve the quality characteristics of the obtained fractions. For example, PEF pretreatment was applied to house crickets (*A. domesticus*), resulting in a ~42% increase in lipid extraction with hexane compared to the extraction without prior treatment (Psarianos et al., 2022). Contrary to these findings, PEF treatment of *H. illucens* larvae did not significantly affect lipid extraction yield compared to untreated larvae (Alles et al., 2020). More expensive methods, such as High hydrostatic pressure (HHP), ultrasound-assisted extraction (UAE), pressurized liquid extraction (PLE) techniques, and supercritical CO_2 lipid extraction, were also applied to assist organic solvent fat extraction from edible insects (Bolat et al., 2021; Boonmee et al., 2024; Fornari et al., 2023; Laroche et al., 2019; Otero et al., 2020). The application of HHP-assisted hexane extraction to *A. domesticus* and *T. molitor* larvae improved fat removal efficiency and influenced the properties of both the extracted lipids and protein powder (Bolat et al., 2021). UAE and PLE techniques combined with ethanol and water-ethanol mixture extraction improved the fatty acid profile of the *A. domesticus* and *T. molitor* larvae fat fraction and lowered its cholesterol level (Otero et al., 2020). Laroche et al. (2019) reported that supercritical CO_2 lipid extraction from *A. domesticus* and *T. molitor* meals was not as effective as conventional ethanol extraction, and they suggested that further optimization should be performed to become cost-effective, particularly in the case of *A. domesticus*. However, Fornari et al. (2023) observed that fat extracted with supercritical CO_2 exhibited improved oxidative stability. Boonmee et al. (2024) investigated the effects of advanced pretreatments, including HPP and ultrasonication (UL), on lipid extraction from commercially farmed edible insects in Thailand, demonstrating that these pretreatments improved lipid yield and reduced microbial loads to varying degrees.

8.3.3 Protein Extraction and Recovery Methods

Edible insects are mostly valued for their high protein content and amino acid profile. Therefore, numerous studies focus on protein extraction and recovery from different edible insect species by applying various processing approaches aimed at increasing extraction yield, protein preparation purity, and protein nutritional and techno-functional quality characteristics (Gkinali, Matsakidou, & Paraskevopoulou, 2022).

Ma et al. (2024) thoroughly investigated protein extraction yield and recovery from waxworm (*Galleria mellonella*) across a range of pH values and ionic strengths. The extractability was maximized at a pH value of 8.0, and was also sensitive to the ionic strength, exceeding 95%. Following the conventional protein extraction approach, alkaline extraction followed by isoelectric precipitation was suggested as a valuable method to obtain a protein-rich fraction (~88% w/w). As expected, variations in pH values and ionic strength during extraction induced changes in the secondary structure of the proteins, affecting their digestibility (Ma et al., 2024). In another case, salt-assisted extraction enhanced the *T. molitor* larvae protein's anti-inflammatory and antihyperglycemia properties (Oh et al., 2024). On the other hand, Mba et al. (2021) observed an increasing trend in protein extraction yield from *R. phoenicis* larvae with increasing pH values, with the highest found at a pH value of 10.0.

Protein preparations from defatted *T. molitor* larvae meal obtained through alkaline extraction, alkaline extraction followed by isoelectric precipitation, and sodium chloride treatment resulted in varying extraction yields (12.5–62.5%) and protein contents (62–75% w/w d.b.). These extraction treatments induced changes in the structure of the obtained protein along with changes in the physicochemical and techno-functional properties (Gkinali, Matsakidou, & Paraskevopoulou, 2022; Gkinali et al., 2023). While all emulsions formed from these proteins were stable under thermal treatment and freeze-thaw cycles, they differed in emulsion droplet size and storage stability (Gkinali et al., 2023). Furthermore, protein extraction under alkaline conditions with or without isoelectric precipitation influenced the protein content of *H. illucens* protein fractions and affected its physicochemical and techno-functional properties (Mshayisa et al., 2022). Alkaline protein extraction followed by isoelectric precipitation and UAE from Honeybee (*Apis mellifera*) pupae and adult grasshopper (*Schistocerca gregaria*) were effective, inducing structural and techno-functional changes to the protein concentrates influencing their surface charge, solubility, and emulsifying ability (Mishyna et al., 2019). Acidic extraction was found to be more effective in isolating protein from defatted cricket, reaching ~62% w/w protein content compared to the content of the protein resulting from the alkaline method (~58% w/w) (Amarender et al., 2020). A similar alkaline extraction procedure (pH 12.0) from *Gryllus assimilis* combined with isoelectric precipitation (pH 4.0) resulted in ~73% protein extraction yield and a protein content of ~71% w/w. The pH values during the extraction and isoelectric precipitation influenced the physicochemical and antioxidant properties of the protein concentrate (Quinteros et al., 2022).

In addition to the effects of pH values and ionic strength on protein extraction yield, extraction time, temperature, and substrate-to-solvent ratio also affected it. The protein extraction yield from defatted *H. illucens* larvae was maximized (64.44%) when the extraction time was set for 60 min, using a ratio of alkaline solution-to-substrate ~25, at 52 °C (Mintah et al., 2020). Zhang, Yue et al. (2023) studied the effects of ultrasound-assisted alkaline protein extraction time, followed by isoelectric precipitation on the secondary/tertiary structure of *T. molitor* larvae proteins. Moderate ultrasound treatment time (30 min) not only enhanced protein digestibility (Zhang, Yue et al., 2023) but also favored protein extraction yield and protein content and improved its water-/oil-holding capacities and emulsifying properties (Zhang, Sun et al., 2023). On the contrary, prolonged treatment resulted in opposite effects due to temperature increase, which caused partial protein denaturation and aggregation and, thus, solubility decrease (Zhang, Yue et al., 2023).

Novel methods, such as PEF, enzymatic-assisted extraction, and high-pressure enzymatic hydrolysis, have been employed to maximize protein extraction yield (Kim et al., 2020; Oh et al., 2024; Psarianos et al., 2022). For example, PEF was significantly successful in boosting protein extraction yield from *T. molitor* larvae up to ~40% higher compared to the conventional method (Psarianos et al., 2022). Enzyme-assisted extraction of proteins from *T. molitor* produced protein fractions with enhanced nutritional value and antioxidant activity (Oh et al., 2024). Kim et al. (2020) reported the beneficial role of high-pressure (100 MPa) enzymatic hydrolysis of *P. brevitarsis seulensis* (Kolbe) larvae on the protein extraction yield, which was found to be ~73% compared to the conventional aqueous extraction at 40 or 95°C and enzymatic hydrolysis at atmospheric conditions. To that end, the adjustment of the solvent-to-substrate ratio and enzyme content could further enhance the protein extraction yield.

8.3.4 Methods for Chitin Isolation

The common methodology for chitin extraction involves a triple-stage procedure: demineralization occurs while the substrate is treated with HCl, deproteinization with NaOH treatment followed by neutralization with HCl, and finally, depigmentation using a solution of ethanol and chloroform (Chalghaf et al., 2023; Machado et al., 2024; Song et al., 2022). Chalghaf et al. (2023), aiming to increase the chitin extraction yield from *T. molitor* adults and larvae and improve the resulting biopolymer properties and the derived chitosan investigated the effect of a reversed sequence of demineralization and deproteinization. When demineralization preceded deproteinization, the yield, the viscosity, and the crystallinity were improved due to pore formation during mineral transfer to the aqueous phase. Yuan et al. (2024) compared different sequences of demineralization and deproteinization followed by depigmentation of *H. illucens* chitin isolation procedure and a protocol skipping completely the depigmentation stage. The sequence of demineralization, deproteinization, and depigmentation maximized both chitin yield and purity and additionally supported chitosan production.

Researchers have investigated chitin extraction from various insect species, comparing conventional procedures with novel methods aimed at increasing yield and improving functionality. Chitin extracted from *Holotrichia parallela* reached a yield value of 15% by a process of deproteinization, demineralization, and depigmentation (Liu et al., 2012). Additionally, enzymatic deproteinization with alcalase was proposed as a potential method for chitin recovery to effectively remove proteins from *T. molitor*, reaching a deproteinization yield of 85% (da Silva Lucas et al., 2021). Chitin isolation yield demonstrated insignificant variations when PEF was applied on *A. domesticus* in contrast to the favorable increase of oil and protein yield. However, PEF treatment induced morphological changes in the isolated chitin. Specifically, PEF-treated chitin appeared smoother because the absence of chemical treatment eliminated cracking on the chitin and limited the formation of the fibril network (Psarianos et al., 2022). Several types of natural deep eutectic solvents (NADES) were used to obtain chitin from *H. illucens* under different conditions. The acidic NADESs consist of a basic hydrogen bond acceptor (HBA) and an acidic hydrogen bond donor (HBD), or the alkali NADESs composed of acid HBA and alkali HBD were found to maximize the recovery rate of chitin. The process involved heating and filtration, followed by washing with hot water to remove residual NADES and oven-drying. By this methodology, simultaneous desalinization and deproteinization were achieved (Zhou et al., 2019). A similar procedure was applied as part of a biorefinery for house crickets (Psarianos et al., 2023).

The aim of obtaining rich-in-chitin fractions from insects is to produce chitosan, a biopolymer that is valued for its biocompatibility and biodegradability and finds applications within the food, cosmetic, and pharmaceutical industries. Chitosan is produced by the deacetylation of chitin by chemical methods involving alkali-NaOH treatment. One-step deacetylation is usually applied. However, a two-step treatment with 20% and 50% w/w NaOH enhanced the deacetylation rate of isolated fly pupae (Sheng et al., 2022). Chitosan of *T. molitor* larvae was prepared by chitin deacetylation with a 50% yield. The acid hydrolysis of chitin and chitosan produced a chitin-oligosaccharide and chitosan-oligosaccharide (Sheng et al., 2022). Another research group tested the competence of choline chloride combined with lactic acid as DES and ionic liquids such as 1-ethyl-3-methyl imidazolium as a pretreatment phase before the chitin deacetylation using chemical means on *H. illucens* and *Bombyx mori*. The extraction process involved the conventional sequence of the 3-stage treatment (Huet et al., 2021).

8.3.5 Processing of Insect Meal

As previously discussed, the processing steps applied during insect fractionation can affect the functional properties of the recovered protein-rich preparations. Additionally, defatting pretreatment can influence their characteristics. For example, the solvent used for defatting *H. illucens* resulted in differences in amino acid profile, surface hydrophobicity, protein solubility, and emulsifying capacity and stability (Kim, Lee et al., 2022). Additionally, the pH value or added salt concentration had an impact on the physicochemical properties of *P. brevitarsis* larvae, such as surface hydrophobicity, solubility, and their techno-functional properties, such as emulsifying capacity and stability, attributed to the chitin content of the extracts (Kim, Yong et al., 2022).

Furthermore, thermally treated at 85°C *H. illucens* larvae exhibited improved protein foaming properties compared to untreated larvae (Queiroz et al., 2021). The authors suggested that heat treatment decreased the surface charge and increased particle size due to protein aggregation contributing to the air–water interface stabilization. On the other hand, Mintah et al. (2020) reported the maximum emulsifying capacity and stability of the *H. illucens* larvae extract at pH values of 6.0 and 4.0, respectively.

On top of that, PEF treatment of adult *A. domesticus* flour led to a remarkable increase in their emulsifying capacity as a result of the increased protein extractability. Moreover, the antioxidant activity of the treated insects was found to be higher than that of the untreated. Additionally, PEF favored the oil-binding capacity of the proteins but did not significantly affect their water-binding capacity (Psarianos et al., 2022).

Another group of investigators examined the impact of a combined treatment of plasma-activated water (PAW) and ultrasound (US) on silkworm pupae, aiming to enhance their microbial stability and nutritional value and improve their physicochemical properties. Although microbial growth was controlled, further optimization is needed to preserve pupal nutritional quality, particularly the fatty acid profile. The combined PAW and US treatment increased phenolic compound levels due to effective cell disruption and release into the aqueous phase. This synergistic effect resulted in remarkable antioxidant activity attributed to the increased phenolic compound release and the presence of other bioactive peptides and amino acids (Ni et al., 2024).

Rivas-Vela et al. (2023) explored the application of subcritical water hydrolysis of grasshoppers *Sphenarium purpurascens* under the effect of various modifiers, such as bicarbonate and citric acid, in combination with changes in temperature and pressure of the process to obtain hydrolysates. The use of citric acid has proven to be more effective in obtaining a lower molecular weight of the hydrolysates, also exhibiting high antioxidant activity (Rivas-Vela et al., 2023). A hydrothermal treatment of silkworm (*Bombyx mori*), bamboo caterpillar (*Omphisa fuscidentalis*), and field cricket (*G. bimaculatus*) at different temperatures and for different durations amplified the antioxidant properties of the defatted insects due to protein hydrolyzation and phenolic compound release. Nevertheless, hydrothermal treatment at the highest temperature and duration (135 °C, 30 min) had the opposite result for uncertain reasons. The investigators speculated that other chemical compounds from chitin may have interfered with the antioxidants (Kurdi et al., 2021). Similar trends were reported for the enzymatic hydrolyzation of *H. illucens* larvae, demonstrating higher foaming and emulsifying properties under acidic and alkaline pH values. This observation was attributed to the effect of hydrolysis on the protein structure (Kim et al., 2023).

In essence, edible insects can serve as rich-in-edible compounds (proteins, lipids, chitin, bioactive), biomass that can be processed further in refinery facilities to separate them and utilize them in different directions (Psarianos et al., 2023). Either ultrasound-assisted or high-pressure processing was applied to a mixture of water and edible house crickets. Organic solvents were then used for fat extraction, while pellets and aqueous phases were treated for chitin protein recovery and phenolic compounds, respectively (Psarianos et al., 2023). The successful exploitation of edible insect biomass is strongly influenced by the processing techniques employed due to their impact on both extraction/recovery yield and the properties of the resulting purified compounds. These factors are essential for the development of safe and commercially viable edible insect products.

8.3.6 Emerging and Innovative Extraction Methods

The methods employed to obtain ingredients from insects are heterogeneous and continuously developing. UAE, supercritical fluid extraction, and enzymatic hydrolysis are some of the emerging, innovative techniques used, offering greater efficiency and higher quality of extracted ingredients. With the increasing demand for sustainable food sources, such methods are becoming vital for the integration of edible insects into mainstream food products. In addition to these methods, traditional extraction techniques are also being re-evaluated and optimized, as presented above.

Coupling enzymatic hydrolysis with fermentation has the potential to produce insect hydrolysates that exert bioactive properties, having the potential to both increase the nutritional value of food products, enhancing the digestibility and functionality of ingredients derived from insects like Hermetia illucens and Tenebrio molitor (Dion-Poulin et al., 2020). UAE is gaining interest due to its efficiency and ability to enhance the extraction of bioactive compounds from insects. This extraction approach shortens the process time, reduces solvent use, and achieves better yields for antioxidants and antimicrobial compounds (Otero et al., 2020). Similar advancements are observed with HPP, which not only aids extraction but also improves the safety of the final products by inactivating pathogens without substantially affecting the nutritional value of the insect-derived ingredients (Dion-Poulin et al., 2020). Supercritical fluid extraction (SFE) is capable of selectively extracting and isolating high-purity compounds from insect biomass. Using supercritical carbon dioxide as a solvent, SFE allows for the isolation of lipids and bioactive compounds while minimizing the thermal degradation often associated with traditional extraction methods. This technique has proven effective for isolating essential fatty acids and antioxidants from insect sources, which may serve functional roles in food products (Franco et al., 2021).

While innovative techniques are gaining popularity, traditional methods of extraction have not been entirely set aside. Researchers are re-evaluating traditional practices to optimize them for insect-derived ingredients (Rahman et al., 2024). This enhancement may involve adapting these techniques to achieve higher yields and better extraction quality, fitting the production of nutrient-rich powders that increase consumer acceptance and usability in diverse culinary applications (Laroche et al., 2019).

8.3.7 Impact of Processing Parameters on Ingredient Performance

Different extraction methods, including HPP and UL, can affect the solubility together with the functional properties of insect-derived proteins (Mintah et al., 2019; Mshayisa et al., 2022). The application of HHP treatment enhances protein solubility while improving their emulsifying properties, making them more suitable for use in multiple food products (Boukil et al., 2022; Kim et al., 2021). Research demonstrates that sonication enhances the antioxidant properties of hydrolyzed proteins from insect meals, indicating that processing conditions can be tailored to enhance specific functional properties (Di Mattia et al., 2019; Mintah et al., 2019). The changes in processing conditions relating to pH and temperature levels play a vital role in the outcome of the process. The solutions containing insect proteins show substantial changes in their solubility properties through pH modifications, thus enabling better food application functionalities (Kim, Yong, Jeong et al., 2019). The solubility behavior of *T. molitor* proteins changes across different pH levels, making it possible to improve their incorporation into food formulations (Kim, Yong, Jeong et al., 2019). Thermal treatment of proteins leads to protein denaturation while simultaneously improving the digestibility and bioavailability of nutrients (Kim et al., 2021; Mshayisa et al., 2022). The substrate on which insects are reared also plays a critical role in determining their nutritional composition and functional properties. The nutrient content and fatty acid profiles of edible insects change substantially when their feed enters different diets, thus affecting ingredient quality (Meyer-Rochow et al., 2021; Riekkinen et al., 2022).

8.3.8 Industrial Scale-Up Considerations

When scaling up the industrial production of insect-derived ingredients, several factors must be carefully considered, particularly in terms of safety, environmental sustainability, efficiency, and consumer acceptance.

From a food safety perspective, it has been reported that certain microbiological species can proliferate in insect-rearing environments. If not properly managed, these microorganisms could end up in the final food ingredient and may pose risks to human health (Kooh et al., 2020; Kooh et al., 2019; Schlüter et al., 2017; Vandeweyer et al., 2020). Additionally, the substrates on which the insects are reared can have a significant role in the safety of the final product, both from a chemical and biological point of view (EFSA NDA Panel et al., 2021a, 2021b, 2021c, 2021d, 2022a, 2022b, 2023, 2024b, 2024c, 2025; EFSA Scientific Committee, 2015; Hoffmans et al., 2024; Mézes, 2018). The safety and quality evaluation of feed, together with robust control procedures throughout processing stages, will contribute to the compliance of insect-derived ingredients with human consumption safety standards (Traynor et al., 2024).

The economically viable scale-up of these ingredients requires evaluation of production expenses alongside market demand and raw material availability. Feeding insects with agricultural by-products and food waste could reduce costs and increase the sustainability profile of insect farming (Elleby et al., 2021; Vrontaki et al., 2024). Minimizing costs and increasing profitability require effective supply chains and production processes (Veldkamp et al., 2022).

As already mentioned, the functional properties and yield of insect-derived ingredients can increase through the employment of techniques like enzymatic hydrolysis and SFE and HPP (Cámara-Ruiz et al., 2023; Kozlu et al., 2024). However, the implementation of these approaches at an industrial production scale demands both infrastructure development and training programs (Cámara-Ruiz et al., 2023). Halonen et al. (2022) argued that automation advancements in insect food production will contribute to price reductions, which will improve product reach and market reception. Moreover, upcycling agricultural by-products and food waste as feed for insects could reduce the production costs and increase the sustainability profile of insect-derived ingredients (Elleby et al., 2021; Vrontaki et al., 2024). Moreover, creating effective supply chains and production processes is essential for minimizing costs and boosting profitability (Veldkamp et al., 2022).

8.4 REGULATORY CONSIDERATIONS AND SAFETY IMPLICATIONS

Adherence to regulatory frameworks and food safety standards is important both to safeguard consumers' health (Ververis et al., 2020) and to help foster public trust in these products (Mancini et al., 2019; Motoki et al., 2020). As the edible insect industry grows, proper labeling is essential to inform consumers of the inclusion of such ingredients in foods to address also the allergenicity potential of insect-derived ingredients.

In the European Union (EU), for example, insects and products thereof are classified as novel foods under Regulation (EU) 2015/2283, and their safety has to be assessed before pre-market authorization may be granted. The safety assessment is conducted by the EU risk assessor, that is, the European Food Safety Authority (EFSA), based on the evidence submitted in an application dossier by the respective food business operator. Aspects such as the identity of the novel food, its production process, its composition, product specifications, and history of use are considered. Moreover, the body of evidence under assessment includes toxicological and nutritional information, the proposed uses, and allergenicity aspects (Precup et al., 2022; Ververis et al., 2020). The guidance provided by EFSA on the data needed to be submitted has been recently updated (EFSA NDA Panel et al., 2024a). Table 8.1 presents the insect-derived ingredients that have been assessed or are being assessed for their safety by EFSA. The assessed novel ingredients are derived from *T. molitor* larvae, *A. domesticus adults*, *A. diaperinus larvae*, and *L. migratoria adults*. Other ingredients from the aforementioned species are under assessment too, alongside ingredients derived from other insect species, such as *H. illucens* and *Apis mellifera*. The forms span from whole frozen and dried insects to powders, protein concentrates, and lipids.

The development of insect farming policies in Africa along with international trade policies focuses on food safety and quality control measures, but comprehensive regulations remain absent in many geographic regions (Musundire et al., 2021; Nakimbugwe et al., 2021; Niassy et al., 2022; Odongo et al., 2018). Safety measures served as the foundations of the recently established Kenyan and Ugandan standards for edible insects (KBS, 2020; UNBS, 2020). Countries of Southeast Asia such as Thailand continue to develop new policies about safe insect consumption, as they extend their historical practice of eating insects (Preteseille et al., 2018). In China, where insects are mainly collected from the wild and consumed on an occasional basis, methods for assessing their food safety have been provided (Feng et al., 2018; Lin et al., 2023). North America controls edible insects through the safety regulations enforced by the US Food and Drug Administration as well as Health Canada (Larouche et al., 2023). While South and Central America maintain a tradition of eating insects, they do not have standardization policies regarding hygiene procedures and health risks or production control, thus potentially obstructing local trading networks (Espitia Buitrago et al., 2021). The implementation of better regulations in those areas shows potential to enhance rural household incomes as well as food distribution stability (Granados-Echegoyen et al., 2024). Native cultures in Oceania regulate insect eating practices and Australia along with New Zealand have allowed restricted insect species as safe food options (Caparros Megido et al., 2024).

8.5 PROCESSING TECHNIQUES VERSUS ALLERGENICITY

8.5.1 Allergenicity Potential of Insect-Derived Ingredients

Similar to other foods and food ingredients containing protein, insects have the potential to elicit allergic reactions in susceptible individuals through primary sensitization and/or cross-reactivity. Although the exact mechanisms of allergenicity remain elusive,

TABLE 8.1 Overview of Insect-Derived Ingredients Assessed under Assessment as Novel Foods by EFSA.

INSECT SPECIES	*FORMS/INGREDIENTS ASSESSED*	*ASSESSMENT STATUS/OUTCOME*	*EFSA JOURNAL REFERENCE*
Tenebrio molitor larvae (yellow mealworm)	Whole dried, powder from whole	Completed/ positive	EFSA NDA Panel et al. (2021a)
Locusta migratoria adults (migratory locusts)	whole frozen, whole dried, powder from whole	Completed/ positive	EFSA NDA Panel et al. (2021b)
Acheta domesticus adults (house crickets)	Whole frozen, whole dried, powder from whole	Completed/ positive	EFSA NDA Panel et al. (2021c)
Tenebrio molitor larvae (yellow mealworm)	Whole frozen, whole dried, powder from whole	Completed/ positive	EFSA NDA Panel et al. (2021d)
Acheta domesticus adults (house crickets)	Partially defatted powder	Completed/ positive	EFSA NDA Panel et al. (2022a)
Alphitobius diaperinus larvae *(lesser mealworm)*	Whole frozen, whole dried, paste from whole, powder from whole	Completed/ positive	EFSA NDA Panel et al. (2022b)
Tenebrio molitor larvae (yellow mealworm)	UV-treated powder from whole	Completed/ positive	EFSA NDA Panel et al. (2023)
Acheta domesticus adults (house crickets)	Powder from whole	Completed/ positive	EFSA NDA Panel et al. (2024b)
Acheta domesticus adults (house crickets)	Whole frozen, whole dried, powder from whole	Completed/ positive	EFSA NDA Panel et al. (2024c)
Tenebrio molitor larvae (yellow mealworm)	Whole frozen, whole dried, powder from whole	Completed/ positive	EFSA NDA Panel et al. (2025)
Acheta domesticus adults (house crickets)	Spray-dried powder	Ongoing/n.a.	EFSA (2025a)
Hermetia illucens larvae (black soldier fly larvae)	Defatted powder	Ongoing/n.a.	EFSA (2025b)
Hermetia illucens larvae (black soldier fly larvae)	Refined fat	Ongoing/n.a.	EFSA (2025c)
Tenebrio molitor larvae (yellow mealworm)	UV-treated mealworm oil	Ongoing/n.a.	EFSA (2025d)
Apis mellifera male pupae (honeybee drone brood)	Whole frozen	Ongoing/n.a.	EFSA (2025e)
Locusta migratoria adults (migratory locusts)	Whole frozen, whole dried, powder from whole	Ongoing/n.a.	EFSA (2025f)
Tenebrio molitor larvae (yellow mealworm)	Protein concentrate	Ongoing/n.a.	EFSA (2025g)
Alphitobius diaperinus larvae *(lesser mealworm)*	Protein powder	Ongoing/n.a.	EFSA (2025h)

pan-allergens, such as arginine kinase and tropomyosin, have been identified in specific edible insect species (Ribeiro et al., 2018, 2021). Recent risk assessments of novel insect-derived ingredients have also highlighted concerns regarding the possible presence of allergens originating from the insects' feed in the final food product (EFSA NDA Panel et al., 2021a, 2021b, 2021c, 2021d, 2022a, 2022b, 2023, 2024b, 2024c, 2025).

8.5.2 Effects of Processing on Allergenicity Profiles

As with other allergenic foods, the allergenicity profile of insect-derived ingredients is not solely dependent on the source but may also be affected by the food processing methods employed. While food processing has been proposed as a strategy to mitigate food allergies, its effect on allergenicity remains inconclusive, with some findings even indicating a potential increase in allergenicity (Pekar et al., 2018; Verhoeckx et al., 2015).

Various studies reported that allergenicity properties can be influenced when proteins are subject to processes such as blanching, boiling, frying, and enzymatic hydrolysis, with complex and species-dependent effects. Blanching decreased tropomyosin and arginine kinase in *H. illucens* (Bessa et al., 2021), while microwave-assisted enzymatic hydrolysis reduced tropomyosin in G. sigillatus (Hall & Liceaga, 2020). The allergenicity of *B. mori* increased following boiling (Jeong et al., 2016). High-degree enzymatic hydrolysis appears to be the most powerful method in allergenicity reduction since it decreased IgE binding capacity in processed insect samples (Hall et al., 2018; Linnemann et al., 2019). Heat treatment, together with enzymatic processes, can mitigate up to an extent allergens' action, but the allergenicity reduction rates vary based on specific insect proteins alongside processing parameters and insect species. Additional research is required to investigate further and optimize processing techniques for reducing allergenic potential while preserving nutritional quality in insect-derived ingredients.

8.6 PRODUCT DEVELOPMENT USING INSECT-DERIVED INGREDIENTS

Scientific research on the utilization of insect-derived ingredients for the production of, for example, bakery products, meat substitutes, and hybrid products shows promising but yet complicated outcomes. Food manufacturers can use insect-derived ingredients to enhance the nutrient profile, environmental performance, and functionality of their products. In general, the interest in food product reformulation continues to increase, considering also the fact that the World Health Organization has highlighted components of dietary concern in current foodstuffs (WHO, 2022). Careful selection and incorporation of insect-derived ingredients could potentially contribute to achieving such issues (Ververis et al., 2024). Moreover, insect-derived ingredients are considered as a means to combat malnutrition in developing countries, where access to nutritious food and animal-derived protein is rather limited (Ledbetter et al., 2024).

8.6.1 Insect-Derived Ingredients in Bakery Products

To date, research on utilizing insect-derived ingredients in food production has primarily focused on reformulating bakery products such as bread, pasta, cakes, muffins, and other pastries. The incorporation of insect-derived ingredients can alter both rheological properties and the nutritional value of the final products (González et al., 2019). Certain of these products, due to the way they are currently designed and produced, are considered of a low nutritional value, not promoting health (e.g., by contributing to a high glycemic index), making their reformulation with insect-derived ingredients an appealing strategy for improvement. Literature on insect-enriched flours is increasing, highlighting interest in their functional and nutritional impact. Most of these studies aim to assess the effect of insect-derived ingredients on food systems and to optimize their integration. For example, mealworm powder in wheat flour was evaluated for pasting and rheological properties (Xie et al., 2024). Flour fortification with black soldier fly prepupae powder was assessed for its effects on dough and baked products (Montevecchi et al., 2021). Moreover, *T. molitor* larvae powder in durum wheat semolina was studied for its techno-functional properties (Carpentieri et al., 2024).

Rice crackers containing 10–30% cricket powder exhibited enhanced antioxidant properties while becoming softer and obtaining reddish-yellow coloration, compared to the control samples. In terms of appearance, the results were less favorable compared to control samples, but the taste and texture of crackers with 20% cricket powder were preferred among the study participants (Suga et al., 2023). Similarly, the addition of increasing quantities (10–30%) of *A. domesticus* powder resulted in denser pancakes with higher consistency, hardness, and chewiness, along with darker coloration. Additionally, the functional changes in batter properties resulted in a minor reduction of cohesiveness, according to Bruttomesso et al. (2024). The use of cricket-derived protein powder at up to 20% in bread and dough products reduced the dough extensibility, which resulted in reduced loaf volume and increased hardness, according to Perez-Fajardo et al. (2023). Flatbread recipes containing cricket powder demonstrated reduced bacterial spore counts and improved stability in microbial growth during shelf life. The textural properties showed decreased hardness, probably because of high chitin content, and these insect-containing flatbreads displayed whole-grain product-like visual appearance (Belleggia et al., 2023).

The quality attributes of product formulations remain greatly influenced by the high fat content of particular insect species, besides their protein composition. Bottle et al. (2024) investigated the use of *T. molitor* and *A. domesticus* meals of three different fat contents (full-fat, partially defatted, fully defatted) as 10% (w/w) wheat flour substitutes in bread preparation. Meals with reduced fat content enabled better water retention in the mixed flour formulations, and T. molitor meals improved dough elasticity. Bread produced with full-fat T. molitor developed the maximum volume and obtained the lowest density alongside a tender texture. The Protein Digestibility Corrected Amino Acid Scores (PDCAAS) of the defatted insect meals exceeded those measured for the control bread samples. The results demonstrate that food producers need to consider fat composition together with flour replacement values in food developing processes.

Phenolic compounds along with other minor non-nutrient components present in insects are under investigation because these components can affect both product flavor and consequently consumers' acceptance. Zhang, Chen et al. (2024) determined how

Protaetia brevitarsis larvae phenolics, amino acids, and fatty acids can influence dough's flavor. The study revealed 11 different phenolic compounds, with the main ones being gentisic acid and vanillic acid. Adding *P. brevitarsis* powder to the dough decreased protein weakening, recovery, brightness, and specific volume, while increasing the water absorption rate. The volatile flavor analysis showed that the addition of powder increased the levels of esters, acids, and ketones while decreasing aldehyde content, thereby changing the organoleptic experience. Sponge cake dough enriched with buffalo worm and cricket and mealworm powders showed increased levels of phenolics composed of protocatechuic acid and its aldehyde, syringic and ferulic acids, and sinapic acids (Kowalski et al., 2023). The omega-6/omega-3 ratio of these products increased though, which diminishes their nutritional value (Simopoulos, 2004; Ververis et al., 2022).

Bakery production requires formulation adjustments to achieve acceptable textures when using insect powders as an ingredient. To address reduced hardness in insect-enriched bakery products, increasing binding agents (e.g., gluten, xanthan gum) can enhance texture without compromising nutrition. Partial replacement of insect powder with chickpea or pea protein can balance fibre-induced textural changes. Additionally, color stabilizers can improve visual appeal, while mild-tasting ingredients or flavor enhancers may enhance consumer acceptance.

The processing of insect-derived ingredients before using them for food formulation can help improve their functionality and taste. The work of Pyo et al. (2024) tested the chitin hydrolysis using lysozymes for improving defatted mealworm powder functionality in fortified breads. The water- and oil-binding capacity and antioxidant activity along with bread volume improved in the cases of enzymatic hydrolysis of insect powder during a 4-hour treatment which resulted in 10% hydrolyzed flour, maintaining the highest moisture content for five days. Moreover, the sensory evaluation established that chitin hydrolysis led to higher acceptability of the final products (Pyo et al., 2024). Pérez-Rodríguez et al. (2023) investigated how the solid-state fermentation of *T. molitor* and *Sphenarium purpurascens* powders by *Aspergillus oryzae* can affect dough properties. The incorporation of 5% S. purpurascens powder into bread enhanced rheological properties and texture alongside improved sensory evaluation of the product. The addition of *T. molitor* powder caused a nonsignificant change in consumer preferences. The study showed that *A. oryzae* can improve insect-containing dough fermentation through its production of amylases, xylanases, and cellulases (Chen et al., 2021).

In addition to food compositional profile and consumer acceptance, protein digestibility of insect-fortified products is an important aspect that remains underexplored. Among biscuits enriched with yellow mealworm (0–20%), those containing 15% showed maximum acceptability levels, enhanced nutrition profile, and reduced 5-hydroxymethylfurfural levels. The protein digestion of fortified biscuits improved, yet lipid digestion remained stable and starch digestion along with glycemic index values decreased, indicating better nutritional features (Mihaly Cozmuta et al., 2023). *Ruspolia differens*-fortified cereal-based cookies had higher content in isoleucine, leucine, and omega-3 fatty acids. The insect-fortified biscuits met microbiological safety standards. The acceptability rate among sensory panelists reached 80%, but the authors suggested that additional research on the flavor is required (Ochieng et al., 2023).

Overall, with regard to bakery products manufactured with insect-derived ingredients, the most studied one is bread, followed by muffins, cookies, and crackers. Reportedly, the most studied insects are crickets, mealworms, palm weevil larvae, grasshoppers, and African emperor moth caterpillars. Products containing up to 10–15% insect powder are generally considered acceptable, depending on the insect powder composition and other ingredients. Future research should focus on comparing water activity, shelf life, cost, and nutritional properties of various insect powder pre-treatment methods for the production of insect-containing bakery products (Amoah et al., 2023).

8.6.2 Insect-Derived Ingredients in Meat Products

Cavalheiro et al. (2023) investigated how house cricket powder (*A. domesticus*) affected compositional, structural, technological, nutritional, and sensory aspects of frankfurters when replacing meat at 2.5%, 5%, and 7.5%. The protein content, saturated fatty acids, and polyunsaturated fatty acids increased when the higher insect powder amount was used. The levels of zinc along with calcium and manganese increased but sodium remained constant. The incorporation of cricket powder at higher percentages negatively affected several technological aspects of frankfurters including their color and texture. The use of yellow mealworm powder (*T. molitor*) as a lean meat substitute in frankfurters (up to 25%) was investigated, with the result showing that moisture, protein, and fat content remained unchanged, but emulsion stability decreased, and cooking loss increased at 20–25% substitution. In vitro digestibility improved, though texture deteriorated due to cavity formation and weak gel networks (Zhang, Li et al., 2024). Another study reported that the utilization of yellow mealworm as meat replacer in sausages led to stable content of biogenic amines, nitrates, nitrites, and peroxide levels, increased fat and omega-3 fatty acid content, and decreased protein, cholesterol, and collagen content (Jankauskienė et al., 2024).

Moreover, the use of cricket powder (*A. domesticus*) was investigated as a partial meat replacer in beef patties, up to 10%. The insect-containing patties had higher protein content, yet they contained greater numbers of total viable and lactic acid bacteria with *Enterobacteriaceae* levels remaining within safety limits according to the authors. The texture of the product remained unaffected by the cricket powder addition up to 5% substitution although darkening and densification occurred. Once cooked, no substantial differences between control and test samples were found in pH or texture, but test samples were harder and more brownish (Cavalheiro et al., 2024). In another similar study, the substitution (up to 20%) of beef in patties by cricket powder (*Gryllus assimilis*) was assessed in terms of physicochemical, technological, and sensory parameters. It was concluded that cricket powder could replace beef by a fraction of up to 10%, obtaining a hybrid meat product with acceptable technological and sensory characteristics (Gomes Martins et al., 2024).

Interestingly, a recent case study using the risk-benefit assessment (RBA) methodology evaluated the impact of replacing red meat with house cricket in beef patties (Ververis et al., 2024), considering nutritional, microbiological, and toxicological aspects. One key finding was that the naturally present high sodium content in cricket powder significantly affected the health impact. Thus, while cricket powder can replace beef in patties, the sodium content of cricket powder must be carefully considered when designing such products to avoid negating the health impact. The study also highlighted that factors like cricket powder hydration ratio influence the outcome.

8.6.3 Insect-Derived Ingredients in Dairy and Fermented Products

Dairy products are being explored as matrices for insect-derived ingredients, mainly to enhance protein content. Andrzejczyk et al. (2024) examined the addition of yellow mealworm powder in yogurts, up to 5%. Analyses included chemical composition, color, texture, viscosity, microstructure, acidity, syneresis, sensory properties, and microbiological aspects. The higher the mealworm powder used, the more increased were protein and fat content. Moreover, it was reported that the presence of the insect powder darkened the color and introduced bitterness, reducing the final product's appeal. A key technological challenge in insect-containing yogurts is achieving uniform blending, especially when additional ingredients are used, for example, flavor masking (e.g., fruits, flavorings) or texture enhancement (e.g., stabilizers, emulsifiers) to maintain a creamy consistency. David-Birman et al. (2022) investigated the impact of silkworm pupae powder on the textural, rheological, and organoleptic properties of whipping and ice-cream. It was reported that the addition of insect powder could enhance ice-cream techno-functional properties and maintain palatability when used at concentrations of ≤1%. Ice-creams containing insect powder at 2.5% and 4% had reduced visual appeal.

With regard to fermented products, a probiotic-fermented drink with cricket protein hydrolysates (CPH) as the primary ingredient was studied (Dridi et al., 2022). The probiotic viability in CPH-containing beverage remained at 8.45 log/mL while the viscosity increased and pH decreased during 5 months of 4°C storage. The results demonstrated that the CPH-containing beverage contained more small peptides than other nonenriched and nonfermented beverages and achieved an 83% survival rate for probiotics. Sensory analysis results were favorable for the CPH-containing beverage due to its texture and color characteristics (Dridi et al., 2022).

While these few studies highlight the potential of insect-derived ingredients in dairy and fermented products, their scope remains limited. Additional research is needed to address technological challenges, optimize formulations, and assess the sensory and nutritional properties of insect-derived ingredients in these products (Diaz-Bustamante et al., 2023).

8.6.4 Other Food Products Produced with Insect-Derived Ingredients

In addition to their use in bakery, meat, and dairy products, the incorporation of insect-derived ingredients is being explored in other food matrices too. For instance, Chamoun et al. (2023) investigated the composition and consumer acceptance of tomato paste containing palm weevil larvae powder (*Rhynchophorus phoenicis* Fabricius) at different levels (8, 15, and 30%). The tomato paste containing 30% of insect powder had the highest protein, fat, and total solids content. Sensory evaluation showed a high overall acceptance and willingness to purchase the insect-containing tomato paste, which improved as consumers learned more about the potential nutritional benefits of palm weevil larvae (Chamoun et al., 2023). Moreover, Gkinali, Matsakidou, Moschakis et al. (2024) assessed yellow mealworm protein extracts for developing egg-yolk-free mayonnaise-type emulsions as a substitute protein source. Different protein extraction processes were used, and the produced emulsion products, alongside standard egg-yolk mayonnaise, were analyzed in terms of droplet size, texture, and flow properties. The results suggested that yellow mealworm protein represents a promising ingredient for producing egg-yolk-free mayonnaise-like emulsions.

There are also several other studies investigating the use of incest-derived ingredients in other foods, with the main focus of these studies being the assessment of consumer acceptance. Examples comprise cream-type soups made from tomatoes and white vegetables with the addition of 20% powder from yellow mealworm (*T. molitor*), house cricket (*A. domesticus*), buffalo worm (*A. diaperinus*), and grasshopper (*R. differens*) (Skotnicka et al., 2023), and cereal bars containing yellow mealworm, banded cricket (*G. sigillatus*), and *house cricket* undergone different processing such as de-fattening (Ribeiro et al., 2019; Ribeiro et al., 2022).

8.6.5 Insect-Derived Ingredients in 3D Printing for Food Products

Research is currently evaluating how insect-derived ingredients can be integrated into food matrices through three-dimensional (3D) printing techniques. Zhang, Jia et al. (2024) examined how the viscosity characteristics and gel combinations of various insects such as honeybee pupae, grasshoppers, and crickets operate in 3D printing in terms of composition and rheological properties. Results confirmed that the printing process caused visual and structural instability for honeybee pupa gels, despite they exhibited good fluidity. Grasshopper gels displayed increased structural stability, yet their fluidity was reduced. Cricket gels showed a balance between fluidity and structural resistance because the protein-to-chitin mass ratio proved significant (Zhang, Jia et al., 2024). Furthermore, the production of snacks incorporating up to 10% *yellow mealworm* and *lesser mealworm* using 3D printing technology has been reported (Herdeiro et al., 2024). Rheological tests were conducted to understand the ink-doughs' properties and their suitability for 3D printing. The nutritional analysis of the resulting snacks demonstrated a substantial protein content, sufficient to qualify them as a "source of protein," along with an enhanced mineral profile, compared to the control snack. The sensory analysis indicated a preference for snacks with lesser mealworm and the combination of both insects (Herdeiro et al., 2024). Additionally, four-dimensional (4D) printing, which involves incorporating stimuli-responsive materials that change shape, size, odor, and color after production, is being explored with insect-derived ingredients (Noree et al., 2023).

8.7 FUTURE DEVELOPMENTS, CHALLENGES, AND OPPORTUNITIES

The potential of insect-derived ingredients to contribute to global food system's transformation is promising, yet underexplored. Multiple challenges, including technological one, must be overcome for these ingredients to become mainstream in food production. With no doubt, the next frontier of insect-derived ingredient development lies in overcoming the barriers of scalability, techno-functional efficiency, nutritional optimization, and sensory enhancement. Beyond these, the intersection of insect farming with other emerging sectors such as circular economy, though the valorization of food waste and food industry by-products could offer exciting prospects for the future.

8.7.1 Further Exploring Insect Species Diversity

A development opportunity in insect-derived ingredients lies in the limited number of insect species that have been extensively studied so far as a source of novel food ingredients. Species such as yellow and lesser mealworm, house and banded crickets, and locusts dominate the research and development efforts in the field (Naska et al., 2022). Identifying and studying additional insect species with promising nutritional and techno-functional properties, such as higher concentrations of specific amino acids, bioactive compounds, or desirable lipid profiles, could offer new opportunities for ingredient development, offering a greater potential and diversity in final product formulations (de Carvalho et al., 2020). This exploration should also address the feasibility of rearing these species on a large scale to ensure stable production volumes.

8.7.2 Scaling-Up Insect Production

Scaling-up the production of insect-derived food ingredients is perhaps one of the most critical challenges in the field, and besides the potential to achieve high rearing volumes, additional aspects are to be considered. Current insect farming methods are still relatively labor-intensive, with large margins for improving efficiency and reducing environmental impact. Innovative approaches in, for example, biotechnology, microbiological interventions, and fermentation could help develop insects optimized for faster growth, enhanced nutrient profile, or more robust resilience to diseases (Carpentier et al., 2024; Vilcinskas, 2020). Insect species and rearing practices could be specifically tailored for achieving particular nutrient compositions, making them more suitable for the production of specific food ingredients (Carpentier et al., 2024). Furthermore, smart farming systems assisted by artificial intelligence could automate the monitoring of insect colonies in real-time, optimizing rearing conditions and monitoring insect health, behavior, and development stages with higher precision (Kröncke et al., 2020; Majewski et al., 2022).

8.7.3 Expanding the Nutritional Diversity of Insect-Derived Ingredients

The use of insects to produce food ingredients extends beyond their protein and fat content since insects contain a vast array of micronutrients, as well as non-nutrient compounds with bioactive properties. For instance, chitin and chitosan, polysaccharides found in the exoskeletons of insects, offer more than just a source of fibre. They have shown potential as ingredients with functional properties, including antioxidant activity (Ma et al., 2023).

Advances in chitin extraction and processing could enable more efficient use in nutraceuticals and functional food formulations (Mohan et al., 2020). Furthermore, insect-derived polyunsaturated fatty acids, in particular omega-3 fatty acids, which are often underrepresented in the current market, could contribute to healthy diets. Manipulation of insects' fatty acids through the feed during rearing is an aspect not to be disregarded when considering the development of new insect-derived ingredients (Riekkinen et al., 2022; Rutaro et al., 2018; Ververis et al., 2022). Insect-derived ingredients could contain or consist of peptides with unique bioactive properties (de la Luz Sánchez-Estrada et al., 2024; Rivero-Pino, Gonzalez-de la Rosa et al., 2024; Rivero-Pino, Ververis et al., 2024). Research into these insect peptides, which are often produced during digestion or fermentation, could unlock applications in immunomodulation, weight management, and gut health. Cutting-edge protein sequencing and metabolomics could contribute to this direction by identifying, extracting, and enhancing such peptides.

8.7.4 Circular Economy and Waste Valorization in Insect Farming

The intersection of insect-derived ingredients and circular economy represents an important opportunity for both environmental sustainability and cost-efficiency. Certain insects thrive on organic waste, such as agricultural by-products, and even food processing waste (Hamam et al., 2024; Sokame et al., 2024; Veldkamp et al., 2022). Technologies in waste-to-feed systems, including high-throughput microbial fermentation processes, can convert waste into high-value insect feed more efficiently, reducing reliance on external feedstocks such as grains (Varelas, 2019).

Innovations in biorefinery processes, where multiple insect-derived ingredients (e.g., protein, fat, chitin, and bioactive compounds) are simultaneously extracted and optimized for various uses, could increase the overall efficiency of these production systems, providing a model for zero-waste food production.

8.7.5 Advanced Processing and Functionalization Technologies

As the demand for insect-derived ingredients rises, so too does the need for advanced processing technologies that can enhance the quality and functionality of these ingredients. Traditional methods such as drying, grinding, and fat extraction have already been optimized for certain applications, but they may not be sufficient for unlocking the full potential of insect-derived ingredients across all food categories.

One important development lies in the use of HPP and PEF to enhance the functionality of insect-derived ingredients. These nonthermal processing techniques can improve the bioavailability of insect-derived nutrients, increase digestibility, and preserve or even enhance the nutritional quality of the products (Cho et al., 2019). Furthermore, functional food technologies, such as microencapsulation and controlled-release systems, could offer opportunities to preserve the bioactive compounds in insect-containing foods, allowing for a more sustained release of nutrients (Hendrawati et al., 2019). Additionally, the combination of different processing techniques should be explored to achieve more efficient extraction of insect-derived ingredients. By integrating multiple methods such as enzymatic hydrolysis and UAE, manufacturers may be able to maximize yield, improve purity, and enhance the functional properties of these ingredients for broader food applications.

8.7.6 Enhancing Storage Stability and Shelf-Life of Insect-Derived Ingredients

Finally, storage stability and shelf-life are often overlooked in insect-derived ingredient research, yet they are critical for product commercialization. As insect-containing foods become more widely available, the contribution of insect-derived ingredients to maintain the nutritional quality and safety of the final products over time will be essential for consumer acceptance. Research on optimal storage conditions, packaging solutions, and the impact of insect-derived ingredients on the shelf-life of processed foods is needed to ensure the long-term viability of insect-containing products in the market (Bernardo & Conte-Junior, 2024; Madau et al., 2020). Advances in active packaging, including the use of biodegradable materials and controlled atmosphere packaging, will help preserve the freshness of insect-derived ingredients, ensuring that they retain their nutritional integrity and safety during transportation and shelf storage (Bernardo & Conte-Junior, 2024; Gan et al., 2022).

8.8 CONCLUSION

Insect-derived ingredients hold a promising opportunity to create a more resilient, sustainable, diverse, and nutritionally rich food system. While considerable challenges remain in terms of scalability, nutritional enhancement, processing innovation, and consumer acceptance, the ongoing advances in food science and technology, circular economy practices, and organoleptic research offer a considerable space for innovation and development of new products. By embracing emerging technologies and interdisciplinary research, the food industry can unlock new potential in insect-derived ingredients.

8.9 DISCLAIMER

The authors F. Rivero-Pino is seconded to and E. Ververis is employed with the European Food Safety Authority (EFSA) in the Nutrition and Food Innovation Unit that provides scientific and administrative support to the Panel on Nutrition, Novel Foods, and Food Allergens, in the area of Novel Foods. The present work is published under the sole responsibility of the authors and may not be considered as an EFSA scientific output. The positions and opinions presented in this work are those of the authors alone and do not necessarily represent the views or scientific work of EFSA. To learn about the views or scientific outputs of EFSA, please consult its website at http://www.efsa.europa.eu.

REFERENCES

Adámek, M., Mlček, J., Adámková, A., Borkovcová, M., Bednářová, M., Juríková, T., Musilová, Z., & Faměra, O. (2019). Is edible insect as a novel food digestible? *Potravinarstvo Slovak Journal of Food Sciences*, 13(1), 470–476.

Adedeji, O. E., Lee, H. E., Kim, Y., Kang, H. J., Kang, M. D., Kim, J. Y., Kim, J. S., Ezekiel, O. O., Kim, W.-C., & Lee, S.-J. (2022). Three-dimensional printing of wheat flour and Acheta domesticus powder blends. *International Journal of Food Science and Technology*, *57*(10), 6279–6285.

Aguilar-Toalá, J. E., Vidal-Limón, A. M., & Liceaga, A. M. (2025). Advancing food security with farmed edible insects: Economic, social, and environmental aspects. *Insects*, *16*(1), 67.

Alles, M. C., Smetana, S., Parniakov, O., Shorstkii, I., Toepfl, S., Aganovic, K., & Heinz, V. (2020). Bio-refinery of insects with pulsed electric field pre-treatment. *Innovative Food Science & Emerging Technologies*, *64*, 102403.

Amarender, R. V., Bhargava, K., Dossey, A. T., & Gamagedara, S. (2020). Lipid and protein extraction from edible insects–Crickets (Gryllidae). *LWT*, *125*, 109222.

Amoah, I., Cobbinah, J. C., Yeboah, J. A., Essiam, F. A., Lim, J. J., Tandoh, M. A., & Rush, E. (2023). Edible insect powder for enrichment of bakery products–a review of nutritional, physical characteristics and acceptability of bakery products to consumers. *Future Foods*, *8*, 100251.

Andrzejczyk, B., Łobacz, A., Ziajka, J., Lis, A., & Małkowska-Kowalczyk, M. (2024). Comprehensive analysis of yoghurt made with the addition of Yellow Mealworm powder (Tenebrio Molitor). *Foods*, *13*(15), 2416.

Ayieko, I. A., Onyango, M., Ngadze, R. T., & Ayieko, M. A. (2021). Edible insects as new food frontier in the hospitality industry. *Frontiers in Sustainable Food Systems*, *5*, 693990.

Azzollini, D., van Iwaarden, A., Lakemond, C. M., & Fogliano, V. (2020). Mechanical and enzyme assisted fractionation process for a sustainable production of black soldier fly (Hermetia illucens) ingredients. *Frontiers in Sustainable Food Systems*, *4*, 80.

Batat, W., & Peter, P. C. (2020). The healthy and sustainable bugs appetite: Factors affecting entomophagy acceptance and adoption in western food cultures. *Journal of Consumer Marketing*, *37*(3), 291–303. https://doi.org/10.1108/jcm-10-2018-2906

Belleggia, L., Foligni, R., Ferrocino, I., Biolcati, F., Mozzon, M., Aquilanti, L., Osimani, A., & Harasym, J. (2023). Morphotextural, microbiological, and volatile characterization of flatbread containing cricket (Acheta domesticus) powder and buckwheat (Fagopyrum esculentum) flour. *European Food Research and Technology*, *249*(11), 2777–2795.

Bernardo, Y. A., & Conte-Junior, C. A. (2024). Oxidative stability in edible insects: Where is the knowledge frontier? *Trends in Food Science & Technology*, 104518.

Bessa, L. W., Pieterse, E., Marais, J., Dhanani, K., & Hoffman, L. C. (2021). Food safety of consuming black soldier fly (Hermetia illucens) larvae: Microbial, heavy metal and cross-reactive allergen risks. *Foods*, *10*(8), 1934.

Boehm, E., Borzekowski, D., Ververis, E., Lohmann, M., & Böl, G.-F. (2021). Communicating food risk-benefit assessments: Edible insects as red meat replacers. *Frontiers in Nutrition*, *8*, 749696.

Bogusz, R., Pobiega, K., Kowalczewski, P. Ł., Onopiuk, A., Szulc, K., Wiktor, A., Rybak, K., & Nowacka, M. (2024). Nutritional value and microbiological aspects of dried yellow mealworm (Tenebrio molitor L.) larvae pretreated with a pulsed electric field. *Applied Sciences*, *14*(3), 968.

Bolat, B., Ugur, A. E., Oztop, M. H., & Alpas, H. (2021). Effects of high hydrostatic pressure assisted degreasing on the technological properties of insect powders obtained from Acheta domesticus & Tenebrio molitor. *Journal of Food Engineering*, *292*, 110359.

Boonmee, N., Chittrakorn, S., Detyothin, S., Tochampa, W., Sriphannam, C., & Ruttarattanamongkol, K. (2024). Effects of high pressure and ultrasonication pretreatments and supercritical carbon dioxide extraction on physico-chemical properties of edible insect oils. *Journal of Food Process Engineering*, *47*(11), e14781.

Bottle, E., Espinosa-Ramírez, J., Serna-Saldívar, S. O., & Tejada-Ortigoza, V. (2024). Effect of full fat and defatted insect meals in breadmaking quality. *LWT*, *191*, 115602.

Boukil, A., Marciniak, A., Mezdour, S., Pouliot, Y., & Doyen, A. (2022). Effect of high hydrostatic pressure intensity on structural modifications in mealworm (Tenebrio molitor) proteins. *Foods*, *11*(7), 956.

Bruttomesso, M., Bianchi, F., Pasqualoni, I., Rizzi, C., & Simonato, B. (2024). Evaluation of the technological and compositional features of pancakes fortified with Acheta domesticus. *LWT*, *199*, 116073.

Cámara-Ruiz, M., Sánchez-Venegas, A., Blasco-Lavilla, N., Hernández, M. D., Sánchez-Liarte, F., Fernández-Gutiérrez, D., & Lara-Guillén, A. J. (2023). Comparative assessment of insect processing technologies for sustainable insect protein production. *Sustainability*, *15*(18), 13735.

Caparros Megido, R., Francis, F., Haubruge, E., Le Gall, P., Tomberlin, J., Miranda, C., Jordan, H., Picard, C., Pino, M., & Ramos-Elordy, J. (2024). A worldwide overview of the status and prospects of edible insect production. *Entomologia Generalis*, *44*(1).

Carpentier, J., Abenaim, L., Luttenschlager, H., Dessauvages, K., Liu, Y., Samoah, P., Francis, F., & Caparros Megido, R. (2024). Microorganism contribution to mass-reared edible insects: Opportunities and challenges. *Insects*, *15*(8), 611.

Carpentieri, S., Orkusz, A., Ferrari, G., & Harasym, J. (2024). Effect of replacing durum wheat semolina with Tenebrio Molitor larvae powder on the techno-functional properties of the binary blends. *Current Research in Food Science*, *8*, 100672.

Castro Delgado, M., Chambers IV, E., Carbonell-Barrachina, A., Noguera Artiaga, L., Vidal Quintanar, R., & Burgos Hernandez, A. (2020). Consumer acceptability in the USA, Mexico, and Spain of chocolate chip cookies made with partial insect powder replacement. *Journal of Food Science*, *85*(6), 1621–1628.

Cavalheiro, C. P., Ruiz-Capillas, C., Herrero, A. M., Pintado, T., Avelar de Sousa, C. C., Sant'Ana Falcão Leite, J., & Costa Alves da Silva, M. (2024). Potential of cricket (Acheta domesticus) flour as a lean meat replacer in the development of beef patties. *Foods*, *13*(2), 286.

Cavalheiro, C. P., Ruiz-Capillas, C., Herrero, A. M., Pintado, T., Cruz, T. D. M. P., & da Silva, M. C. A. (2023). Cricket (Acheta domesticus) flour as meat replacer in frankfurters: Nutritional, technological, structural, and sensory characteristics. *Innovative Food Science & Emerging Technologies*, *83*, 103245.

Chalghaf, M., Charradi, K., Ksouri, R., Alsulami, Q. A., Jaouani, A., Keshk, S. M., & Hayouni, E. A. (2023). Physicochemical characterization of chitin extracted by different treatment sequences from an edible insect. *International Journal of Biological Macromolecules*, *253*, 127156.

Chamoun, L., Karboune, S., Lutterodt, H. E., & Melgar-Quinonez, H. (2023). Nutritional composition and consumer acceptance of tomato paste fortified with palm weevil larvae (Rhynchophorus phoenicis Fabricius) in the Ashanti region, Ghana. *Food Science & Nutrition*, *11*(8), 4583–4595.

Chantakun, K., Petcharat, T., Wattanachant, S., Ab Karim, M. S. B., & Kaewthong, P. (2024). Fatty acid profile and thermal behavior of fat-rich edible insect oils compared to commonly consumed animal and plant oils. *Food Science of Animal Resources*, *44*(4), 790.

Chen, Y., Eder, S., Schubert, S., Gorgerat, S., Boschet, E., Baltensperger, L., Boschet, E., Städeli, C., Kuster, S., & Fischer, P. (2021). Influence of amylase addition on bread quality and bread staling. *Acs Food Science & Technology*, *1*(6), 1143–1150.

Cho, K. H., Kang, S. W., Yoo, J. S., Song, D. K., Chung, Y. H., Kwon, G. T., & Kim, Y. Y. (2019). Effects of mealworm (Tenebrio Molitor) larvae hydrolysate on nutrient ileal digestibility in growing pigs compared to those of defatted mealworm larvae meal, fermented poultry by-product, and hydrolyzed fish soluble. *Asian-Australasian Journal of Animal Sciences*, *33*(3), 490.

Collins, C. M., Vaskou, P., & Kountouris, Y. (2019). Insect food products in the western world: Assessing the potential of a new 'green' market. *Annals of the Entomological Society of America*, *112*(6), 518–528. https://doi.org/10.1093/aesa/saz015

Cruz-López, S. O., Escalona-Buendía, H. B., Martinez-Arellano, I., Domínguez-Soberanes, J., & Alvarez-Cisneros, Y. M. (2024). Physicochemical and techno-functional characterization of soluble proteins extracted by ultrasound from the cricket Acheta domesticus. *Heliyon*, *10*(23).

da Costa Rocha, A. C., de Andrade, C. J., & de Oliveira, D. (2021). Perspective on integrated biorefinery for valorization of biomass from the edible insect Tenebrio Molitor. *Trends in Food Science & Technology*, *116*, 480–491.

da Silva Lucas, A. J., Oreste, E. Q., Costa, H. L. G., López, H. M., Saad, C. D. M., & Prentice, C. (2021). Extraction, physicochemical characterization, and morphological properties of chitin and chitosan from cuticles of edible insects. *Food chemistry*, *343*, 128550.

Dalle Zotte, A., Singh, Y., Michiels, J., & Cullere, M. (2019). Black soldier fly (Hermetia illucens) as dietary source for laying quails: live performance, and egg physico-chemical quality, sensory profile and storage stability. *Animals*, *9*(3), 115.

David-Birman, T., Romano, A., Aga, A., Pascoviche, D., Davidovich-Pinhas, M., & Lesmes, U. (2022). Impact of silkworm pupae (Bombyx mori) powder on cream foaming, ice cream properties and palatability. *Innovative Food Science & Emerging Technologies*, *75*, 102874.

de Carvalho, N. M., Madureira, A. R., & Pintado, M. E. (2020). The potential of insects as food sources–a review. *Critical Reviews in Food Science and Nutrition*, *60*(21), 3642–3652.

de la Luz Sánchez-Estrada, M., Aguirre-Becerra, H., & Feregrino-Pérez, A. A. (2024). Bioactive compounds and biological activity in edible insects: A review. *Heliyon*, *10*(2).

del Hierro, J. N., Cantero-Bahillo, E., Fornari, T., & Martin, D. (2021). Effect of defatting and extraction solvent on the antioxidant and pancreatic lipase inhibitory activities of extracts from Hermetia illucens and Tenebrio Molitor. *Insects*, *12*(9), 789.

Deroy, O., Reade, B., & Spence, C. (2015). The insectivore's dilemma, and how to take the West out of it. *Food Quality and Preference*, *44*, 44–55.

Detilleux, L., Bayendi Loudit, S., Le Gall, P., Francis, F., Caparros Megido, R., & Dogot, T. (2024). Consumers of insect-based foods: A cross-cultural study between Belgium and Gabon. *Journal of Insect Science*, *24*(3), 2.

Di Mattia, C., Battista, N., Sacchetti, G., & Serafini, M. (2019). Antioxidant activities in vitro of water and liposoluble extracts obtained by different species of edible insects and invertebrates. *Frontiers in Nutrition*, *6*, 438996.

Diaz-Bustamante, M. L., Keppler, J. K., Reyes, L. H., & Solano, O. A. A. (2023). Trends and prospects in dairy protein replacement in yogurt and cheese. *Heliyon*, *9*(6).

Dion-Poulin, A., Laroche, M., Doyen, A., & Turgeon, S. L. (2020). Functionality of cricket and mealworm hydrolysates generated after pretreatment of meals with high hydrostatic pressures. *Molecules*, *25*(22), 5366.

Dridi, C., Millette, M., Aguilar, B., Salmieri, S., & Lacroix, M. (2022). Storage stability of a fermented probiotic beverage enriched with cricket protein hydrolysates. *Food and Bioprocess Technology*, *15*(11), 2587–2600.

EFSA. (2025a, March 1). *Application for authorisation of Acheta domesticus (house cricket) spray-dried powder as a novel food*. EFSA. https://open.efsa.europa.eu/questions/EFSA-Q-2023–00374

EFSA. (2025b, March 1). *Application for authorisation of Dried defatted powder of Hermetia Illucens larvae as a novel food*. EFSA. https://open.efsa.europa.eu/questions/EFSA-Q-2023–00703

EFSA. (2025c, March 1). *Application for authorisation of Hermetia illucens larvae refined fat (Refined Black soldier fly larvae fat) as a novel food*. EFSA. https://open.efsa.europa.eu/questions/EFSA-Q-2023–00847

EFSA. (2025d, March 1). *Application for authorisation of Vitamin D3 containing UV-treated mealworm oil as a novel food.* EFSA. https://open.efsa.europa.eu/questions/EFSA-Q-2022–00534

EFSA. (2025e, March 1). *Request for a scientific opinion on Apis mellifera male pupae as a novel food.* EFSA. https://open.efsa.europa.eu/questions/EFSA-Q-2019–00201

EFSA. (2025f, March 1). *Request for a scientific opinion on Locusta Migratoria as a novel food.* EFSA. https://open.efsa.europa.eu/questions/EFSA-Q-2018–00513

EFSA. (2025g, March 1). *Request for a scientific opinion on Protein-rich flour from fresh larvae of mealworm (Tenebrio Molitor) as a novel food.* EFSA. https://open.efsa.europa.eu/questions/EFSA-Q-2021–00105

EFSA. (2025h, March 1). *Request for a scientific opinion on Protein powders from the Alphitobius diaperinus larva as a novel food.* EFSA. https://open.efsa.europa.eu/questions/EFSA-Q-2019–00690

EFSA NDA Panel on Nutrition, Novel Foods and Food Allergens (NDA), Turck, D., Bohn, T., Cámara, M., Castenmiller, J., De Henauw, S., Hirsch-Ernst, K. I., Jos, Á., Maciuk, A., Mangelsdorf, I., McNulty, B., Naska, A., Pentieva, K., Siani, A., Thies, F., Aguilera-Gómez, M., Cubadda, F., Frenzel, T., Heinonen, M.,. . . . McArdle, H. J. (2024c). Safety of frozen, dried, and powder forms of house crickets (Acheta domesticus) as a novel food pursuant to Regulation (EU) 2015/2283. *EFSA Journal*, *22*(12). https://doi.org/https://doi.org/10.2903/j.efsa.2024.9101

EFSA NDA Panel on Nutrition, Novel Foods and Food Allergens (NDA), Turck, D., Bohn, T., Cámara, M., Castenmiller, J., De Henauw, S., Hirsch-Ernst, K. I., Jos, Á., Maciuk, A., Mangelsdorf, I., McNulty, B., Naska, A., Pentieva, K., Siani, A., Thies, F., Aguilera-Gómez, M., Cubadda, F., Frenzel, T., Heinonen, M.,. . . . McArdle, H. J. (2025). Safety of frozen and dried forms of whole yellow mealworm (Tenebrio Molitor larva) as a novel food pursuant to Regulation (EU) 2015/2283. *EFSA Journal*, *23*(1). https://doi.org/https://doi.org/10.2903/j.efsa.2025.9155

EFSA NDA Panel on Nutrition, Novel Foods and Food Allergens (NDA), Turck, D., Castenmiller, J., De Henauw, S., Hirsch-Ernst, K. I., Kearney, J., Maciuk, A., Mangelsdorf, I., McArdle, H. J., Naska, A., Pelaez, C., Pentieva, K., Siani, A., Thies, F., Tsabouri, S., Vinceti, M., Cubadda, F., Frenzel, T., Heinonen, M.,. . . . Knutsen, H. K. (2021a). Safety of dried yellow mealworm (Tenebrio Molitor larva) as a novel food pursuant to Regulation (EU) 2015/2283. *EFSA Journal*, *19*(1). https://doi.org/https://doi.org/10.2903/j.efsa.2021.6343

EFSA NDA Panel on Nutrition, Novel Foods and Food Allergens (NDA), Turck, D., Castenmiller, J., De Henauw, S., Hirsch-Ernst, K. I., Kearney, J., Maciuk, A., Mangelsdorf, I., McArdle, H. J., Naska, A., Pelaez, C., Pentieva, K., Siani, A., Thies, F., Tsabouri, S., Vinceti, M., Cubadda, F., Frenzel, T., Heinonen, M.,. . . . Knutsen, H. K. (2021b). Safety of frozen and dried formulations from migratory locust (Locusta migratoria) as a novel food pursuant to Regulation (EU) 2015/2283. *EFSA Journal*, *19*(7). https://doi.org/https://doi.org/10.2903/j.efsa.2021.6667

EFSA NDA Panel on Nutrition, Novel Foods and Food Allergens (NDA), Turck, D., Bohn, T., Castenmiller, J., De Henauw, S., Hirsch-Ernst, K. I., Maciuk, A., Mangelsdorf, I., McArdle, H. J., Naska, A., Pelaez, C., Pentieva, K., Siani, A., Thies, F., Tsabouri, S., Vinceti, M., Aguilera-Gómez, M., Cubadda, F., Frenzel, T.,. . . . Knutsen, H. K. (2023). Safety of UV-treated powder of whole yellow mealworm (Tenebrio Molitor larva) as a novel food pursuant to Regulation (EU) 2015/2283. *EFSA Journal*, *21*(6). https://doi.org/https://doi.org/10.2903/j.efsa.2023.8009

EFSA NDA Panel on Nutrition, Novel Foods and Food Allergens (NDA), Turck, D., Bohn, T., Castenmiller, J., De Henauw, S., Hirsch-Ernst, K. I., Maciuk, A., Mangelsdorf, I., McArdle, H. J., Naska, A., Pelaez, C., Pentieva, K., Siani, A., Thies, F., Tsabouri, S., Vinceti, M., Cubadda, F., Frenzel, T., Heinonen, M.,. . . . Knutsen, H. K. (2021c). Safety of frozen and dried formulations from whole house crickets (Acheta domesticus) as a novel food pursuant to Regulation (EU) 2015/2283. *EFSA Journal*, *19*(8), e06779. https://doi.org/10.2903/j.efsa.2021.6779

EFSA NDA Panel on Nutrition, Novel Foods and Food Allergens (NDA), Turck, D., Bohn, T., Castenmiller, J., De Henauw, S., Hirsch-Ernst, K. I., Maciuk, A., Mangelsdorf, I., McArdle, H.

J., Naska, A., Pelaez, C., Pentieva, K., Siani, A., Thies, F., Tsabouri, S., Vinceti, M., Cubadda, F., Frenzel, T., Heinonen, M.,. . . . Knutsen, H. K. (2021d). Safety of frozen and dried formulations from whole yellow mealworm (Tenebrio Molitor larva) as a novel food pursuant to Regulation (EU) 2015/2283. *EFSA Journal*, *19*(9). https://doi.org/https://doi.org/10.2903/j.efsa.2021.6778

EFSA NDA Panel on Nutrition, Novel Foods and Food Allergens (NDA), Turck, D., Bohn, T., Castenmiller, J., De Henauw, S., Hirsch-Ernst, K. I., Maciuk, A., Mangelsdorf, I., McArdle, H. J., Naska, A., Pelaez, C., Pentieva, K., Siani, A., Thies, F., Tsabouri, S., Vinceti, M., Cubadda, F., Frenzel, T., Heinonen, M.,. . . . Knutsen, H. K. (2022a). Safety of partially defatted house cricket (Acheta domesticus) powder as a novel food pursuant to Regulation (EU) 2015/2283. *EFSA Journal*, *20*(5). https://doi.org/https://doi.org/10.2903/j.efsa.2022.7258

EFSA NDA Panel on Nutrition, Novel Foods and Food Allergens (NDA), Turck, D., Bohn, T., Castenmiller, J., De Henauw, S., Hirsch-Ernst, K. I., Maciuk, A., Mangelsdorf, I., McArdle, H. J., Naska, A., Pelaez, C., Pentieva, K., Siani, A., Thies, F., Tsabouri, S., Vinceti, M., Cubadda, F., Frenzel, T., Heinonen, M.,. . . . Knutsen, H. K. (2022b). Safety of frozen and freeze-dried formulations of the lesser mealworm (Alphitobius diaperinus larva) as a novel food pursuant to Regulation (EU) 2015/2283. *EFSA Journal*, *20*(7). https://doi.org/https://doi.org/10.2903/j.efsa.2022.7325

EFSA NDA Panel on Nutrition, Novel Foods and Food Allergens (NDA), Turck, D., Bohn, T., Castenmiller, J., de Henauw, S., Hirsch-Ernst, K. I., Maciuk, A., Mangelsdorf, I., McArdle, H. J., Naska, A., Pentieva, K., Siani, A., Thies, F., Tsabouri, S., Vinceti, M., Aguilera Gómez, M., Cubadda, F., Frenzel, T., Heinonen, M.,. . . . Knutsen, H. K. (2024a). Guidance on the scientific requirements for an application for authorisation of a novel food in the context of Regulation (EU) 2015/2283. *EFSA Journal*, *22*(9), e8961.

EFSA NDA Panel on Nutrition, Novel Foods and Food Allergens (NDA), Turck, D., Bohn, T., Castenmiller, J., De Henauw, S., Hirsch-Ernst, K. I., Maciuk, A., Mangelsdorf, I., McArdle, H. J., Naska, A., Pentieva, K., Siani, A., Thies, F., Tsabouri, S., Vinceti, M., Aguilera-Gómez, M., Cubadda, F., Frenzel, T., Heinonen, M.,. . . . Knutsen, H. K. (2024b). Safety of Acheta domesticus powder as a novel food pursuant to Regulation (EU) 2015/2283. *EFSA Journal*, *22*(7). https://doi.org/https://doi.org/10.2903/j.efsa.2024.8919

EFSA Scientific Committee. (2015). Risk profile related to production and consumption of insects as food and feed. *EFSA Journal*, *13*(10), 4257.

Elleby, C., Jensen, H. G., Domínguez, I. P., Chatzopoulos, T., & Charlebois, P. (2021). Insects reared on food waste: A game changer for global agricultural feed markets? *EuroChoices*, *20*(3), 56–62.

Espinosa-Solís, A., Velázquez-Segura, A., Lara-Rodríguez, C., Martínez, L. M., Chuck-Hernández, C., & Rodríguez-Sifuentes, L. (2024). Optimizing chitin extraction and chitosan production from house cricket flour. *Processes*, *12*(3), 464.

Espitia Buitrago, P. A., Hernández, L. M., Burkart, S., Palmer, N., & Cardoso Arango, J. A. (2021). Forage-fed insects as food and feed source: Opportunities and constraints of edible insects in the tropics. *Frontiers in Sustainable Food Systems*, *5*, 724628.

Feng, Y., Chen, X. M., Zhao, M., He, Z., Sun, L., Wang, C. Y., & Ding, W. F. (2018). Edible insects in China: Utilization and prospects. *Insect Science*, *25*(2), 184–198.

Fornari, T., Vázquez, L., Villanueva-Bermejo, D., Hurtado-Ribeira, R., Martín Hernández, D., & Martin, D. (2023). Effect of moisture and oil content in the supercritical CO2 defatting of Hermetia illucens larvae. *Foods*, *12*(3), 490.

Franco, A., Salvia, R., Scieuzo, C., Schmitt, E., Russo, A., & Falabella, P. (2021). Lipids from insects in cosmetics and for personal care products. *Insects*, *13*(1), 41.

Gan, J., Zhao, M., He, Z., Sun, L., Li, X., & Feng, Y. (2022). The effects of antioxidants and packaging methods on inhibiting lipid oxidation in deep fried crickets (Gryllus bimaculatus) during storage. *Foods*, *11*(3), 326.

Gharibzahedi, S. M. T., & Altintas, Z. (2023). Lesser mealworm (Alphitobius diaperinus L.) larvae oils extracted by pure and binary mixed organic solvents: Physicochemical and antioxidant properties, fatty acid composition, and lipid quality indices. *Food Chemistry*, *408*, 135209.

Gkinali, A.-A., Matsakidou, A., Michailidis, A., & Paraskevopoulou, A. (2024). How do Greeks feel about eating insects? A study of consumer perceptions and preferences. *Foods*, *13*(19), 3199.

Gkinali, A.-A., Matsakidou, A., Moschakis, T., & Paraskevopoulou, A. (2024). Egg-free mayonnaise-type emulsions stabilized with proteins derived from the larvae of Tenebrio molitor. *Food Hydrocolloids*, *156*, 110249.

Gkinali, A.-A., Matsakidou, A., & Paraskevopoulou, A. (2022). Characterization of Tenebrio Molitor larvae protein preparations obtained by different extraction approaches. *Foods*, *11*(23), 3852.

Gkinali, A.-A., Matsakidou, A., & Paraskevopoulou, A. (2023). Assessing the emulsifying properties of Tenebrio Molitor larvae protein preparations: Impact of storage, thermal, and freeze-thaw treatments on o/w emulsion stability. *International Journal of Biological Macromolecules*, *250*, 126165.

Gkinali, A.-A., Matsakidou, A., Vasileiou, E., & Paraskevopoulou, A. (2022). Potentiality of Tenebrio Molitor larva-based ingredients for the food industry: A review. *Trends in Food Science & Technology*, *119*, 495–507.

Gomes Martins, V. M., Milano, P., Rodrigues Pollonio, M. A., dos Santos, M., de Oliveira, A. P., Savay-da-Silva, L. K., Ferreira Ignácio Câmara, A. K., & de Souza Paglarini, C. (2024). Adding cricket (Gryllus assimilis) flour in hybrid beef patties: Physicochemical, technological and sensory challenges. *International Journal of Food Science and Technology*, *59*(1), 450–459.

González, C. M., Garzón, R., & Rosell, C. M. (2019). Insects as ingredients for bakery goods. A comparison study of H. illucens, A. domestica and T. Molitor flours. *Innovative Food Science & Emerging Technologies*, *51*, 205–210.

Gould, J., & Wolf, B. (2018). Interfacial and emulsifying properties of mealworm protein at the oil/water interface. *Food Hydrocolloids*, *77*, 57–65.

Granados-Echegoyen, C., Vásquez-López, A., Calderón-Cortés, N., Gallego-Ocampo, H. L., Gómez-Rodríguez, C. H., Rodríguez-Vélez, J. M., Sarmiento-Cordero, M. A., Salamanca-Canizales, L. J., Rodríguez-Vélez, B., & Arroyo-Balán, F. (2024). Brief overview of edible insects: Exploring consumption and promising sustainable uses in Latin America. *Frontiers in Sustainable Food Systems*, *8*, 1385081.

Hall, F., Johnson, P. E., & Liceaga, A. (2018). Effect of enzymatic hydrolysis on bioactive properties and allergenicity of cricket (Gryllodes sigillatus) protein. *Food Chemistry*, *262*, 39–47.

Hall, F., & Liceaga, A. (2020). Effect of microwave-assisted enzymatic hydrolysis of cricket (Gryllodes sigillatus) protein on ACE and DPP-IV inhibition and tropomyosin-IgG binding. *Journal of Functional Foods*, *64*, 103634.

Hall, F. G. (2020). *Bioactive and allergenic properties of edible cricket (Gryllodes sigillatus) peptides*. Purdue University.

Halloran, A., Münke, C., Vantomme, P., Reade, B., & Evans, J. (2015). Broadening insect gastronomy. In *The Routledge handbook of sustainable food and gastronomy* (pp. 199–205). Routledge.

Halonen, V., Uusitalo, V., Levänen, J., Sillman, J., Leppäkoski, L., & Claudelin, A. (2022). Recognizing potential pathways to increasing the consumption of edible insects from the perspective of consumer acceptance: Case study from Finland. *Sustainability*, *14*(3), 1439. https://doi.org/10.3390/su14031439

Hamam, M., D'Amico, M., & Di Vita, G. (2024). Advances in the insect industry within a circular bioeconomy context: A research agenda. *Environmental Sciences Europe*, *36*(1), 29.

Harris, K. J. (2016). Culture and theory: Considerations for the ethnic restaurant and food safety culture. *Athens Journal of Tourism*, *3*(4), 263–272.

Hartmann, C., Shi, J., Giusto, A., & Siegrist, M. (2015). The psychology of eating insects: A cross-cultural comparison between Germany and China. *Food Quality and Preference*, *44*, 148–156.

Hartmann, C., & Siegrist, M. (2016). Becoming an insectivore: Results of an experiment. *Food Quality and Preference*, *51*, 118–122.

Hendrawati, T. Y., Sari, A. M., Rahman, M. I. S., Nugrahani, R. A., & Siswahyu, A. (2019). Microencapsulation techniques of herbal compounds for raw materials in food industry, cosmetics and pharmaceuticals. In *Microencapsulation-processes, technologies and industrial applications*. IntechOpen.

Herdeiro, F. M., Carvalho, M. O., Nunes, M. C., & Raymundo, A. (2024). Development of healthy snacks incorporating meal from Tenebrio Molitor and Alphitobius diaperinus using 3D printing technology. *Foods*, *13*(2), 179.

Hlongwane, Z., Siwela, M., Slotow, R., & Munyai, T. (2022). Effect of geographical location, insect type and cooking method on the nutritional composition of insects consumed in South Africa. *Journal of Insects as Food and Feed*, *8*(5), 537–556.

Hoffmans, Y., Veldkamp, T., Meijer, N., Brust, G., van der Schans, M., Prins, T., van Rozen, K., Elissen, H., van Wikselaar, P., & van der Weide, R. (2024). Can black soldier fly larvae (Hermetia illucens) be reared on waste streams for food and feed?–A safety perspective. *Journal of Insects as Food and Feed*, *10*(7), 1211–1221.

Huet, G., Hadad, C., Gonzalez-Dominguez, J. M., Courty, M., Jamali, A., Cailleu, D., & van Nhien, A. N. (2021). IL versus DES: Impact on chitin pretreatment to afford high quality and highly functionalizable chitosan. *Carbohydrate Polymers*, *269*, 118332.

Huis, A. V., Itterbeeck, J. V., Klunder, H., Mertens, E., Halloran, A., Muir, G., & Vantomme, P. (2013). *Edible insects: Future prospects for food and feed security*. Food and Agriculture Organization of the United Nations.

Huis, A. V., Rumpold, B. A., Maya, C., & Roos, N. (2021). Nutritional qualities and enhancement of edible insects. *Annual Review of Nutrition*, *41*(1), 551–576. https://doi.org/10.1146/annurev-nutr-041520-010856

Hurtado-Ribeira, R., Hernández, D. M., Villanueva-Bermejo, D., García-Risco, M. R., Hernández, M. D., Vázquez, L., Fornari, T., & Martin, D. (2023). The interaction of slaughtering, drying, and defatting methods differently affects oxidative quality of the fat from black soldier fly (Hermetia illucens) larvae. *Insects*, *14*(4), 368.

Hurtado-Ribeira, R., Villanueva-Bermejo, D., García-Risco, M. R., Hernández, M. D., Sánchez-Muros, M. J., Fornari, T., Vázquez, L., & Martin, D. (2023). Evaluation of the interrelated effects of slaughtering, drying, and defatting methods on the composition and properties of black soldier fly (Hermetia illucens) larvae fat. *Current Research in Food Science*, *7*, 100633.

Igwe, C., Ujowundu, C., Nwaogu, L., & Okwu, G. (2011). Chemical analysis of an edible African termite Macrotermes nigeriensis, a potential antidote to food security problem. *Biochemistry and Analytical Biochemistry*, *1*(105), 2161–1009.

Jankauskienė, A., Kiselioviene, S., Aleknavičius, D., Miliūnaitė, I., Kerzienė, S., Gaižauskaitė, Ž., Jukniené, I., Zaviztanavičiūtė, P., & Kabašinskienė, A. (2024). Innovative applications of Tenebrio Molitor larvae in the production of sustainable meat sausages: Quality and safety aspects. *Foods*, *13*(10), 1451.

Jeong, K. Y., Son, M., Lee, J. Y., Park, K. H., Lee, J.-H., & Park, J.-W. (2016). Allergenic characterization of 27-kDa glycoprotein, a novel heat stable allergen, from the pupa of silkworm, Bombyx mori. *Journal of Korean medical science*, *31*(1), 18–24.

Jeong, M.-S., Lee, S.-D., & Cho, S.-J. (2021). Effect of three defatting solvents on the techno-functional properties of an edible insect (Gryllus bimaculatus) protein concentrate. *Molecules*, *26*(17), 5307.

KBS. (2020). KS 2922–1:2020 Edible insects – specification Part 1: Edible insects products. In *Kenya Bureau of Standards*. Kenya Bureau of Standards.

Khatun, H., Lievens, S., Smets, R., Akhtaruzzaman, M., Van Der Borght, M., & Claes, J. (2022). Optimizing the rheological and textural properties of chapatti enriched with house crickets (Acheta domesticus) flour using hydrocolloids by an I-optimal design. *Foods*, *11*(21), 3467.

Kim, T.-E., Kim, C.-T., Sim, H.-J., Lee, H.-Y., Kim, S.-O., Kim, D.-K., Jo, M.-N., Lee, S.-J., Jeon, Y.-D., & Song, Y.-J. (2020). Production of protein hydrolysate from Protaetia brevitarsis seulensis (Kolbe) larvae by enzyme treatment under high pressure. *Food Science and Biotechnology*, *29*, 1187–1194.

Kim, T.-K., Lee, J.-H., Yong, H. I., Kang, M.-C., Cha, J. Y., Chun, J. Y., & Choi, Y.-S. (2022). Effects of defatting methods on the physicochemical properties of proteins extracted from Hermetia illucens larvae. *Foods*, *11*(10), 1400.

Kim, T.-K., Yong, H., Jang, H., Jung, S., & Choi, Y. (2022). Effect of extraction condition on technological properties of protein fromProtaetia brevitarsis larvae. *Journal of Insects as Food and Feed*, *8*(2), 147–156.

Kim, T.-K., Yong, H. I., Jeong, C. H., Han, S. G., Kim, Y.-B., Paik, H.-D., & Choi, Y.-S. (2019). Technical functional properties of water-and salt-soluble proteins extracted from edible insects. *Food Science of Animal Resources*, *39*(4), 643.

Kim, T.-K., Yong, H. I., Kim, Y.-B., Jung, S., Kim, H.-W., & Choi, Y.-S. (2021). Effects of organic solvent on functional properties of defatted proteins extracted from Protaetia brevitarsis larvae. *Food Chemistry*, *336*, 127679.

Kim, T.-K., Yong, H. I., Kim, Y.-B., Kim, H.-W., & Choi, Y.-S. (2019). Edible insects as a protein source: A review of public perception, processing technology, and research trends. *Food Science of Animal Resources*, *39*(4), 521.

Kim, Y., Kim, T., Choi, Y., Lee, J., Jang, H., Kang, M., & Choi, Y. (2023). Effects of hydrolysis and swelling on structural and functional properties ofHermetia illucens L.: insoluble protein in residue fraction. *Journal of Insects as Food and Feed*, *9*(6), 799–808.

Kooh, P., Jury, V., Laurent, S., Audiat-Perrin, F., Sanaa, M., Tesson, V., Federighi, M., & Boué, G. (2020). Control of biological hazards in insect processing: Application of HACCP method for yellow mealworm (Tenebrio Molitor) powders. *Foods*, *9*(11), 1528.

Kooh, P., Ververis, E., Tesson, V., Boué, G., & Federighi, M. (2019). Entomophagy and public health: A review of microbiological hazards. *Health*, *11*(10), 1272–1290.

Kowalski, S., Gumul, D., Oracz, J., Rosicka-Kaczmarek, J., Mikulec, A., Mickowska, B., Skotnicka, M., & Zborowski, M. (2023). Chemical composition, antioxidant properties and sensory aspects of sponge cakes supplemented with edible insect flours. *Antioxidants*, *12*(11), 1912.

Kozlu, A., Ngasakul, N., Klojdová, I., & Baigts-Allende, D. K. (2024). Edible insect-processing techniques: A strategy to develop nutritional food products and novelty food analogs. *European Food Research and Technology*, *250*(5), 1253–1267.

Krawczyk, A., Fernández-López, J., & Zimoch-Korzycka, A. (2024). Insect protein as a component of meat analogue burger. *Foods*, *13*(12), 1806.

Kröger, T., Dupont, J., Büsing, L., & Fiebelkorn, F. (2022). Acceptance of insect-based food products in western societies: A systematic review. *Frontiers in nutrition*, *8*, 759885.

Kröncke, N., Baur, A., Böschen, V., Demtröder, S., Benning, R., & Delgado, A. (2020). Automation of insect mass rearing and processing technologies of mealworms (Tenebrio Molitor). In *African edible insects as alternative source of food, oil, protein and bioactive components* (pp. 123–139). Springer International Publishing.

Kurdi, P., Chaowiwat, P., Weston, J., & Hansawasdi, C. (2021). Studies on microbial quality, protein yield, and antioxidant properties of some frozen edible insects. *International Journal of Food Science*, *2021*(1), 5580976.

Lange, K. W., & Nakamura, Y. (2023). Potential contribution of edible insects to sustainable consumption and production. *Frontiers in Sustainability*, *4*, 1112950

Laroche, M., Perreault, V., Marciniak, A., Gravel, A., Chamberland, J., & Doyen, A. (2019). Comparison of conventional and sustainable lipid extraction methods for the production of oil and protein isolate from edible insect meal. *Foods*, *8*(11), 572.

Larouche, J., Campbell, B., Hénault-Éthier, L., Banks, I. J., Tomberlin, J. K., Preyer, C., Deschamps, M.-H., & Vandenberg, G. W. (2023). The edible insect sector in Canada and the United States. *Animal Frontiers*, *13*(4), 16–25.

Ledbetter, M., Wilkin, J. D., Mubaiwa, J., Manditsera, F. A., Macheka, L., Ruzengwe, F. M., Madimutsa, O. N., Chopera, P., Matsungo, T. M., & Cottin, S. C. (2024). Development of a nutritious cereal-based instant porridge by the incorporation of protein-rich insect powder–An example from Zimbabwe. *Journal of Functional Foods*, *112*, 105957.

Lee, M. H., Kim, T.-K., Cha, J. Y., Yong, H. I., Kang, M.-C., Jang, H. W., & Choi, Y.-S. (2022). Physicochemical characteristics and aroma patterns of oils prepared from edible insects. *LWT*, *167*, 113888.

Liang, Z., Zhu, Y., Leonard, W., & Fang, Z. (2024). Recent advances in edible insect processing technologies. *Food Research International*, 114137.

Lin, X., Wang, F., Lu, Y., Wang, J., Chen, J., Yu, Y., Tao, X., Xiao, Y., & Peng, Y. (2023). A review on edible insects in China: Nutritional supply, environmental benefits, and potential applications. *Current Research in Food Science*, *7*, 100596.

Linnemann, D. E. L., Meinlschmidt, P., Purschke, B., Hofstetter, G., Monroy, F. A. R., Einhorn, L., Mothes-Luksch, N., Jaeger, H., Jensen-Jarolim, E., & Pali-Schoell, I. (2019). Novel foods: Enzymatic and thermal food processing make edible insects non-allergenic. *Journal of Allergy and Clinical Immunology*, *143*(2), AB35.

Liu, S., Sun, J., Yu, L., Zhang, C., Bi, J., Zhu, F., Qu, M., Jiang, C., & Yang, Q. (2012). Extraction and characterization of chitin from the beetle Holotrichia parallela Motschulsky. *Molecules*, *17*(4), 4604–4611.

Lorini, C., Ricotta, L., Vettori, V., Riccio, M. D., Biamonte, M. A., & Bonaccorsi, G. (2021). Insights into the predictors of attitude toward entomophagy: the potential role of health literacy: A cross-sectional study conducted in a sample of students of the University of Florence. *International Journal of Environmental Research and Public Health*, *18*(10), 5306. https://doi.org/10.3390/ijerph18105306

Lorrette, B., & Sánchez, L. E. (2022). New lipid sources in the insect industry, regulatory aspects and applications. *OCL*, *29*, 22. https://doi.org/10.1051/ocl/2022017

Lu, J., Guo, Y., Muhmood, A., Zeng, B., Qiu, Y., Wang, P., & Ren, L. (2022). Probing the antioxidant activity of functional proteins and bioactive peptides in Hermetia illucens larvae fed with food wastes. *Scientific Reports*, *12*(1), 2799.

Lucas-González, R., Fernández-López, J., Pérez-Álvarez, J. A., & Viuda-Martos, M. (2019). Effect of drying processes in the chemical, physico-chemical, techno-functional and antioxidant properties of flours obtained from house cricket (Acheta domesticus). *European Food Research and Technology*, *245*(7), 1451–1458.

Ma, Z., Mondor, M., Goycoolea, F. M., Ganji, S. R., & Hernández-Álvarez, A. J. (2024). Unlocking the potential of waxworm (Galleria mellonella) proteins: Extraction, fractionation, and protein quality assessment. *Food Bioscience*, *59*, 103955.

Ma, Z., Mondor, M., Valencia, F. G., & Hernández-Álvarez, A. J. (2023). Current state of insect proteins: extraction technologies, bioactive peptides and allergenicity of edible insect proteins. *Food & Function*, *14*(18), 8129–8156.

Machado, S. S. N., da Silva, J. B. A., Nascimento, R. Q., Lemos, P. V. F., de Jesus Assis, D., Marcelino, H. R., de Souza Ferreira, E., Cardoso, L. G., Pereira, J. D., & Santana, J. S. (2024). Insect residues as an alternative and promising source for the extraction of chitin and chitosan. *International Journal of Biological Macromolecules*, *254*, 127773.

Madau, F. A., Arru, B., Furesi, R., & Pulina, P. (2020). Insect farming for feed and food production from a circular business model perspective. *Sustainability*, *12*(13), 5418.

Mahmoudi-Kordi, F., Balvardi, M., Akhavan, H.-R., & Jafari, S. M. (2022). Optimization of ethanol-assisted aqueous oil extraction from Cicadatra querula. *Journal of Food Measurement and Characterization*, *16*(2), 1426–1436.

Majewski, P., Zapotoczny, P., Lampa, P., Burduk, R., & Reiner, J. (2022). Multipurpose monitoring system for edible insect breeding based on machine learning. *Scientific Reports*, *12*(1), 7892.

Mancini, S., Moruzzo, R., Riccioli, F., & Paci, G. (2019). European consumers' readiness to adopt insects as food. A review. *Food Research International*, *122*, 661–678.

Mastore, M., Quadroni, S., Caramella, S., & Brivio, M. F. (2021). The silkworm as a source of natural antimicrobial preparations: Efficacy on various bacterial strains. *Antibiotics*, *10*(11), 1339.

Matandirotya, N. R., Filho, W. L., Mahed, G., Maseko, B., & Murandu, C. V. (2022). Edible insects consumption in Africa towards environmental health and sustainable food systems: A bibliometric study. *International Journal of Environmental Research and Public Health*, *19*(22), 14823.

Mba, A. R. F., David-Briand, E., Viau, M., Riaublanc, A., Kansci, G., & Genot, C. (2021). Protein extraction yield, lipid composition, and emulsifying properties of aqueous extracts of Rhynchophorus phoenicis larvae extracted at pH 3.0 to 10.0. *Future Foods*, *4*, 100037.

Megido, R. C., Gierts, C., Blecker, C., Brostaux, Y., Haubruge, É., Alabi, T., & Francis, F. (2016). Consumer acceptance of insect-based alternative meat products in Western countries. *Food Quality and Preference*, *52*, 237–243.

Megido, R. C., Poelaert, C., Ernens, M., Liotta, M., Blecker, C., Danthine, S., Tyteca, E., Haubruge, É., Alabi, T., & Bindelle, J. (2018). Effect of household cooking techniques on the microbiological load and the nutritional quality of mealworms (Tenebrio Molitor L. 1758). *Food Research International*, *106*, 503–508.

Melgar-Lalanne, G., Hernández-Álvarez, A. J., & Salinas-Castro, A. (2019). Edible insects processing: Traditional and innovative technologies. *Comprehensive Reviews in Food Science and Food Safety*, *18*(4), 1166–1191.

Mendoza-Salazar, A., Santiago-López, L., Torres-Llanez, M. J., Hernández-Mendoza, A., Vallejo-Cordoba, B., Liceaga, A. M., & González-Córdova, A. F. (2021). In vitro antioxidant and antihypertensive activity of edible insects flours (mealworm and grasshopper) fermented with Lactococcus lactis strains. *Fermentation*, *7*(3), 153.

Messina, C. M., Gaglio, R., Morghese, M., Tolone, M., Arena, R., Moschetti, G., Santulli, A., Francesca, N., & Settanni, L. (2019). Microbiological profile and bioactive properties of insect powders used in food and feed formulations. *Foods*, *8*(9), 400.

Meyer-Rochow, V. B., Gahukar, R. T., Ghosh, S., & Jung, C. (2021). Chemical composition, nutrient quality and acceptability of edible insects are affected by species, developmental stage, gender, diet, and processing method. *Foods*, *10*(5), 1036.

Meyer-Rochow, V. B., & Jung, C. (2020). Insects used as food and feed: Isn't that what we all need? *Foods*, *9*(8), 1003. https://doi.org/10.3390/foods9081003

Mézes, M. (2018). Food safety aspect of insects: A review. *Acta Alimentaria*, *47*(4), 513–522.

Mihaly Cozmuta, A., Uivarasan, A., Peter, A., Nicula, C., Kovacs, D. E., & Mihaly Cozmuta, L. (2023). Yellow mealworm (Tenebrio Molitor) powder promotes a high bioaccessible protein fraction and low glycaemic index in biscuits. *Nutrients*, *15*(4), 997.

Mintah, B. K., He, R., Agyekum, A. A., Dabbour, M., Golly, M. K., & Ma, H. (2020). Edible insect protein for food applications: Extraction, composition, and functional properties. *Journal of Food Process Engineering*, *43*(4), e13362.

Mintah, B. K., He, R., Dabbour, M., Xiang, J., Agyekum, A. A., & Ma, H. (2019). Techno-functional attribute and antioxidative capacity of edible insect protein preparations and hydrolysates thereof: Effect of multiple mode sonochemical action. *Ultrasonics Sonochemistry*, *58*, 104676.

Mishyna, M., Chen, J., & Benjamin, O. (2020). Sensory attributes of edible insects and insect-based foods–Future outlooks for enhancing consumer appeal. *Trends in Food Science & Technology*, *95*, 141–148.

Mishyna, M., Martinez, J.-J. I., Chen, J., & Benjamin, O. (2019). Extraction, characterization and functional properties of soluble proteins from edible grasshopper (Schistocerca gregaria) and honey bee (Apis mellifera). *Food Research International*, *116*, 697–706.

Mohan, K., Ganesan, A. R., Muralisankar, T., Jayakumar, R., Sathishkumar, P., Uthayakumar, V., Chandirasekar, R., & Revathi, N. (2020). Recent insights into the extraction, characterization, and bioactivities of chitin and chitosan from insects. *Trends in Food Science & Technology*, *105*, 17–42.

Montevecchi, G., Licciardello, F., Masino, F., Miron, L. T., & Antonelli, A. (2021). Fortification of wheat flour with black soldier fly prepupae. Evaluation of technological and nutritional parameters of the intermediate doughs and final baked products. *Innovative Food Science & Emerging Technologies*, *69*, 102666.

Moruzzo, R., Mancini, S., & Гуиди, A. (2021). Edible insects and sustainable development goals. *Insects*, *12*(6), 557. https://doi.org/10.3390/insects12060557

Motoki, K., Ishikawa, S.-i., Spence, C., & Velasco, C. (2020). Contextual acceptance of insect-based foods. *Food Quality and Preference*, *85*, 103982.

Mozhui, L., Kakati, L., Kiewhuo, P., & Changkija, S. (2020). Traditional knowledge of the utilization of edible insects in Nagaland, North-East India. *Foods*, *9*(7), 852.

Mshayisa, V. V., Van Wyk, J., & Zozo, B. (2022). Nutritional, techno-functional and structural properties of black soldier fly (Hermetia illucens) larvae flours and protein concentrates. *Foods*, *11*(5), 724.

Musundire, R., Ngonyama, D., Chemura, A., Ngadze, R. T., Jackson, J., Matanda, M. J., Tarakini, T., Langton, M., & Chiwona-Karltun, L. (2021). Stewardship of wild and farmed edible insects as food and feed in Sub-Saharan Africa: A perspective. *Frontiers in Veterinary Science*, *8*, 601386.

Nakimbugwe, D., Ssepuuya, G., Male, D., Lutwama, V., Mukisa, I., & Fiaboe, K. (2021). Status of the regulatory environment for utilization of insects as food and feed in Sub-Saharan Africa-a review. *Critical Reviews in Food Science and Nutrition*, *61*(8), 1269–1278.

Naska, A., Ververis, E., Niforou, A., Pires, S. M., Poulsen, M., Jakobsen, L. S., Becker, N., Lohmann, M., Tesson, V., & Federighi, M. (2022). Novel foods as red meat replacers–an insight using Risk Benefit Assessment methods (the NovRBA project). *EFSA Supporting Publications*, *19*(5), 7316E.

Ndiritu, A. K., Kinyuru, J. N., Kenji, G. M., & Gichuhi, P. N. (2017). Extraction technique influences the physico-chemical characteristics and functional properties of edible crickets (Acheta domesticus) protein concentrate. *Journal of Food Measurement and Characterization*, *11*, 2013–2021.

Ni, J.-B., Luo, S.-Y., Bi, Y.-X., Zielinska, S., Ding, C.-J., Tao, J.-L., Ning, Z., Tian, W.-L., Peng, W.-J., & Fang, X.-M. (2024). The combined effects of ultrasound and plasma-activated water on silkworm pupae: Physicochemical properties, microbiological diversity and ultrastructure. *Ultrasonics Sonochemistry*, *107*, 106927.

Niassy, S., Omuse, E., Roos, N., Halloran, A., Eilenberg, J., Egonyu, J., Tanga, C. M., Meutchieye, F., Mwangi, R., & Subramanian, S. (2022). *Safety, regulatory and environmental issues related to breeding and international trade of edible insects in Africa*. OIE Revue Scientifique et Technique, *41*(1), 117–131.

Nino, M. C., Reddivari, L., Ferruzzi, M. G., & Liceaga, A. M. (2021). Targeted phenolic characterization and antioxidant bioactivity of extracts from edible Acheta domesticus. *Foods*, *10*(10), 2295.

Nongonierma, A. B., Lamoureux, C., & FitzGerald, R. J. (2018). Generation of dipeptidyl peptidase IV (DPP-IV) inhibitory peptides during the enzymatic hydrolysis of tropical banded cricket (Gryllodes sigillatus) proteins. *Food & function*, *9*(1), 407–416.

Noree, S., Pinyakit, Y., Tungkijanansin, N., Kulsing, C., & Hoven, V. P. (2023). Shape transformation of 4D printed edible insects triggered by thermal dehydration. *Journal of Food Engineering*, *358*, 111666.

Ochieng, B. O., Anyango, J. O., Khamis, F. M., Ekesi, S., Egonyu, J. P., Subramanian, S., Nduko, J. M., Nakimbugwe, D., Cheseto, X., & Tanga, C. M. (2023). Nutritional characteristics, microbial loads and consumer acceptability of cookies enriched with insect (Ruspolia differens) meal. *LWT*, *184*, 115012.

Odongo, W., Okia, C., Nalika, N., Nzabamwita, P., Ndimubandi, J., & Nyeko, P. (2018). Marketing of edible insects in Lake Victoria basin: The case of Uganda and Burundi. *Journal of Insects as Food and Feed*, *4*(4), 285–294.

Oh, E., Kang, Y., Cho, I., Koh, J., Park, W. J., & Kim, Y. (2024). Anti-inflammatory and anti-hyperglycemia effects of mealworm (Tenebrio Molitor larvae) protein extracted by four methods: Alkali, salt, enzyme, and screw press. *Journal of Food Science*, *89*(9), 5280–5292.

Olivadese, M., & Dindo, M. L. (2023). Edible insects: A historical and cultural perspective on entomophagy with a focus on western societies. *Insects*, *14*(8), 690.

Otero, P., Gutierrez-Docio, A., del Hierro, J. N., Reglero, G., & Martin, D. (2020). Extracts from the edible insects Acheta domesticus and Tenebrio Molitor with improved fatty acid profile due to ultrasound assisted or pressurized liquid extraction. *Food Chemistry*, *314*, 126200.

Pan, J., Xu, H., Cheng, Y., Mintah, B. K., Dabbour, M., Yang, F., Chen, W., Zhang, Z., Dai, C., He, R., & Ma, H. (2022). Recent insight on edible insect protein: Extraction, functional properties, allergenicity, bioactivity, and applications. *Foods*, *11*(19), 2931. https://doi.org/10.3390/foods11192931

Pekar, J., Ret, D., & Untersmayr, E. (2018). Stability of allergens. *Molecular Immunology*, *100*, 14–20.

Perez-Fajardo, M., Bean, S. R., & Dogan, H. (2023). Effect of cricket protein powders on dough functionality and bread quality. *Cereal Chemistry*, *100*(3), 587–600.

Pérez-Rodríguez, E., Ibarra-Herrera, C. C., & Pérez-Carrillo, E. (2023). Effect of incorporation of solid-state fermented edible insects Tenebrio Molitor and Sphenarium purpurascens with Aspergillus oryzae in the elaboration of bread. *LWT*, *184*, 115003.

Perez-Santaescolastica, C., De Winne, A., Devaere, J., & Fraeye, I. (2022). The flavour of edible insects: A comprehensive review on volatile compounds and their analytical assessment. *Trends in Food Science & Technology*, *127*, 352–367.

Pippinato, L., Gasco, L., Vita, G. D., & Mancuso, T. (2020). Current scenario in the European edible-insect industry: A preliminary study. *Journal of Insects as Food and Feed*, *6*(4), 371–381. https://doi.org/10.3920/jiff2020.0008

Pranil, T., Moontree, T., Moongngarm, A., & Loypimai, P. (2024). Ultrasound-assisted extraction of edible cricket oils from Gryllus bimaculatus and Acheta domesticus. *Journal of Food Measurement and Characterization*, *18*(7), 6145–6155.

Precup, G., Ververis, E., Azzollini, D., Rivero-Pino, F., Zakidou, P., & Germini, A. (2022). The safety assessment of insects and products thereof as novel foods in the European Union. *Novel Foods and Edible Insects in the European Union*, *123*.

Preteseille, N., Deguerry, A., Reverberi, M., & Weigel, T. (2018). Insects in Thailand: National leadership and regional development, from standards to regulations through association. *Edible Insects in Sustainable Food Systems*, 435–442.

Psarianos, M., Aghababaei, F., & Schlüter, O. K. (2025). Bioactive compounds in edible insects: Aspects of cultivation, processing and nutrition. *Food Research International*, 115802.

Psarianos, M., Dimopoulos, G., Ojha, S., Cavini, A. C. M., Bußler, S., Taoukis, P., & Schlüter, O. K. (2022). Effect of pulsed electric fields on cricket (Acheta domesticus) flour: Extraction yield (protein, fat and chitin) and techno-functional properties. *Innovative Food Science & Emerging Technologies*, *76*, 102908.

Psarianos, M., Marzban, N., Ojha, S., Schneider, R., & Schlüter, O. K. (2025). Functional and bioactive properties of chitosan produced from Acheta domesticus with fermentation, enzymatic and microwave-assisted extraction. *Sustainable Food Technology*, *3*(1), 277–285.

Psarianos, M., Ojha, S., & Schlüter, O. K. (2023). Evaluating an emerging technology-based biorefinery for edible house crickets. *Frontiers in Nutrition*, *10*, 1185612.

Purschke, B., Tanzmeister, H., Meinlschmidt, P., Baumgartner, S., Lauter, K., & Jäger, H. (2018). Recovery of soluble proteins from migratory locust (Locusta migratoria) and characterisation of their compositional and techno-functional properties. *Food Research International*, *106*, 271–279.

Pyo, S.-H., Moon, C.-R., Park, S.-W., Choi, J.-y., Park, J.-D., Sung, J. M., Choi, E.-J., & Son, Y.-J. (2024). Quality and staling characteristics of white bread fortified with lysozyme-hydrolyzed mealworm powder (Tenebrio molitor L.). *Current Research in Food Science*, *8*, 100685.

Queiroz, L. S., Regnard, M., Jessen, F., Mohammadifar, M. A., Sloth, J. J., Petersen, H. O., Ajalloueian, F., Brouzes, C. M. C., Fraihi, W., & Fallquist, H. (2021). Physico-chemical and colloidal properties of protein extracted from black soldier fly (Hermetia illucens) larvae. *International Journal of Biological Macromolecules*, *186*, 714–723.

Queiroz, L. S., Silva, N. F. N., Jessen, F., Mohammadifar, M. A., Stephani, R., de Carvalho, A. F., Perrone, Í. T., & Casanova, F. (2023). Edible insect as an alternative protein source: A review on the chemistry and functionalities of proteins under different processing methods. *Heliyon*, *9*(4).

Quinteros, M. F., Martínez, J., Barrionuevo, A., Rojas, M., & Carrillo, W. (2022). Functional, antioxidant, and anti-Inflammatory properties of cricket protein concentrate (Gryllus assimilis). *Biology*, *11*(5), 776.

Rahman, M. M., Byanju, B., & Lamsal, B. P. (2024). Protein, lipid, and chitin fractions from insects: Method of extraction, functional properties, and potential applications. *Critical Reviews in Food Science and Nutrition*, *64*(18), 6415–6431.

Ramos-Elorduy, J. (1998). *Creepy crawly cuisine: The gourmet guide to edible insects*. Inner Traditions/Bear & Co.

Ranga, L., Panagiotou, M., Noci, F., Charalampidou, M., Gkatzionis, K., & Dermiki, M. (2025). Cross-cultural perspectives on insect-based foods: insights from consumers in Greece and Ireland. *Foods*, *14*(3), 490.

Ranga, L., Vishnumurthy, P., & Dermiki, M. (2024). Willingness to consume insects among students in France and Ireland. *Irish Journal of Agricultural and Food Research*, *62*(1). https://doi.org/10.15212/ijafr-2023–0106

Ribeiro, J., Sousa-Pinto, B., Fonseca, J., Fonseca, S. C., & Cunha, L. (2021). Edible insects and food safety: Allergy. *Journal of Insects as Food and Feed*, *7*(5), 833–848.

Ribeiro, J. C., Cunha, L. M., Sousa-Pinto, B., & Fonseca, J. (2018). Allergic risks of consuming edible insects: A systematic review. *Molecular Nutrition & Food Research*, *62*(1), 1700030.

Ribeiro, J. C., Lima, R. C., Maia, M. R., Almeida, A. A., Fonseca, A. J., Cabrita, A. R. J., & Cunha, L. M. (2019). Impact of defatting freeze-dried edible crickets (Acheta domesticus and Gryllodes sigillatus) on the nutritive value, overall liking and sensory profile of cereal bars. *LWT*, *113*, 108335.

Ribeiro, J. C., Pintado, M. E., & Cunha, L. M. (2024). Consumption of edible insects and insect-based foods: A systematic review of sensory properties and evoked emotional response. *Comprehensive Reviews in Food Science and Food Safety*, *23*(1), e13247.

Ribeiro, J. C., Santos, C., Lima, R. C., Pintado, M. E., & Cunha, L. M. (2022). Impact of defatting and drying methods on the overall liking and sensory profile of a cereal bar incorporating edible insect species. *Future Foods*, *6*, 100190.

Riekkinen, K., Väkeväinen, K., & Korhonen, J. (2022). The effect of substrate on the nutrient content and fatty acid composition of edible insects. *Insects*, *13*(7), 590.

Rivas-Vela, C. I., Castaño-Tostado, E., Cardador-Martínez, A., Amaya-Llano, S. L., & Castillo-Herrera, G. A. (2023). Subcritical water hydrolysis for the obtention of bioactive peptides from a grasshopper Sphenarium purpurascens protein concentrate. *The Journal of Supercritical Fluids*, *197*, 105893.

Rivero-Pino, F., Gonzalez-de la Rosa, T., & Montserrat-de la Paz, S. (2024). Edible insects as a source of biopeptides and their role in immunonutrition. *Food & Function*, *15*, 2789–2798.

Rivero-Pino, F., Ververis, E., Pérez-Gálvez, R., & Naska, A. (2024). Biological properties and safety aspects of edible insects. *Insects as Food and Food Ingredients*, 173–190.

Rumpold, B. A., & Schlüter, O. (2013). Nutritional composition and safety aspects of edible insects. *Molecular Nutrition & Food Research*, *57*(5), 802–823. https://doi.org/10.1002/mnfr.201200735

Rutaro, K., Malinga, G. M., Opoke, R., Lehtovaara, V. J., Omujal, F., Nyeko, P., Roininen, H., & Valtonen, A. (2018). Artificial diets determine fatty acid composition in edible Ruspolia differens (Orthoptera: Tettigoniidae). *Journal of Asia-Pacific Entomology*, *21*(4), 1342–1349.

Schlüter, O., Rumpold, B., Holzhauser, T., Roth, A., Vogel, R. F., Quasigroch, W., Vogel, S., Heinz, V., Jäger, H., & Bandick, N. (2017). Safety aspects of the production of foods and food ingredients from insects. *Molecular Nutrition & Food Research*, *61*(6), 1600520.

Sheng, Z., Guo, A., Wang, J., & Chen, X. (2022). Preparation, physicochemical properties and antimicrobial activity of chitosan from fly pupae. *Heliyon*, *8*(10).

Sikora, D., & Rzymski, P. (2025). Disgust or curiosity? Acceptance of edible insects as food alternative in Poland. *European Food Research and Technology*, 1–12.

Simopoulos, A. P. (2004). Omega-6/omega-3 essential fatty acid ratio and chronic diseases. *Food Reviews International*, *20*(1), 77–90.

Singh, Y., Cullere, M., Kovitvadhi, A., Chundang, P., & Dalle Zotte, A. (2020). Effect of different killing methods on physicochemical traits, nutritional characteristics, in vitro human digestibility and oxidative stability during storage of the house cricket (Acheta domesticus L.). *Innovative Food Science & Emerging Technologies*, *65*, 102444.

Skotnicka, M., Mazurek, A., & Kowalski, S. (2023). The acceptance of cream soups with the addition of edible insects (mealworm, T. Molitor; house cricket, A. domesticus; buffalo worm, A. diaperinus; grasshopper, R. differens) among young people and seniors in Poland. *Nutrients*, *15*(24), 5047.

Sokame, B. M., Runyu, J. C., & Tonnang, H. E. (2024). Integrating edible insect into circular agriculture for sustainable production. *Sustainable Production and Consumption*, *52*, 80–94.

Song, Y. S., Jo, Y. H., Han, Y. S., & Jung, W. J. (2022). Production of chitin-and chitosan-oligosaccharide using the edible insect, Tenebrio Molitor. *Entomological Research*, *52*(4), 207–213.

Suga, N., Tsumura, E., Naito, Y., Hamaguchi, I., Matsuda, S., Kawabata, K., & Sakamoto, K. (2023). Thermal stability of cricket powder and its effects on antioxidant activity, physical, and sensory properties of rice crackers. *LWT*, *186*, 115267.

Sweers, L., Mishyna, M., Boom, R., Fogliano, V., Keppler, J., & Lakemond, C. (2023). Microfiltration for effective microbiological decontamination of edible insects–Protein hydrolysis, aggregation and pH are critical for protein recovery. *Food and Bioproducts Processing*, *141*, 128–138.

Tagawa, K., Hosoya, T., Hyakumura, K., Suzuki, D., Yoshizawa, S., & Praxaysombath, B. (2022). The effects of season, geography, and urbanization on the diversity of edible insects at food markets in Laos. *PLoS One*, *17*(4), e0267307.

Tan, H. S. G., Fischer, A. R., Tinchan, P., Stieger, M., Steenbekkers, L., & Van Trijp, H. C. (2015). Insects as food: Exploring cultural exposure and individual experience as determinants of acceptance. *Food Quality and Preference*, *42*, 78–89.

Tan, Y. N., Chin, Y. L., & Chen, W. N. (2021). Comparison of sustainable lipid and protein removal methods for the isolation of insect chitin from black soldier fly exoskeleton. *Acs Food Science & Technology*, *1*(4), 698–706. https://doi.org/10.1021/acsfoodscitech.0c00104

Teixeira, C. S., Villa, C., Costa, J., Ferreira, I. M., & Mafra, I. (2023). Edible insects as a novel source of bioactive peptides: A systematic review. *Foods*, *12*(10), 2026.

Tejada, L., Buendía-Moreno, L., Hernández, I., Abellán, A., Cayuela, J. M., Salazar, E., & Bueno-Gavilá, E. (2022). Bioactivities of mealworm (Alphitobius diaperinus L.) larvae hydrolysates obtained from artichoke (Cynara scolymus L.) proteases. *Biology*, *11*(5), 631.

Traynor, A., Burns, D. T., Wu, D., Karoonuthaisiri, N., Petchkongkaew, A., & Elliott, C. (2024). An analysis of emerging food safety and fraud risks of novel insect proteins within complex supply chains. *NPJ Science of Food*, *8*(1), 7.

Tzompa-Sosa, D. A., Moruzzo, R., Mancini, S., Schouteten, J. J., Liu, A., Li, J., & Sogari, G. (2023). Consumers' acceptance toward whole and processed mealworms: A cross-country study in Belgium, China, Italy, Mexico, and the US. *PLoS One*, *18*(1), e0279530.

Tzompa-Sosa, D. A., Sogari, G., Copelotti, E., Andreani, G., Schouteten, J. J., Moruzzo, R., Liu, A., Li, J., & Mancini, S. (2023). What motivates consumers to accept whole and processed mealworms in their diets? A five-country study. *Future Foods*, *7*, 100225.

UNBS. (2020). *US 2146:2020 edible insects – Specification*. Uganda National Bureau of Standards.

Van Huis, A., & Oonincx, D. G. (2017). The environmental sustainability of insects as food and feed. A review. *Agronomy for Sustainable Development*, *37*, 1–14.

Van Moll, L., De Smet, J., Paas, A., Tegtmeier, D., Vilcinskas, A., Cos, P., & Van Campenhout, L. (2022). In vitro evaluation of antimicrobial peptides from the black soldier fly (Hermetia illucens) against a selection of human pathogens. *Microbiology Spectrum*, *10*(1), e01664–01621.

Vandeweyer, D., Lievens, B., & Van Campenhout, L. (2020). Microbiological safety of industrially reared insects for food: Identification of bacterial endospores and targeted detection of foodborne viruses. *bioRxiv*, 2020.2004. 2022.055236.

Varelas, V. (2019). Food wastes as a potential new source for edible insect mass production for food and feed: A review. *Fermentation*, *5*(3), 81.

Veldkamp, T., Meijer, N., Alleweldt, F., Deruytter, D., Van Campenhout, L., Gasco, L., Roos, N., Smetana, S., Fernandes, A., & Van der Fels-Klerx, H. (2022). Overcoming technical and market barriers to enable sustainable large-scale production and consumption of insect proteins in Europe: A SUSINCHAIN perspective. *Insects*, *13*(3), 281.

Verhoeckx, K. C., Vissers, Y. M., Baumert, J. L., Faludi, R., Feys, M., Flanagan, S., Herouet-Guicheney, C., Holzhauser, T., Shimojo, R., & van der Bolt, N. (2015). Food processing and allergenicity. *Food and Chemical Toxicology*, *80*, 223–240.

Ververis, E., Ackerl, R., Azzollini, D., Colombo, P. A., de Sesmaisons, A., Dumas, C., Fernandez-Dumont, A., da Costa, L. F., Germini, A., & Goumperis, T. (2020). Novel foods in the European Union: Scientific requirements and challenges of the risk assessment process by the European food safety authority. *Food Research International*, *137*, 109515.

Ververis, E., Boue, G., Poulsen, M., Pires, S. M., Niforou, A., Thomsen, S. T., Tesson, V., Federighi, M., & Naska, A. (2022). A systematic review of the nutrient composition, microbiological and toxicological profile of Acheta domesticus (house cricket). *Journal of Food Composition and Analysis*, *114*, 104859.

Ververis, E., Niforou, A., Poulsen, M., Pires, S. M., Federighi, M., Samoli, E., Naska, A., & Boué, G. (2024). Substituting red meat with insects in burgers: Estimating the public health impact using risk-benefit assessment. *Food and Chemical Toxicology*, *189*, 114764.

Vilcinskas, A. (2020). Insect biotechnology: Insects as a resource. In *Biological transformation* (pp. 247–260). Springer.

Vrontaki, M., Adamaki-Sotiraki, C., Rumbos, C. I., Anastasiadis, A., & Athanassiou, C. G. (2024). Valorization of local agricultural by-products as nutritional substrates for Tenebrio Molitor larvae: A sustainable approach to alternative protein production. *Environmental Science and Pollution Research*, *31*(24), 35760–35768.

Wade, M., & Hoelle, J. (2020). A review of edible insect industrialization: Scales of production and implications for sustainability. *Environmental Research Letters*, *15*(12), 123013.

Wang, J.-Z., Wu, C.-K., Wang, X.-Y., Xu, Y.-Z., Herman, R. A., Sheng, S., You, S., Wang, J., & Wu, F.-A. (2023). One positive part of copper ion on biocatalysts: CNPs@ Cu-Trypsin-assisted aqueous extraction technique for simultaneous exploration of edible insect oil and protein. *Chemical Engineering Journal*, *464*, 142560.

Wendin, K., Olsson, V., & Langton, M. (2019). Mealworms as food ingredient – Sensory investigation of a model system. *Foods*, *8*(8), 319. https://doi.org/10.3390/foods8080319

WHO. (2022). *Reformulation of food and beverage products for healthier diets: Policy brief*. World Health Organization.

Wu, Q., Patočka, J., & Kuča, K. (2018). Insect antimicrobial peptides, a mini review. *Toxins*, *10*(11), 461.

Xie, X., Zhang, J., Yuan, Z., Wang, H., An, J., & Deng, L. (2024). Characterization of wheat dough and Chinese steamed bread using mealworm powder formulated with medium-gluten and whole wheat flour. *Journal of the Science of Food and Agriculture*, *104*(3), 1732–1740.

Yuan, B.-Q., Yu, T.-H., Chen, S.-C., Zhang, Z.-Q., Guo, Z.-K., Huang, G.-X., Xiao, J.-h., & Huang, D.-W. (2024). Physical and chemical characterization of chitin and chitosan extracted under different treatments from black soldier fly. *International Journal of Biological Macromolecules*, *279*, 135228.

Zhang, F., Li, X., Liang, X., Kong, B., Sun, F., Cao, C., Gong, H., Zhang, H., & Liu, Q. (2024). Feasibility of Tenebrio Molitor larvae protein to partially replace lean meat in the processing of hybrid frankfurters: Perspectives on quality profiles and in vitro digestibility. *Food Research International*, *176*, 113846.

Zhang, F., Sun, Z., Li, X., Kong, B., Sun, F., Cao, C., Chen, Q., Zhang, H., & Liu, Q. (2023). Ultrasound-assisted alkaline extraction of protein from Tenebrio Molitor larvae: Extraction kinetics, physiochemical, and functional traits. *Ultrasonics Sonochemistry*, *95*, 106379.

Zhang, F., Xu, Y., Kong, B., Chen, Q., Sun, F., Zhang, H., & Liu, Q. (2022). Comparative study of two types of pre-extraction treatment (drying or non-drying) on physicochemical, structural and functional properties of extracted insect proteins from Tenebrio Molitor larvae. *Current Research in Food Science*, *5*, 1570–1580.

Zhang, F., Yue, Q., Li, X., Kong, B., Sun, F., Cao, C., Zhang, H., & Liu, Q. (2023). Mechanisms underlying the effects of ultrasound-assisted alkaline extraction on the structural properties and in vitro digestibility of Tenebrio Molitor larvae protein. *Ultrasonics Sonochemistry*, *94*, 106335.

Zhang, J., Li, M., Han, Z., & Shao, J.-H. (2025). A review of insect peptides: From extraction and preparation to bioactivities. *Current Opinion in Food Science*, 101301.

Zhang, W., Jia, Y., Guo, C., Devahastin, S., Hu, X., & Yi, J. (2024). Effect of compositions and physical properties on 3D printability of gels from selected commercial edible insects: Role of protein and chitin. *Food Chemistry*, *433*, 137349.

Zhang, Z.-Q., Chen, S.-C., Wang, Q.-L., Liu, C.-Q., Xiao, J.-H., & Huang, D.-W. (2024). Nutritional value of Protaetia brevitarsis powder and its effect on the flavor and antioxidant capacity of dough. *LWT*, *195*, 115849.

Zhou, P., Li, J., Yan, T., Wang, X., Huang, J., Kuang, Z., Ye, M., & Pan, M. (2019). Selectivity of deproteinization and demineralization using natural deep eutectic solvents for production of insect chitin (Hermetia illucens). *Carbohydrate Polymers*, *225*, 115255.

Zhou, Y., Wang, D., Zhou, S., Duan, H., Guo, J., & Yan, W. (2022). Nutritional composition, health benefits, and application value of edible insects: A review. *Foods*, *11*(24), 3961.

Zielińska, E. (2022). Evaluating the functional characteristics of certain insect flours (non-defatted/defatted flour) and their protein preparations. *Molecules*, *27*(19), 6339.

Zielińska, E., Baraniak, B., & Karaś, M. (2018). Identification of antioxidant and anti-inflammatory peptides obtained by simulated gastrointestinal digestion of three edible insects species (Gryllodes sigillatus, Tenebrio Molitor, Schistocerca gragaria). *International Journal of Food Science and Technology*, *53*(11), 2542–2551.

Index

Note: Page numbers in *italics* indicate a figure, and page numbers in **bold** indicate a table on the corresponding page.

U

V

W

For Product Safety Concerns and Information please contact our EU representative GPSR@taylorandfrancis.com Taylor & Francis Verlag GmbH, Kaufingerstraße 24, 80331 München, Germany